REGIONAL OCEANOGRAPHY:

AN INTRODUCTION

Some Other Related Pergamon Titles of Interest

LALLI & PARSONS
Biological Oceanography: An Introduction

PARSONS et al
A Manual of Chemical and Biological Methods for Seawater Analysis

PARSONS et al
Biological Oceanographic Processes, 3rd Edition

PICKARD & EMERY
Descriptive Physical Oceanography, 5th Enlarged Edition

POND & PICKARD
Introductory Dynamical Oceanography, 2nd Edition

OPEN UNIVERSITY
The Ocean Basins: Their Structure and Evolution
Seawater: Its Composition, Properties and Behaviour
Ocean Circulation
Waves, Tides and Shallow-Water Processes
Ocean Chemistry and Deep-Sea Sediments
Case Studies in Oceanography and Marine Affairs

WILLIAMS & ELDER
Fluid Physics for Oceanographers and Physicists

Forthcoming Books

EMERY & THOMSON
Data Analysis Methods in Physical Oceanography

Journals

Continental Shelf Research
Deep-Sea Research
Oceanographic Literature Review
Progress in Oceanography

All the above books are available under the terms of the Pergamon textbook inspection copy service. Full details of all Pergamon publications/free specimen copy of any Pergamon journal available on request from your nearest Pergamon office.

REGIONAL OCEANOGRAPHY:
AN INTRODUCTION

MATTHIAS TOMCZAK

School of Earth Sciences
The Flinders University of South Australia

and

J. STUART GODFREY

CSIRO Division of Oceanography, Tasmania, Australia

PERGAMON

U.K.	Elsevier Science Ltd, Headington Hill Hall, Oxford OX3 0BW, England
U.S.A.	Elsevier Science Inc, 660 White Plains Road, Tarrytown, New York 10591-5153, U.S.A.
JAPAN	Elsevier Science Japan, Tsunashima Building Annex, 3-20-12 Yushima, Bunkyo-ku, Tokyo 113, Japan

First edition 1994

British Library Cataloguing in Publication Data
A catalogue record for this book is available from the British Library.

Library of Congress Cataloguing-in-Publication Data
Tomczak, M., 1941
Regional oceanography: an introduction/by Matthias Tomczak and J. Stuart Godfrey.
p. cm.
Includes index.
1. Oceanography.
I. Godfrey, J. Stuart. II. Title.
GC11.2.T66 1993
551.46—dc20

0 08 041021 9 Hardcover
0 08 041020 0 Flexicover

Printed and bound in Great Britain by
Butler & Tanner Ltd, Frome and London

Preface

This book developed from lectures for undergraduate students of marine sciences in Sydney. Following a good tradition, the curriculum at the university began with an interdisciplinary introduction into all sciences of the oceans. A similar curriculum operates at the Flinders University of South Australia, where the introductory course on regional oceanography brings together students of marine geology, biology, chemistry, geography, and physical oceanography. The choice of textbook for such a course usually follows the rule of the lowest common denominator. This eliminates most textbooks commonly used in physical oceanography because they require some understanding of mathematics and theoretical physics which not all students bring to the course.

When looking around for alternatives we were dissatisfied with the available material. Non-mathematical texts were either written for high school or college use and considered too elementary for a university curriculum, or they were outdated, leaving too large a gap between what students were taught and what they find in research publications. We decided that it was time to produce a set of lectures which would take into account modern findings and modern ideas and present them on a level suitable for an introductory undergraduate course.

This book is the result of our efforts. Originally the material presented to our students was covered in 21 one-hour lectures. When we decided to develop it into a book the material grew in response to comments and suggestions from colleagues who were asked by Pergamon Press to comment on our proposal. In writing this book we have been surprised to learn how much of the ocean's behaviour as a component of climate - the particular reasons why it absorbs heat in one region, or restores it to the atmosphere in another - can be understood by combining an understanding of simple physical principles with knowledge of the ocean's geographical features. We therefore expanded on some aspects of the ocean's role in climate in this book; however, we have tried to give a simple account of that role, which should prove useful in more advanced studies, aimed towards actual prediction of possible climate changes. The division into chapters has been retained, but in its present form it is unrealistic to cover the material in 21 hours.

Most horizontal property distributions are shown in the Peters projection which combines fidelity of area with a rectangular coordinate grid (Peters, 1989). Although the problem which projection is appropriate for a given task is not a trivial one, oceanographers usually do not even realize that there is a problem. A basic requirement for a regional description of the world ocean is fidelity of area. The Mercator projection, which will remain the ideal charting tool at sea, grossly overemphasizes the temperate and subpolar regions, at the expense of the tropics and subtropics to which most of the world ocean belongs. Most commonly used projections with fidelity of area are based on curved coordinates and therefore require a latitude/longitude grid across the map for the location of features. The Peters projection keeps the map surface free for the information of interest while doing justice to the relative roles of all climatic regions. Distortions of distance are severe only in the polar regions. To rectify this, pole-centred stereographic projections are used in the discussion of the Arctic and Antarctic oceans.

Although the list of references is long, it is obvious that an introductory text is not the place for a bibliography on regional oceanographic studies. Readers have to be aware of the eclectic character of our reference list. The fact that a certain paper is not quoted does not mean that we did not consider it important, nor can it be concluded that inclusion in the list of references puts a paper into the "very important" category. Our approach to the vexed question of references in elementary textbooks is that we follow accepted procedures and generally acknowledge sources of figures, preferring those which highlight the essentials of a situation to those with more detail or priority of thought. Beyond that, we restrict documentation with references to those parts of the text where new information leads us to modify or contradict earlier work. Where reference is not made to the original source it can always be found by following up the references given in the quoted literature.

A substantial part of this book was prepared while one of us (M.T.) was on study leave at the Institut für Meereskunde an der Universität Kiel. The hospitality and facilities of that institution are gratefully acknowledged. Various colleagues from Kiel commented on early drafts of the text, particularly the chapters on the Southern and Atlantic Oceans. We thank in particular Ray Peterson, Lothar Stramma, and Walter Zenk for helpful suggestions. We are indebted to Birgit Klein for information from her work not published at the time. Among our Sydney colleagues we thank John Luick for his attentive reading of early drafts. Janet Sprintall prepared most of the computer-generated figures, and she and You Yuzhu gave us generous access to work in progress at the time. Cesar Villanoy provided valuable assistance with chapter 13 and assisted in the preparation of many computer-generated figures. Brenda Durie skilfully converted the GEBCO topography into the topographic charts of this book. Drafting staff at Pergamon Press guaranteed high standards by preparing most figures from our often rough drafts. At Flinders University, Gail Jackson drafted many figures with never-flinching dedication to highest figure standards; without her help this book would not have gone through the printer's presses for another twelve months.

The *Oceanographic Literature Review* section of Deep-Sea Research proved invaluable during the preparation of this book, and we express our sincere appreciation to the people behind this excellent research tool.

Finally, we note a few things which regrettably did not get the attention they deserve. Much can be learned about the oceanic circulation and the life cycle of the various water masses by combining information from CTD and current meter data with information on oxygen, nutrients, and other chemical tracers. Unfortunately marine chemistry has always been among the authors' weak points, and the treatment of the chemical tracers in this book is below acceptable level - it cannot even be called elementary. Should this text prove its usefulness with students and lecturers, to the extent that a revised edition seems justified, a first improvement should be proper coverage of the distribution of all major chemical tracers in the world ocean. Also, our text is clearly a product of what is known as the "Western World": It is based nearly exclusively on research reported in English and published in North America or Europe. While there is no need for apologies in that respect, it is true that in a field such as regional oceanography significant research is sometimes reported in a journal published closer to the regions of interest. It is likely that more accurate descriptions of the oceanography particularly of some of the marginal seas exist in Japanese, Russian, Chinese, or other languages. We welcome the assistance of oceanographers who know of such descriptions and communicate relevant information to us.

Adelaide and Hobart, May 1993

Matthias Tomczak   J. Stuart Godfrey

Regional Oceanography:
An Introduction

Matthias Tomczak and J. Stuart Godfrey

Contents

CHAPTER 1

Introduction: What drives the ocean currents?

SIXTY years ago, this textbook would have been titled "Introductory Geography of the Oceans". Physical oceanography then was a close relative of physical geography and shared its descriptive character. This period culminated in textbooks such as *Geographie des Atlantischen Ozeans* (1912) and *Geographie des Indischen und Stillen Ozeans* (1935) by G. Schott, books which conveyed to the reader through a passioned yet accurate description of its features the fascination which the oceanic environment exerts on the oceanographer. Oceanography has come a long way since then, having concentrated on understanding the physical principles that drive the oceans and using the tools of mathematics and theoretical fluid dynamics to forecast their behaviour. Students of oceanography now spend more time trying to come to grips with vorticity, inverse methods and normal mode analysis than learning about the features of the deep sea basins and marginal seas or about the climatic regions of the oceans. And this is rightly so, for little is learned in science through mere description; analysis and conclusion are required before anyone can claim to understand.

As it turns out, understanding the ocean circulation is impossible without knowledge of geographical details - the depth of certain ocean sills, for example, or the peculiarities of the wind field and its seasonal variation. To separate the facts about the geography of the ocean from acquired knowledge about its dynamics would be like separating the memorizing of the vocabulary of a new language from the learning of its grammar. What we propose to do is describe the features of the world ocean both as a systematic exercise in geography and as examples of physical principles at work.

These physical principles are sufficiently powerful and all-pervasive that it is worth introducing them first. This and the following four chapters are a summary of some of the principles and their consequences, and will serve as a reference for all chapters to come. Students of oceanography who are using this book as their introduction to the discipline will find them essential reading. Advanced students who use the book because of a need to brush up their knowledge on the geography of the oceans can skip the first five chapters and go straight to the chapter of interest to them. Both should take particular notice of the figures which accompany the text. With a bit of guidance, much can be learnt by looking closely at observational data. Our text will provide the guidance, but it will not go into detailed descriptions of what can be seen more easily in figures. The figures are therefore not illustrations of the text; they are an integral part of this book.

Some knowledge of the geography of the oceans is essential in regional oceanography. While we attempted to include as much relevant geographical information as possible, clarity of figures must rank higher than detail in an introductory text. Sometimes the location of a feature can be determined by consulting the index. In other cases the use of an elementary geographical atlas may be required.

One final note on the use of geographic and oceanographic nomenclature, before some readers turn the page and proceed to other chapters. Although the use of geographical names and the rules on naming newly discovered geographic features are regulated by an international advisory body, use of geographical names in oceanography is not uniform. This is particularly true for features such as currents, fronts, or water masses, which are not covered by the international regulations on the use of geographical names. Oceanographers have an unfortunate habit of trying to make their mark by putting names to their liking on features which may or may not have been named before (probably to someone else's liking). In this text we adopt the principle that geographic features are referred to under the names used on the GEBCO charts (IHO/IOC/CHS, 1984). In references to currents and fronts we use generally accepted names where they exist, preferring names which include a reference to geographic features (e.g. Peru/Chile Current, not Humboldt Current). Universally accepted names for water masses exist only for the major oceanic water masses; other water masses can be found under a variety of names in the literature. Our usage of water mass names is based partly on historical use, partly on the systematic approach to water mass analysis described in Chapter 5. Wherever possible we use names already introduced by others and do not invent our own.

We return to our discussion of the most important physical principles. A discussion of what in essence are elements of geophysical fluid dynamics is not to everyone's liking; nevertheless, some of the principles determining ocean flows turn out to be quite simple, and by understanding them it is possible to go a long way towards understanding the role of the ocean in climate variations, both natural or man-made. The ocean is unique in this respect; it can absorb heat in one region, and restore it to the atmosphere (perhaps decades or centuries later) at a quite different place. This has become a topic of widespread interest and intensive research in recent years, and by spending some effort on understanding the underlying principles readers will find that they can gain an understanding of much of the modern literature on this topic.

If we exclude tidal forces, which have little effect on the long-term mean properties of the ocean, the oceanic circulation is driven by three external influences: wind stress, heating and cooling, and evaporation and precipitation - all of which, in turn, are ultimately driven by radiation from the sun. To understand why temperature, salinity and all other properties of the oceans' waters are distributed the way they are, a basic knowledge about these external forces is necessary. We therefore begin our description of the geography of the oceans with a brief look at the atmosphere, which holds the key to the question how the energy received from the sun keeps the ocean circulation going.

We note at the outset that this approach ignores the fact that the circulation of the atmosphere is in turn influenced by the distribution of oceanic properties, such as sea surface temperature (in oceanography often abbreviated as SST) and the distribution of sea ice. In particular, the amount of evaporation from the ocean depends strongly on the sea surface temperature; and when the evaporated water is returned as rain it releases its latent heat into the surrounding air. This heating is probably the strongest driving force for the atmospheric winds. To understand the oceanic and atmospheric circulation fully we should treat them as a single system of two interacting components, coupled at the air-sea interface through the fluxes of momentum, heat, and mass. This of course complicates the task and could not be achieved with traditional oceanographic or meteorological tools. Although the stage has now been reached where treatment of the ocean and the atmosphere as a coupled system is becoming more and more feasible, it seems good advice for an introductory text to follow the traditional approach and consider the state of the atmosphere as determined independently of the state of the ocean. We shall return to the question of interaction between ocean and atmosphere in the last three chapters when we discuss interannual climate fluctuations and long-term climate change.

The amount of heat radiation received by the outer atmosphere varies from the equator to the poles. The difference varies with the seasons, but on average the equatorial regions receive much more heat than the polar regions. The cold air at the poles is denser than the warm air at the equator; and since the air pressure at the sea surface or on land is determined by the weight of the air above the observation point, air pressure at sea level is higher at the poles than at the equator - in other words, a pressure gradient is set up which is directed from the poles toward the equator. The pressure gradient in the upper part of the atmosphere has the opposite sign.

In fluids and gases, pressure gradients produce flow from regions of high pressure to regions of low pressure. If the earth were not rotating, the response to these pressure gradients would be direct and simple. Two circulation cells would be set up, one in either hemisphere, by the differential solar heating. At sea level, winds would blow from the poles to the equator;

the air would then rise and recirculate back to the poles at great height. On a rotating earth this pattern is modified quite strongly, in two ways. Firstly, as air moves towards the equator, the rotation of the earth shifts ocean and land eastward under it. An observer moving with the land experiences the air movement as an "easterly" wind, i.e. a wind blowing from the east, with an equatorward component. In the tropics and subtropics this wind is known as the Trade Wind, in polar latitudes it occurs as the Polar Easterlies. The outcome is that the wind no longer blows from regions of high pressure to regions of low pressure along the most direct route but tends to follow contours of constant pressure (isobars) - hence the usefulness of isobars on the daily weather map in the television news.

Because on a rotating earth the flow of air is more zonal (directed east-west) than meridional (directed north-south), the importance of the vertical component of air movement is reduced: the flow can circle the earth with great speed without need for uplift or sinking. This produces the second, more drastic modification of the simple hemispheric cell arrangement. It turns out that zonal flow of high speed becomes unstable, creating eddies which in turn reshape the air pressure distribution. As a result, an intermediate air pressure maximum is established at mid-latitudes. The reversal of the meridional pressure gradient establishes a band of "westerly" wind, i.e. wind blowing from the west (Figure 1.1). Seafarers know them well and refer to them as the Roaring Forties, thus expressing their experience that between 40° and 50° latitude the winds are usually strong, highly variable and very gusty.

Figure 1.2 gives the winds at sea level in the real world. The features seen in Figure 1.1 come out clearly, but the presence of continental land masses modifies the atmospheric circulation further. Because air heats up faster over the continents than over the oceans during summer, and cools faster during winter, large land masses are characterized by low air pressure in summer - relative to air pressure over the ocean at the same latitude - and high air pressure in winter. This results in a deviation of average wind direction from mainly easterly or mainly westerly over some parts of the oceans. Some ocean regions experience strong seasonal variations in wind direction, including complete reversal. Such wind systems are known as monsoons.

In passing, it should be noted that the convention for indicating the direction of ocean currents differs from the convention used for wind directions. A "westerly" wind is a wind which blows from the west and goes to the east; a "westward" current is a current which comes from the east and flows towards west. This can cause confusion to people who rarely, if ever, go to sea; but it is easily understood and remembered when related to practical experience with winds and ocean currents. On land, it is important to know from where the wind blows: any windbreak must be erected in this direction. Where the wind goes is of no consequence. At sea, the important

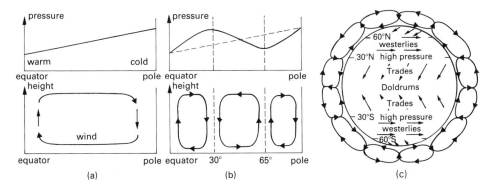

FIG. 1.1. Schematic diagram of the meridional air pressure distribution and associated air movement (a) on a non-rotating earth, (b) on a rotating earth without continents, (c) viewed from above (the Polar Easterlies poleward of 65° are not labelled).

information is where the current goes: a ship exposed to current drift has to stay well clear from obstacles downstream. Where the water comes from is irrelevant.

From the point of view of oceanography, knowledge of the planetary wind field above the sea surface has always been unsatisfactory. It is difficult to obtain quantitatively accurate wind data from the oceans, particularly from regions remote from major shipping routes. Advances in numerical modelling of the atmosphere and the use of drifting buoys equipped with pressure sensors greatly improved our knowledge of winds over the Antarctic ocean, but the data are still not adequate for many oceanographic purposes. What is needed in oceanography is accurate measurement of wind gradients rather than pure wind strength, which places much more stringent quality requirements on the individual data. However, significant progress has been made, and will be made over the next decade.

We include for completeness information on the distribution of air pressure at sea level. Air pressure variations affect the ocean only indirectly, through the associated wind systems, and oceanographers are not usually concerned with air pressure maps. Meteorologists need air pressure maps as a basic tool of their trade; but they prepare them for their own purposes. In contrast to physical oceanography, where (with the exception of the northern Indian Ocean) a discussion of the oceanic circulation starts from a well-defined annual mean, meteorology rarely looks at the annual mean atmospheric circulation. This has to do with the low thermal capacity of air, which results in much larger seasonal changes in the atmosphere than in the ocean (see Chapter 18) and makes the annual mean a rather irrelevant quantity. Figure 1.3 therefore shows only the January and July situation. Nevertheless, it gives some useful and instructive information; a comparison with the corresponding wind fields of Figure 1.2 in later chapters will document that the same rules

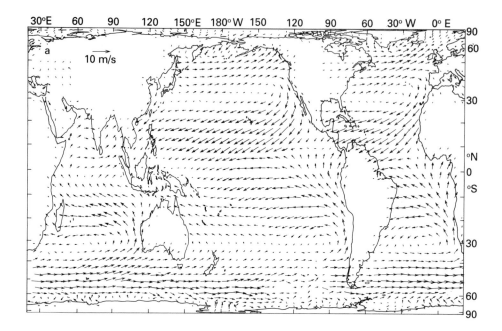

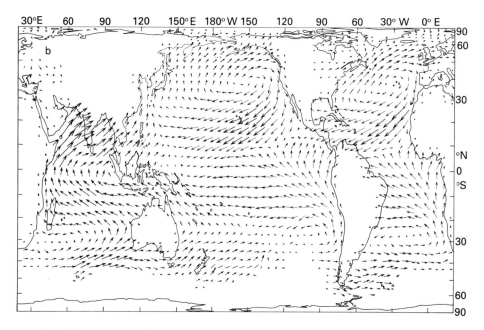

FIG. 1.2. Surface winds over the World Ocean. (a) Annual mean, (b) July mean. Data from Wright (1988). All means computed for the period 1950-1979.

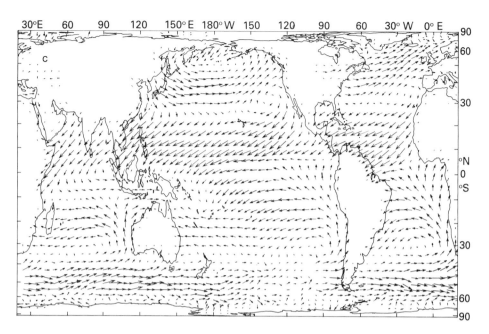

Fɪɢ. 1.2. Surface winds over the World Ocean. (c) January mean. Data from Wright (1988). All means computed for the period 1950-1979.

which will be derived for the ocean in chapters 3 - 5 operate in the atmosphere.

The modifications of the basic air pressure pattern of Figure 1.1 by the distribution of land and water are the outstanding features in the seasonal pressure maps. The zonal arrangement of high and low air pressure is seen most clearly in the southern hemisphere. Alternation between low pressure over continents and high pressure over the oceans during summer is particularly evident in the subtropics, but it is easy to see that the basic pattern is preserved in the zonal average. In the northern hemisphere the zonal distribution is disturbed by the Asian land mass which produces a summer low in northern Pakistan and an intense winter high over Mongolia and is responsible for the monsoon winds which dominate the Indian Ocean.

When the wind field is compared with the pressure field it is seen that the nearly zonal pressure distribution in the southern hemisphere produces strong and persistent Westerlies between 40° and 60°S. The remainder of the ocean is dominated by wind systems characterized by wind movement around centres of high and low pressure. During northern summer, for example, the Trade Wind and the Westerlies over the North Pacific Ocean are elements of a wind system in which air circulates around an atmospheric high in a clockwise manner. It is a general rule that air moves around atmospheric highs in a clockwise direction in the northern hemisphere and in an anti-clockwise direction in the southern hemisphere. Likewise, movement around atmospheric lows is anti-clockwise in the northern hemisphere, clockwise in the southern hemisphere. In meteorology and oceanography, circulation around a centre of low pressure is called cyclonic, circulation around centres of high pressure is called anti-cyclonic.

Winds drive ocean currents by releasing some of their momentum to the oceanic surface layer. The important quantity in this process is the wind stress, which is roughly a quadratic function of wind speed. Our knowledge of the wind stress distribution over the ocean is even less well established than our knowledge of the wind field. Most winds contain a considerable amount of turbulence, experienced as short gusts interspersed with periods of relative calm. Because of the quadratic relationship between wind speed and wind stress, gusty winds create larger stresses than would a steady wind of the same average speed. It is possible that our standard measuring equipment does not resolve all wind gusts adequately and that as a consequence our estimates of oceanic wind stress are too low. Direct measurement of the wind stress is difficult; it requires special equipment and has only been done in a small number of locations. The few direct observations were used to develop a formula useful for routine estimation of wind stress. The formula links the stress τ to routine merchant ship observations of wind speed, air and sea temperatures, wave state and other relevant quantities. τ is a vector with units of force per unit area (kg m^{-1} s^{-2}, or Newton per square metre) that points directly downwind; its magnitude is calculated from

$$| \tau | = C_d\, \rho_a\, U^2 \qquad\qquad (1.1)$$

where ρ_a is air density (about 1.2 kg m^{-3} at mid-latitudes), U is wind speed at 10 m above sea level, and C_d is the dimensionless "drag coefficient". (Here and in the following, we use bold characters to denote vectors and normal italics for scalars and constants.) Appropriate values for C_d are still the subject of active research, and the uncertainty about its value adds to the lack of precise knowledge about the wind stress distribution over the ocean. C_d varies from about 0.001 to 0.0025 depending on the air-sea temperature difference, the water roughness, and on the wind speed itself. A median value is about 0.0013. Figure 1.4 shows a recent representation of the oceanic wind stress field calculated from eqn (1.1) on the basis of merchant ship data and often used in numerical ocean models. Note that the mean wind stress is not necessarily parallel to the mean wind but is determined by the direction of the strongest winds. Around Antarctica, for example, mean winds are westerlies (Figure 1.2) but the mean wind stresses follow the northwesterly direction of the strong winds in the storm systems. In the northern hemisphere the gusty Westerlies produce larger stresses than the strong but less gusty Trades.

We conclude this chapter by briefly reviewing the atmospheric conditions imposed on the fluxes of heat and mass. Figure 1.5 gives annual mean solar radiation as received at the sea surface; 93% of it is absorbed by the ocean. Solar radiation is naturally largest in the tropics and in cloud-free regions.

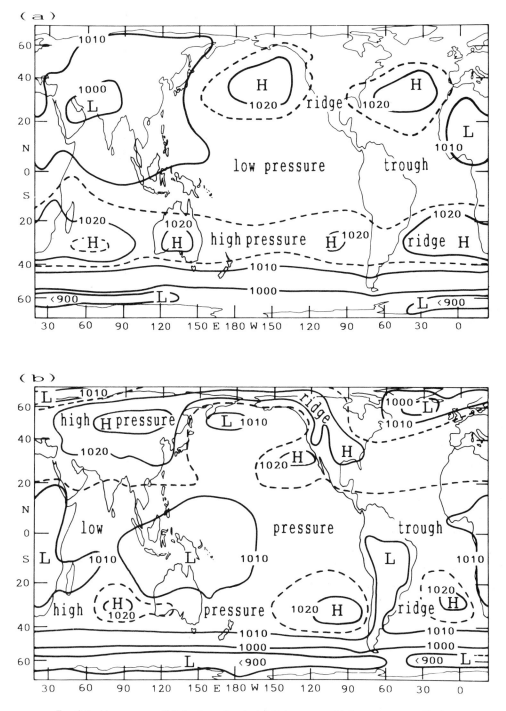

Fɪɢ. 1.3. Air pressure (hPa) at sea level. (a) July mean, (b) January mean. Broken lines show isobars at 5 hPa separation. Adapted from U.S. Navy (1981).

Again, the observed field shows significant departures from a simple zonal pattern as a result of the distribution of land and water, mainly through the effect this distribution has on the distribution of atmospheric water vapour and clouds. As an example, the wind convergence between the northern and southern hemispheres' Trades, known as the Intertropical Convergence Zone or ITCZ, consistently shows strong cloud cover and high rainfall (this will be discussed in more detail in Chapter 18) and is thus characterized by a regional minimum of solar radiation.

The heat flux through the ocean surface is determined by the balance between four components - incoming solar radiation, outgoing back radiation, heat loss from evaporation, and mechanical heat transfer between the ocean and the atmosphere (for details see Dietrich *et al.*, 1980, or Pond and Pickard, 1983), and their sum can be positive or negative. Each of the four contributions are hard to estimate accurately, so their balance is not very accurately established. Nevertheless, the need for heat flux values as input for ocean models caused several researchers to draw world maps of the balance (Figure 1.6). Generally speaking, the ocean gains heat in the tropics (between 20°S and 20°N) and loses heat in the temperate and polar regions. Departures from this simple zonal distribution are, however, so large that this generalization becomes rather meaningless. Cool water must flow into the regions of net ocean heat gain, and the warmed water must flow away from these regions; this advection does not occur uniformly at all longitudes but in currents of limited longitudinal extent, e.g. along the coasts of Peru and Somalia. Similarly the large heat losses in the Kuroshio and Gulf Stream regions along the coasts of Japan and the eastern USA are caused by rapid poleward advection of warm water. These processes will be addressed in detail in the discussion of individual oceans.

The mass or freshwater flux, i.e. the transport of water between the ocean and the atmosphere, is controlled by the difference between rainfall and runoff from land on one hand and evaporation from the ocean surface on the other hand. (Evaporation from land need not be considered here, since it does not represent a gain or loss to the ocean.) Figure 1.7 shows a recent estimate of the annual mean distribution of precipitation minus evaporation (*P-E*). Maximum *P-E* values are found in the ITCZ (known to mariners as the Doldrums) where moist air rises to great hight, releasing its water vapour; values of 500 cm/year and more are observed east of Indonesia. Mean sea surface salinity, shown in Figure 2.5b, clearly reflects the mass flux field, most notably in the generally zonal arrangement of the isohalines: lowest salinities tend to occur in regions of maximum *P-E*, although the relationship is not that simple in detail. Again, modifications are generated by the distribution of land and water and by air and ocean currents. An obvious example is the effect of the limited communication between Indian Ocean and Red Sea waters which produces extreme surface salinities in the latter. Further discussion of these and other aspects is postponed to the appropriate chapters.

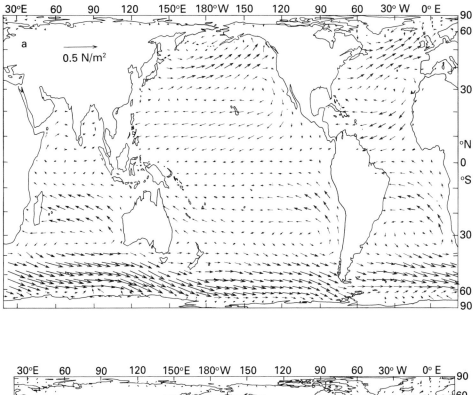

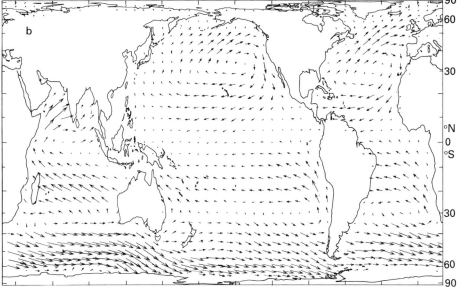

FIG. 1.4. Wind stress over the World Ocean. (a) Annual mean, (b) June - August.
All means computed for the period 1980-1986. Continued page 12.

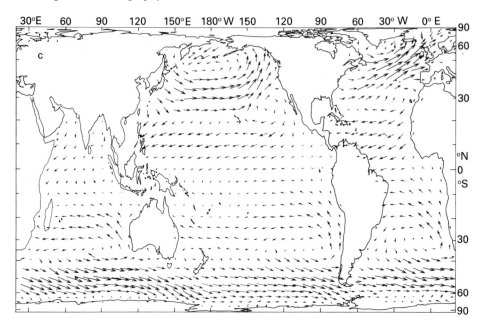

FIG. 1.4. Wind stress over the World Ocean. (c) December - February. Data from Trenberth *et al.* (1989). All means computed for the period 1980-1986.

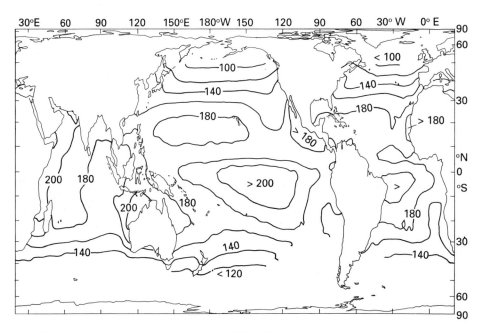

FIG. 1.5. Annual mean solar radiation (W m-2) received at sea level. Data from Oberhuber (1988). 200 W m-2 will warm a layer of water 50 m deep by about 2.5°C per month if unopposed by heat losses from other effects.

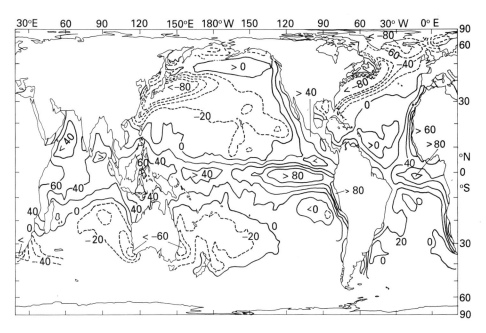

FIG. 1.6 Annual mean heat flux into the ocean (W m^{-2}). Data from Oberhuber (1988).

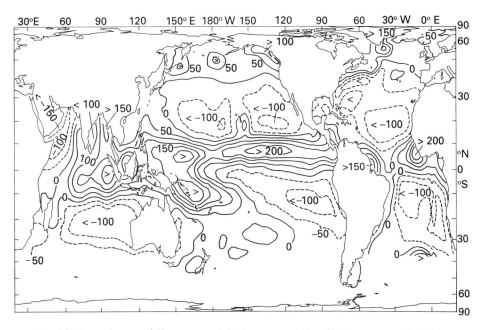

FIG. 1.7. Annual mean difference precipitation - evaporation (*P-E*, cm per year). Data from Oberhuber (1988). Positive values indicate freshwater gain. The quantity *E-P*, often seen in oceanography, is the negative of the quantity displayed here.

CHAPTER 2

Temperature, salinity, density and the oceanic pressure field

The ratios of the many components which make up the salt in the ocean are remarkably constant, and salinity, the total salt content of seawater, is a well-defined quantity. For a water sample of known temperature and pressure it can be determined by only one measurement, that of conductivity.

Today, the single most useful instrument for oceanographic measurements is the CTD, which stands for "Conductivity-Temperature-Depth". It is sometimes also known as the STD, which stands for "Salinity-Temperature-Depth"; but CTD is the more accurate description, because in both systems salinity is not directly measured but determined through a conductivity measurement. Even the term CTD is inaccurate, since depth is a distance, and a CTD does not measure its distance from the sea surface but employs a pressure measurement to indicate depth. But the three most important oceanographic parameters which form the basis of a regional description of the ocean are temperature, salinity, and pressure, which the CTD delivers.

In this text we follow oceanographic convention and express temperature T and potential temperature Θ in degrees Celsius (°C) and pressure p in kiloPascal (kPa, 10 kPa = 1 dbar, 0.1 kPa = 1 mbar; for most applications, pressure is proportional to depth, with 10 kPa equivalent to 1 m). Salinity S is taken to be evaluated on the Practical Salinity Scale (even when data are taken from the older literature) and therefore carries no units. Density ρ is expressed in kg m^{-3} or represented by $\sigma_t = \rho - 1000$. As is common oceanographic practice, σ_t does not carry units (although strictly speaking it should be expressed in kg m^{-3} as well). Readers not familiar with these concepts should consult textbooks such as Pickard and Emery (1990), Pond and Pickard (1983), or Gill (1982); the last two include information on the Practical Salinity Scale and the Equation of State of Seawater which gives density as a function of temperature, salinity, and pressure. We use z for depth (z being the vertical coordinate in a Cartesian xyz coordinate system with x pointing east and y pointing north) and count z positive downward from the undisturbed sea surface $z = 0$.

A CTD typically returns temperature to 0.003°C, salinity to 0.003, and depth to an accuracy of 1 - 2 m. Depth resolution can be much better, and advanced

CTD systems, which produce data triplets at rates of 20 Hz or more and apply data averaging, give very accurate pictures of the structure of the ocean along a vertical line. The basic CTD data set, called a CTD station or cast, consists of continuous profiles of temperature and salinity against depth (Figure 2.1 shows an example). The task of an oceanographic cruise for the purpose of regional oceanography is to obtain sufficient CTD stations over the region of interest to enable the researcher to develop a three-dimensional picture of these parameters and their variations in time. As we shall see later, such a data set generally gives a useful picture of the velocity field as well.

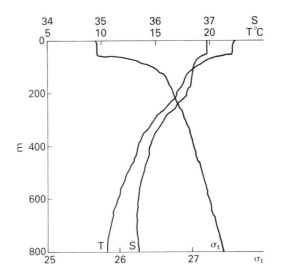

FIG. 2.1. An example of the basic CTD data set. Temperature T and salinity S are shown against pressure converted to depth. Also shown is the derived quantity σ_t.

For a description of the world ocean it is necessary to combine observations from many such cruises, which is only possible if all oceanographic institutions calibrate their instrumentation against the same standard. The electrical sensors employed in CTD systems do not have the long-term stability required for this task and have to be routinely calibrated against measurements obtained with precision reversing thermometers and with salinometers, which compare water samples from CTD stations with a seawater standard of known salinity (for details see Dietrich *et al.*, 1980). A CTD is therefore usually housed inside a frame, with 12 or more bottles around it (Figure 2.2). The water samples collected in the bottles are used for calibration of the CTD sensors. In addition, oxygen and nutrient content of the water can be determined from the samples in the vessel's laboratory.

The CTD developed from a prototype built in Australia in the 1950s and has been a major tool of oceanography at the large research institutions since the 1970s. Two decades are not enough to explore the world ocean fully, and regional oceanography still has to rely on much information gathered through bottle casts, which produce 12 - 24 samples over the entire observation depth and therefore are of much lower vertical resolution. Although bottle data have been collected for nearly 100 years now, significant data gaps still exist, as is evident from the distribution of oceanographic stations shown in Figure 2.3. In the deep basins of the oceans, where variations of temperature and salinity are small, very high data accuracy is required to allow integration of data from different cruises into a single data set. Many cruise data which are quite adequate for an oceanographic study of regional importance turn out to be inadequate for inclusion in a world data set.

Fig. 2.2. A CTD is retrieved after completion of a station. The instrument is mounted in the lower centre, protected by a metal cage to prevent damage in rough weather. Above the CTD are 12 sampling bottles for the collection of water samples. The white plastic frames attached to some of them carry precision reversing thermometers.

To close existing gaps and monitor long-term changes in regions of adequate data coverage, a major experiment, planned for the decade 1990 - 2000, is under way. This World Ocean Circulation Experiment (WOCE) will cover the world ocean with a network of CTD stations, extending from the surface to the ocean floor and including chemical measurements. Figure 2.4 shows the planned global network of cruise tracks along which CTD stations will be made at intervals of 30 nautical miles (half a degree of latitude, or about 55 km). As a result, we can expect to have a very accurate global picture

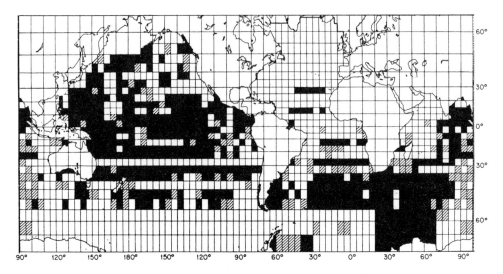

Fig. 2.3. World wide distribution of oceanographic stations of high data quality shortly before 1980. Unshaded 5° squares contain at least one high-quality deep station. Shaded 5° squares contain at least one high-quality station in a shallow area. Black 5° squares contain no high-quality station. FromWorthington (1981)

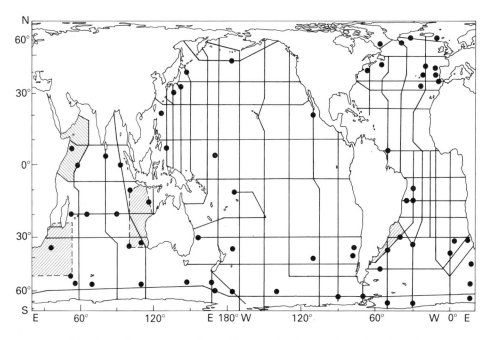

Fig. 2.4. The hydrographic sections of the World Ocean Circulation Experiment (WOCE). Shaded regions indicate intensive study areas. Dots indicate positions of current meter moorings.

of the distribution of the major oceanographic parameters by the turn of the century.

Because of the need for a global description of the oceanic parameter fields, researchers have attempted to extract whatever information they can from the existing data base. Figure 2.5 is an example of a recent and widely used attempt. It includes all available oceanographic data regardless of absolute accuracy and shows that many features of the oceans can be studied without the very high data accuracy required for the analysis of the deep basins. Features such as the large pool of very warm surface water in the equatorial western Pacific and eastern Indian Oceans, the outflow of high salinity water from the Eurafrican Mediterranean Sea into the Atlantic Ocean below 1000 m depth, the formation of cold bottom water in the Weddell and Ross Seas near Antarctica, or the outflow of low salinity water from the Indonesian seas into the Indian Ocean, are all clearly visible in the existing data base. However, it should be remembered that the number of observations available for every 2° square varies considerably over the area and decreases quickly with depth; in the polar regions it is also biased towards summer observations. Detailed interpretation of these and similar maps always has to take into account the actual data distribution.

Of itself, such information is only mildly interesting; but it is surprising what can be deduced from it. These deductions go into much more detail and reach much further than the examples just listed, which follow from simple qualitative arguments about the shape of isotherms or isohalines. More detailed analysis is based on the fact that most ocean currents can be adequately described if the oceanic pressure field is known (just as the atmospheric wind field follows from the air pressure distribution). Pressure at a point in the ocean is determined by the weight of the water above, which depends on the depth of the point and on the density of the water above it. As already noted, seawater density is a function of temperature, salinity and pressure. It is therefore possible - subject to some assumptions - to deduce the pressure distribution in the ocean and thus the current field from observations of temperature and salinity. How this is done is reviewed in the remainder of this chapter.

The first step in an accurate calculation of the oceanic pressure field is the calculation of density ρ from the Equation of State

$$\rho = \rho\,(\,T,\,S,\,p\,)\ . \tag{2.1}$$

Much care has gone into the laboratory measurements of density as a function of temperature T, salinity S and pressure p, and the Equation of State of Seawater now allows the calculation of density to a fractional accuracy of $3 \cdot 10^{-5}$ (0.03 kg m^{-3}) (Unesco, 1981; Millero and Poisson, 1981). We are now able to construct the density field to an accuracy comparable with the best

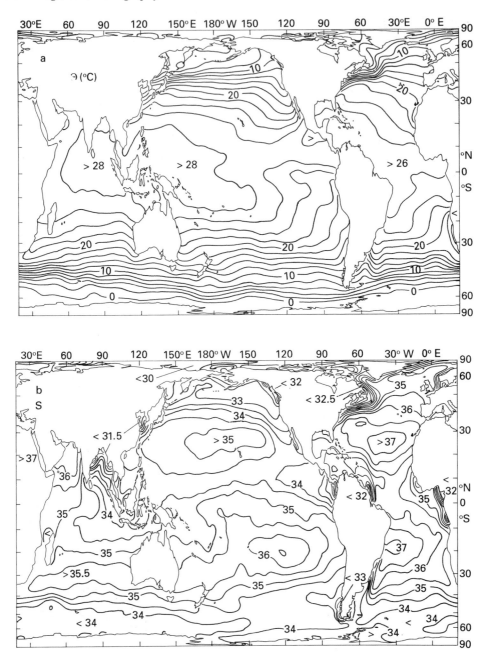

FIG. 2.5. Climatological mean potential temperature Θ (°C) and salinity S for the world ocean. (a) Θ at $z = 0$ m, (b) S at $z = 0$ m. From *Levitus* (1982). The maps were constructed from mean values calculated from all available data for "2° squares", i.e. elements of 2° longitude by 2° latitude, and smoothed over an area of approximately 700 km diameter. Continued on page 21.

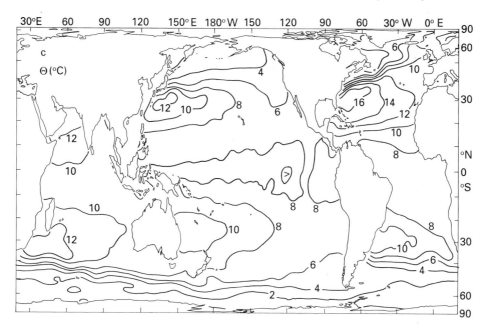

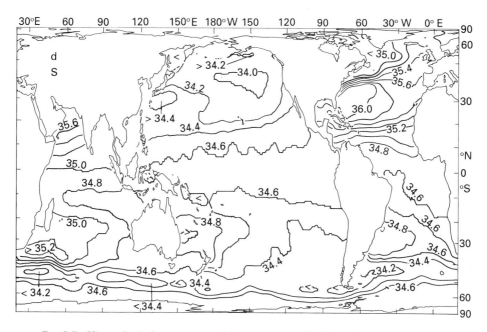

FIG. 2.5. Climatological mean potential temperature Θ (°C) and salinity S for the world ocean. (c) Θ at $z = 500$ m, (d) S at $z = 500$ m. From *Levitus* (1982). The maps were constructed from mean values calculated from all available data for "2° squares", i.e. elements of 2° longitude by 2° latitude, and smoothed over an area of approximately 700 km diameter. Temperatures below 1°C are not plotted.

Continued on page 22.

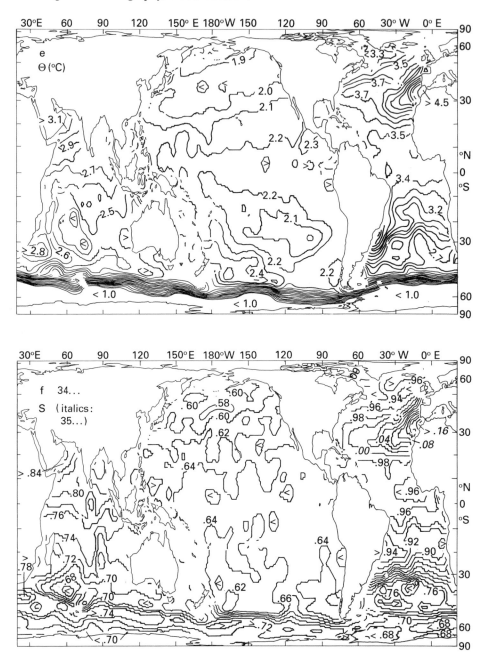

Fig. 2.5. Climatological mean potential temperature Θ (°C) and salinity S for the world ocean. (e) Θ at $z = 2000$ m, (f) S at $z = 2000$ m. From *Levitus* (1982). The maps were constructed from mean values calculated from all available data for "2° squares", i.e. elements of 2° longitude by 2° latitude, and smoothed over an area of approximately 700 km diameter. Temperatures below 1°C are not plotted. The lowest values reached at the 2000 m level are around 0.0°C in the Antarctic and near -0.9°C in the Arctic region.

field measurements of T, S and depth (or pressure p). The lowest accuracy is actually in the determination of depth since the pressure sensor is usually accurate to within 0.5 - 1% of full range, i.e. to 5 - 10 m if the sensor range is 1000 m. However, the oceanic pressure field can be determined with much higher accuracy from the distribution of density, as will be seen in a moment.

The quality of T and S measurements and of the presently-used equation of state can be checked by examining data collected in the high pressures found at great depth, where our measurement techniques are given their severest test. Figure 2.6 shows measurements from the vicinity of the deepest known place in the ocean. The measured temperature T increases with depth over the last six kilometres, while salinity S varies little. In a constant pressure environment this would indicate an apparent static instability, i.e. errors in the Equation of State. However, the same laboratory experiments that gave us the Equation of State allow us to determine very accurately the temperature drop that would occur if a parcel of water were brought to the surface without heat exchange. It would cool on decompression, by approximately 0.035°C per 1 km, and attain its "potential temperature" Θ. Figure 2.6 shows both T and Θ. It is seen that Θ is constant within measurement error. The "potential density" (the density the water would have if it were brought to the surface without changing salinity and potential temperature) is thus constant within measurement error, too. The available oceanic observations of today show that it is extremely rare to find inversions of potential density, i.e. situations where denser water appears to be lying on top of lighter water. Because in reality such inversions are unstable and overturn very quickly,

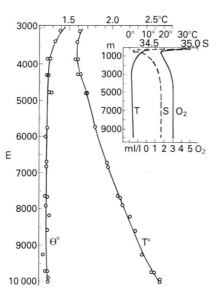

FIG. 2.6. Temperature T and potential temperature Θ in the Philippines Trench. The inset also shows salinity S and oxygen O_2. From Bruun *et al.* (1956)

their absence in observational data implies that eqn (2.1) obtained from laboratory measurements is in fact valid in the ocean.

Precise knowledge of the density field is the basis for the second step, accurate calculation of the pressure field $p(z)$ from the hydrostatic relation

$$\partial p / \partial z = g \rho \, , \qquad (2.2)$$

where g is gravity, $g = 9.8$ m s^{-2}, and depth z increases downwards. This equation is not uniformly valid (it does not hold for wind waves, for example); but it can be shown that it holds very accurately, to the accuracy of eqn (2.1), if it is applied to situations of sufficiently large space and time scales. It forms the basis of regional oceanography.

Evaluation of the pressure field from the hydrostatic equation involves a vertical integration of density. The advantage of an integration is that it eliminates the uncertainty in the measurement of depth as a source for inaccuracy. Its disadvantage is that it requires a reference pressure as a starting point. Without that information, eqn (2.2) can be used to get *differences* between pressures at different depths. An alternative way, which is common practice in oceanography, is to determine the distance, or depth difference, between two surfaces of constant pressure. For this purpose, a quantity called steric height h is introduced and defined as

$$h (z_1 , z_2) = \int_{z_1}^{z_2} \delta \; (T, \; S, \; p) \; \rho_0 dz \qquad (2.3)$$

where ρ_0 is a reference density, and

$$\delta \; (T, \; S, \; p) = \rho \; (T, \; S \; p)^{-1} - \rho \; (0, \; 35, \; p)^{-1} \qquad (2.4)$$

is called the specific volume anomaly; it is the difference in volume between a unit mass of water at temperature T and salinity S and a unit mass at the standard salinity $S = 35.0$ and temperature $T = 0°C$. Steric height h has the dimension of height and is expressed in metres. To sufficient approximation,

$$\delta \; (T, \; S, \; p) = \frac{\rho_0 - \rho \; (T, \; S, \; p)}{\rho_0^2} \qquad (2.4a)$$

so that eqn (2.3) can also be written

$$h \; (z_1 , z_2) = \int_{z_1}^{z_2} \Delta\rho \; (T, \; S, \; p) \; / \; \rho_0(p) \; dz \qquad (2.3a)$$

where $\rho_0(p) = \rho \; (0, \; 35, \; p)$ and $\Delta\rho \; (T, \; S, \; p) = \rho_0 - \rho \; (T, \; S, \; p)$. $h(z_1, z_2)$ measures the height by which a column of water between depths z_1 and z_2 with standard temperature $T = 0°C$ and salinity $S = 35.0$ expands if its temperature and salinity are changed to the observed values. Typically, h is a few tens of centimetres. Because the weight of the water is not changed during expansion, the pressure difference between top and bottom remains the same. It is seen then that $h(z_1, z_2)$ measures variations in the vertical distance between two surfaces of constant pressure.

For oceanographic purposes, the sea surface can always be regarded as an isobaric surface. As any meteorological air pressure map (such as Figure 1.3) tells us, this is only an approximation. In reality, pressure differences between atmospheric highs and lows are of the order of 2 - 3 kPa. However, the ocean reacts to these differences by expanding in regions of low atmospheric pressure and contracting under atmospheric highs. These vertical movements are of the order of 0.2 - 0.3 m; like the tides, they have little effect on the long-term flow field and can be disregarded in our discussion, which is concerned with water movement induced by the oceanic pressure field.

Unfortunately, the sea surface cannot be used as the reference surface for the integration of eqns (2.3) or (2.3a) because it is not necessarily flat, and its exact shape is not known. We may, however, assume for the moment that a constant pressure surface which does not vary in depth can be found somewhere in the ocean and call it, for reasons which will become clear in a moment, a "depth of no motion". It then becomes possible using eqn (2.3) to map the oceanic pressure distribution by measuring, for every isobaric surface, its distance from the position z_0 of this depth of no motion. A possible pressure distribution is sketched in Figure 2.7. Because the weight of the water above the depth of no motion has to be the same everywhere, the sea surface, which is given by the steric height $h(0, z_0)$, is lower above the region of high water density than above the region with low water density. It is seen that a two-dimensional representation of the situation on a horizontal map is possible in two ways. We can either select a constant depth surface $z = z_r$ and map the intersections of the isobars with the depth surface, or we can select a constant pressure surface p_1 and draw contours of constant steric height. The first method is well known from meteorology; daily weather forecasts are based on maps of isobars at sea level (considered flat for the purpose of meteorology). In oceanography the position of the sea surface is unknown and has to be determined by analysis. Oceanographers therefore map the shape of the sea surface by showing contours of equal steric height relative to a depth of no motion, where pressure is assumed to be constant.

It is easy to show - subject to our assumption of a depth of no motion - that at any depth level, contours of steric height coincide with contours of constant pressure. The hydrostatic relation (eqn 2.2) tells us that if pressure is constant at $z = z_0$, the quantity $\rho_0 g h (z, z_0)$ measures the pressure variations along a surface of constant height z. Thus, a contour map of h is an isobar map scaled by the factor $\rho_0 g$ (see Figure 2.7).

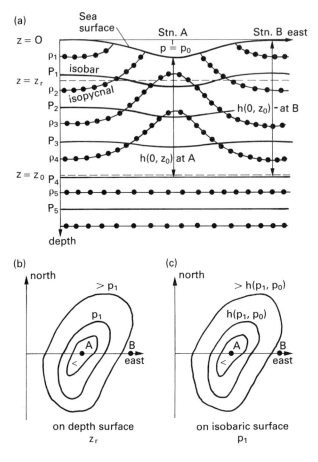

Fig. 2.7. Schematic illustration of steric height as a measure of distance between isobaric surfaces, and of the relationship between maps of isobars at constant height and maps of steric height at constant pressure. (a) Distribution of isobars and isopycnals: at any depth level above $z = z_o$, water at station A is denser than water at station B. As the weight of the water above $z = z_o$ is the same, the water column must be longer at B than at A. The steric height of the sea surface relative to $z = z_o$ is given by $h(p_o, p_4)$, which in oceanographic applications is often given as $h(0, z_o)$, i.e. with reference to depth rather than pressure. The difference is negligible. (b) The corresponding pressure map at constant depth $z = z_r$. (c) The corresponding map of steric height at constant pressure $p = p_1$. The diagram requires some study, but it is well worth it; understanding these principles is the basis for the interpretation of many features found in the ocenic circulation.

A quantity widely found in oceanographic literature is the dynamic height D. It is defined as

$$D\ (p_1,\ p_2)\ =\int_{p_1}^{p_2}\delta\ (T,\ S,\ p)\ dp \tag{2.5}$$

and is equal to gh, i.e. the product of gravity and steric height. Maps of dynamic height, also known as dynamic topography maps, are therefore maps of steric height scaled by the factor g or pressure maps scaled by the factor ρ_0. We prefer the use of steric height because it has the unit of length and therefore can be directly interpreted in terms of, for example, the shape of the sea surface. Other representations do, of course, just as well, as long as we remember that the "dynamic metre" often given as the unit for D is not a length but corresponds to $m^2\ s^{-2}$.

Is it possible to find a flat pressure surface in the ocean, i.e. one where the horizontal pressure gradient vanishes? One consequence of zero horizontal pressure gradient would be the absence of a current at that depth - hence the name "depth of no motion". Observations support the idea that in the deep ocean flow might, indeed, be so slow that the deep pressure map can be treated as flat. They show that below about 1300 m temperature and salinity are rather uniform, at least within a given basin. This comes out clearly in the maps of Figure 2.5, which would not show any structure at 2000 m depth outside the Southern Ocean if the relatively coarse contour interval of the 500 m maps were applied here. Furthermore the T and S gradients contribute roughly equal and opposite amounts to the density, so that density is remarkably uniform at such depths (even in the Southern Ocean). Within a basin, the density field is so horizontally uniform that steric height at 1500 m relative to 2000 m, which is shown in Figure 2.8a, displays horizontal variations of only a centimetre or so, within a given basin - and those variations look

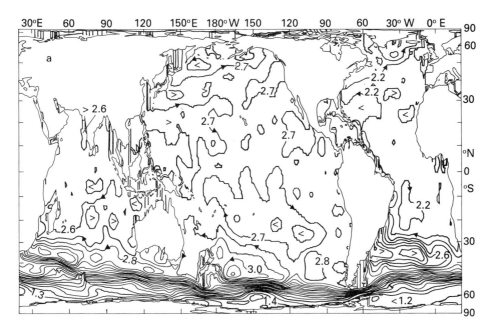

Fig. 2.8. Dynamic height ($m^2\ s^{-2}$), or steric height multiplied by gravity, for the world ocean. (a) at 1500 m relative to 2000 m. Arrows indicate the direction of the implied movement of water, as explained in Chapter 3. (Divide contour values by 10 to obtain approximate steric height in m.) From *Levitus* (1982). Continued page 28.

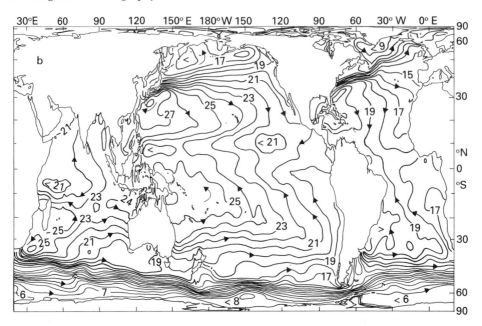

FIG. 2.8. Dynamic height (m² s²), or steric height multiplied by gravity, for the world ocean. (b) at 0 m relative to 2000 m. Arrows indicate the direction of the implied movement of water, as explained in Chapter 3. (Divide contour values by 10 to obtain approximate steric height in m.) From *Levitus* (1982).

so random that they can be just as much a result of noise in the small data base as a real effect. By contrast, the steric height map for the sea surface relative to 2000 m (Figure 2.8b) shows differences of 0.5 m in a single basin and 1.8 m or so from highest to lowest point in the ocean, because the horizontal gradients of density are much greater (by a factor of several hundred) near the surface than at depths of 1500 - 2000 m.

These facts do not prove that the ocean is moving relatively slowly at depths of 1500 m or 2000 m; all they show is that if flow is slow at the one depth, it is slow at the other. However, in most parts of the ocean similar remarks apply for all pairs of depths (z_1, z_2) when both lie at or below about 1500 m, so the observations show that if there is any strong motion at these depths, it must take the form of a nearly vertically uniform flow everywhere below 1500 m. Because of the ocean's rough bottom it seems unlikely that such motions can be very strong or extend over great distances. Unfortunately, direct measurements of deep ocean currents are much harder to make than measurements of temperature and salinity; but in most parts of the ocean, such measurements as have been made support the idea that flow at these depths is very slow.

CHAPTER 3

The Coriolis force, geostrophy, Rossby waves and the westward intensification

The oceanic circulation is the result of a certain balance of forces. Geophysical Fluid Dynamics shows that a very good description of this balance is achieved if the oceans are subdivided into dynamical regions as sketched in Figure 3.1. We note that frictional forces are only important in the vicinity of ocean boundaries; in the vast expanse of the ocean interior below the surface layer they are negligible in comparison to the force set up by the pressure gradient. We know from Chapter 2 how to calculate the pressure field - subject to our choice of the depth of no motion - and should therefore be able to determine the flow in the largest of the dynamic regions.

The pressure gradient force cannot be the only acting force; otherwise the water would accelerate towards the centres of low pressure, as air movement on a non-rotating earth in Figure 1.1. Eventually, bottom friction would limit the growth of the velocity, and the circulation would become steady. Flow in the ocean interior is generally sluggish, and the friction force is no match for the pressure gradient force produced by variations in the density field. However, the pressure gradient force is not unopposed; it is balanced by a force which is known as the

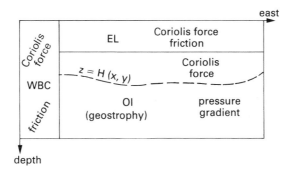

Fig. 3.1. An east-west section through an idealized ocean basin away from the equator, showing the subdivision into three dynamic regions, the ocean interior (*OI*), the surface boundary or Ekman layer (*EL*), and the western boundary current region (*WBC*). In each region the Coriolis force is balanced by a different force. The $1^1/_2$ layer model discussed in the text divides the ocean interior further, into a dynamically active layer above the interface $z = H(x,y)$ and a layer with no motion below. The relative sizes of the various regions are not to scale.

Coriolis force and is the result of the earth's rotation. It is an apparent force, that is, it is only apparent to an observer in a rotating frame of reference. To see this, consider a person standing on a merry-go-round, facing a ball thrown by a person from outside. To follow the ball the person would have to turn and therefore conclude that a force must be acting on the ball to deflect it from the shortest (straight) path. The person throwing the ball sees it follow a straight path and thus does not notice the force, and indeed the force does not exist for any person not on the merry-go-round.

In oceanography currents are always expressed relative to the ocean floor - which rotates with the earth - and can therefore only be described correctly if the Coriolis force is taken into account in the balance of forces. The Coriolis force is proportional in magnitude to the flow speed and directed perpendicular to the direction of the flow. It acts to the left of the flow in the southern hemisphere, and to the right in the northern hemisphere. A somewhat inaccurate but helpful way to see why the direction is different in the two hemispheres is related to the principle of conservation of angular momentum.

A water particle at rest at the equator carries angular momentum from the earth's rotation. When it is moved poleward it retains its angular momentum while its distance from the earth's axis is reduced. To conserve angular momentum it has to increase its rotation around the axis, just as ballet dancers increase their rate of rotation when pulling their arms towards their bodies (bringing them closer to their axis of rotation). The particle therefore starts spinning faster then the earth below it, i.e. it starts moving eastward. This results in a deflection from a straight path towards right in the northern hemisphere and towards left in the southern hemisphere. Likewise, a particle moving toward the equator from higher latitudes increases its distance from the axis of rotation and falls back in the rotation relative to the earth underneath; it starts moving westward, or again to the right in the northern and to the left in the southern hemisphere.

More on the Coriolis force can be found in Pond and Pickard (1983) or other text books. Neumann and Pierson (1966) give a detailed derivation based on Newton's Law of Motion.

The balance between the Coriolis force and the pressure gradient force is called *geostrophic balance,* and the corresponding flow is known as geostrophic flow. Compared to movement on a non-rotating earth, where the flow crosses isobars from high to low pressure, geostrophic flow is characterized by movement *along* isobars. We can see an example of geostrophic motion in the atmosphere if we recall the relation between air pressure (Figure 1.3) and wind (Figure 1.2). As already noted in Chapter 1, the wind direction nearly coincides with the orientation of the isobars. In the upper atmosphere - the analogon to the ocean interior - winds are strictly geostrophic. Winds at sea level are affected by bottom friction and therefore blow at a small angle to the isobars.

A useful quantity for the description of the oceanic circulation is mass transport. The basic definition is

$$M^* = \rho v \qquad (3.1)$$

Here, M^* is the mass transport through an area of unit width (1 m^2) perpendicular to the direction of the flow and v the velocity vector with components (u,v,w) along the (x,y,z) axes. M^* is therefore a vector which points in the same direction as velocity and has units of mass per unit area and unit time, or kg m^{-2} s^{-1}. More commonly, mass transport refers to the total transport of mass in a current, *i.e.* integrated over the width and the depth of the current. It then has dimensions of mass per unit time, or kg s^{-1}. It is also possible to define the mass transport in a layer of water between depths z_1 and z_2:

$$M = \int_{z_1}^{z_2} \rho v \, dz \qquad (3.2)$$

Thus, M represents the transport between depths z_1 and z_2 per unit width (1 m) perpendicular to the flow and has units of mass per unit width and unit time (kg m^{-1} s^{-1}). Likewise, the transport per unit depth (1 m) between two stations A and B is given by

$$M' = \int_A^B \rho v_n \, dl \qquad (3.3)$$

where v_n now is the velocity normal to the line between A and B, and M' is the transport in the direction of v_n, with units of mass per unit depth and unit time (again kg m^{-1} s^{-1}). Unfortunately oceanographers refer to all three quantities (3.1), (3.2), and (3.3) as mass transport, and care has to be taken to verify which quantity is used in any particular study.

A quantity often found in oceanography is volume transport, defined as mass transport integrated over the width and depth of a current, divided by density. It has the unit m^3 s^{-1}. More commonly used is the unit Sverdrup (Sv), defined as 1 Sv = 10^6 m^3 s^{-1}.

The qualitative properties of geostrophic flow can be summarized as

> **Rule 1**: *In geostrophic flow, water moves along isobars, with the higher pressure on its left in the Southern Hemisphere and to its right in the Northern Hemisphere. In the ocean interior away from the equator, the flow of water is geostrophic.*

The magnitude of geostrophic flow, expressed as mass transport per unit depth between two points A and B, is given to considerable accuracy over most of the ocean by:

$$M' = \frac{\rho_0 \, g \, T_d \, \Delta h}{4\pi \, \sin \phi} = \frac{\rho_0 \, g \, \Delta h}{f} \qquad (3.4)$$

where ρ_0 is an average water density; g is the acceleration of gravity, $g = 9.8$ m s^{-2}; T_d is the length of a day $= 86,400$ s; ϕ is the latitude; and Δh is the difference in steric height between two adjacent isobars. $f = (T_d / 4\pi \sin\phi)^{-1}$ is known as the Coriolis parameter; it has the dimension of frequency and is positive north and negative south of the equator. Figure 3.2 is an illustration of Rule 1; it also demonstrates how the transport per unit depth between two streamlines can be evaluated.

Again, we can verify our rule by looking at the atmosphere (Figures 1.2 and 1.3). Whether the circulation is cyclonic or anticyclonic, high air pressure is always on the left of the wind direction in the Southern Hemisphere and to the right

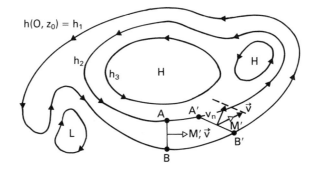

FIG. 3.2: Illustration of the relationship between a map of steric height (dynamic topography), geostrophic flow, and the evaluation of the geostrophic mass transport per unit depth M' between two streamlines (contours of constant steric height). For both station pairs, A and B and A' and B', Δh in eqn (3.4) is given by $h_2 - h_1$. The geostrophic velocity is inversely proportional to the distance between streamlines, or equal to M' divided by density and by the distance between points A and B, because the section AB is perpendicular to the streamlines. If station pair A' and B' is used for the calculation, eqn. 3.4 still produces the correct geostrophic mass transport M' between streamlines h_1 and h_2, but the velocity derived from M' and distance $A'B'$ is only the velocity component v_n perpendicular to the section $A'B'$.

in the Northern Hemisphere. Meteorologists refer to this rule as Buys-Ballot's Law. We note at this stage that the equatorial region has to be considered separately, because the Coriolis parameter vanishes at the equator and another force is needed to balance the pressure gradient force.

Because f varies with latitude, the dependence of M' on f gives rise to waves of very large wavelength known as *Rossby waves*. To understand the mechanism of these waves it is useful to introduce an approximation to the ocean's density structure known as the "$1^1/_2$ layer ocean". In such a model the ocean is divided into a deep layer of contant density ρ_2 and a much shallower layer above it, again of constant density $\rho_1 = \rho_2 - \Delta\rho$. The lower layer is considered motionless on account of its large vertical extent. The thickness of the upper layer or interface $z = H (x,y,t)$ is allowed to vary. In the real ocean a fairly sharp density interface exists outside the polar regions, characterized by a rapid temperature change from near 20°C to below 10°C (the permanent thermocline or pycnocline, see

Figure 6.1). The $1^1/_2$ layer ocean is a somewhat crude approximation of that situation but can describe flow above the pycnocline quite well.

The driving force for geostrophic currents are the horizontal differences in pressure or steric height, so we can add any arbitrary constant to the steric height field without affecting the currents deduced from it. In our $1^1/_2$ layer model, we choose a depth of no motion z_{nm} at an arbitrary depth that lies entirely in the lower layer (see Figure 3.3) and evaluate steric height $h(x,y)$ by integrating from z_{nm} upwards. Being the distance between isobaric surfaces, $h(x,y)$ is then independent of x and y in the lower layer. We conclude from eqn (3.4) that there is no geostrophic flow in the lower layer; this is consistent with the idea that the lower layer is at rest.

When the integral of (2.3a) is carried to some depth z_1 in the upper layer the result is

$$h(x, y, z_1) = \frac{(z_{nm} - z_1)\ (\rho_0 - \rho_2) - (H(x, y) - z_1)\ \Delta\rho}{\rho_0} \tag{3.5}$$

This does vary with position, because the interface depth $H(x,y)$ varies. The constant term $((z_{nm} - z_1)\ (\rho_2 - \rho_0) - \Delta\rho\ z_1)\ /\ \rho_0$ does not affect horizontal differences and can be dropped; so the steric height is then just

$$h(x,y) = -\Delta\rho/\rho_0\ H(x, y) \tag{3.6}$$

Notice that horizontal gradients of steric height are independent of depth in the upper layer, so geostrophic flow in the upper layer will be independent of depth as well (see Figure 3.3).

The factor $\Delta\rho/\rho_0$ is of the order 0.01 or less; so $H(x,y)$ has to be much larger than $h(x,y)$. The negative sign in eqn (3.6) indicates that $H(x,y)$ slopes upward where $h(x,y)$ slopes downward and *vice versa*. At the surface, $h(x,y)$ measures the surface elevation needed to maintain constant weight of water, above every point on a depth level in the lower layer (see Figure 3.3). It follows that in a $1^1/_2$ layer ocean the sea surface is a scaled mirror image of the interface. This result is of sufficient importance to formulate it as

> **Rule 1a:** *In most ocean regions (where the $1^1/_2$ layer model is a good approximation) the thermocline slopes opposite to the sea surface, and at an angle usually 100 - 300 times larger than the sea surface.*

This is an important rule, because in contrast to the slope of the sea surface the slope of the thermocline can be seen in measurements made aboard a ship. This allows oceanographers to get a qualitative idea of currents from inspection of temperature and salinity data. If we now remember that Rule 1 links the direction of geostrophic flow with the sea surface slope we find that Rule 1a links the direction of geostrophic flow with the slope of the thermocline - a result we formulate as

> **Rule 2:** *In a hydrographic section across a current, looking in the direction of flow, in most ocean regions (where the $1^1/_2$ layer model is a good approximation) the thermocline slopes upward to the right of the current in the southern hemisphere, downward to the right in the northern hemisphere.*

This simple rule allows a very easy check on the current direction from hydrographic measurements; readers may want to verify it on Figure 3.3. While Rule 1 expresses the properties of geostrophy and is thus valid wherever geostrophy holds, Rules 1a and 2 are based on the $1^1/_2$-layer model and therefore not as widely applicable. A more complete derivation based only on the assumption of geostrophy, which contains the $1^1/_2$-layer ocean as a special case but applies to the continuously stratified ocean as well, leads to

> **Rule 2a:** *If in the southern (northern) hemisphere isopycnals slope upward to the right (left) across a current when looking in the direction of flow, current speed decreases with depth; if they slope downward, current speed increases with depth.*

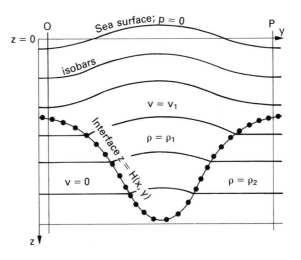

Fig. 3.3. Side view of a $1^1/_2$-layer ocean. The thermocline depth H is variable, but the density difference $\Delta\rho = \rho_2 - \rho_1$ between both layers is constant. Horizontal gradients of steric height $h(x,y)$ are independent of depth in the top layer, so by Rule 1 the currents are also independent of depth in this layer. $h(x,y)$ also measures the surface elevation, and is proportional to the thermocline depth $H(x,y)$ through eqn (3.6). The depth of no motion z_{nm} can be anywhere where it is located entirely in the lower layer.

In most oceanic situations, and in particular in the upper 1500 m of the ocean, the effect of the vertical salinity gradient on density is much smaller than the effect of the vertical temperature gradient, and the word "isopycnals" can be replaced by "isotherms". A vertical temperature section is then sufficient to get an idea of the direction of flow perpendicular to the section. In the atmosphere density is mostly determined by temperature alone, and application of Rule 2a gives a direct relationship between the structure of the temperature and the wind field. Rule 2a is therefore known as the *thermal wind relation*. The term has been adopted in oceanography; a current which is recognized by sloping isopycnals is sometimes called a "thermal wind".

For completeness we note without verification that Rule 2a is valid in western boundary currents and for zonal flow near the equator if the hydrographic section is taken perpendicular across the current axis.

Some consequences of Rule I: Rossby waves and western boundary currents

We consider a $1^{1}/_{2}$ -layer ocean - to define the sign of f, we take it in the southern hemisphere - with the bottom layer at rest. Suppose there is a large region in which the layer depth H is deeper than in surrounding regions, where both layers are at rest (*i.e.* H is constant there). Figure 3.4 shows a map of H for this situation. Appropriately scaled (by the factor $\Delta\rho/\rho_0$; see eqn (3.6) above) it is also a map of steric height, from which the flow at all depths in the upper layer can be deduced through eqn (3.4). It is seen that the feature represents a large anticyclonic eddy.

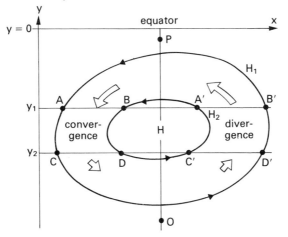

FIG. 3.4. Plan view of the eddy of Fig. 3.3. A and B are two points on the western side of the eddy at latitude y_1, on two isobars separated by an amount $\Delta h = \Delta\rho(H_1 - H_2)/\rho$ in steric height. C and D are two similar points at latitude $y_2 = y + \Delta y$. By eqn (3.4), total southward flow is greater in magnitude between A and B than between C and D because f is smaller in magnitude at A and B than at C and D; the thermocline deepens in $ABCD$. By the same argument, the thermocline shallows in $A'B'C'D'$: the eddy moves west.

Consider now the transport between two isobars corresponding to layer depths of H and $H + \Delta H$, at latitudes y_1 and y_2; ΔH is assumed to be small so the average depth of the layer is H. The total southward transport in the upper layer through the area between A and B is then $M_{tot} = H \cdot M'$ which, from eqns (3.4) and (3.6), is

$$M_{tot} = \frac{gH \, \Delta \rho \, \Delta H}{f(y_1)} = \frac{\rho_0 \, gH \, \Delta h}{f(y_1)} \qquad (3.7)$$

where $\Delta h = \Delta \rho \Delta H / \rho_0$ is the steric height difference between A and B, and $f(y_1)$ is the Coriolis parameter at the latitude y_1 of A and B. Similarly, at latitude y_2 between C and D:

$$M_{tot} = \frac{gH \, \Delta \rho \, \Delta H}{f(y_2)} = \frac{\rho_0 \, gH \, \Delta h}{f(y_2)} \qquad (3.8)$$

The Coriolis parameter varies with latitude, increasing in magnitude with distance from the equator: $| f(y_1) | < | f(y_2) |$. If the depression of the interface illustrated in Figure 3.4 covers a large enough region (typically some hundreds of kilometres across), the southward transport between the two isobars is smaller between C and D than between A and B. As the water which passes between A and B has to go somewhere, some of it is pushed downwards. We conclude that on the *western* side of the eddy the interface is pushed downwards. By contrast, a similar argument shows that on the *eastern* side of the eddy flow which passes between A' and B' is larger than flow between C' and D', and the interface is pulled upwards. The net effect is a westward movement of the interface depression and with it the eddy. The same derivation of westward movement can be made for cyclonic (shallow thermocline) eddies.

The westward movement of such "planetary eddies" is known as Rossby wave propagation. Rossby waves tend to carry energy from the ocean interior into the western boundary current region of Figure 3.1. The accumulation of energy in the west leads to an intensification of the currents on the western side of all oceans; examples are the East Australian Current, the Gulf Stream, or the Agulhas Current. Our Rule 1 becomes invalid in these narrow western boundary currents, where friction and nonlinear effects lead to dissipation of energy. Because the western boundary layer is only about 100 km wide, these currents follow the coast closely and are only poorly resolved by the climatological maps of Figure 2.8, which smooth all data over horizontal distances of about 700 km. It should also be observed that much of the flow in the western boundary layers occurs over the continental slope and shelf, where steric height relative to 2000 m cannot be defined. However, the intense outflows of the western boundary currents can be seen moving eastwards from the western edge of each ocean at 30 - 40° N or S.

We can readily estimate the speed of a Rossby wave. This is done most easily and to sufficient accuracy on the so-called β-plane, which approximates the Coriolis parameter by $f = f_0 + \beta y$, i.e. a function which varies linearly with latitude. Between the two latitudes y_1 and y_2, the Coriolis parameter then changes by an amount $\beta \Delta y$, where $\Delta y = y_1 - y_2$. For small Δy, the net mass convergence between the streamlines through A and B or C and D is, from eqns (3.7) and (3.8) and noting that f is negative in the southern hemisphere,

$$g\,H\,\Delta\rho\,\Delta H\; \frac{1}{f(y_1)} - \frac{1}{f(y_1)} = g\,H\,\Delta\rho\,\Delta H\; \frac{\beta\,\Delta y}{f^2(y_1)} \tag{3.9}$$

where ΔH is the difference in thermocline depth between A and B (or between C and D). This mass convergence will force the interface down over the area $\Delta x \Delta y$ defined by A, B, C, and D. For small Δx and Δy we find:

$$\rho_0\,\frac{H}{t} = \frac{gH\,\Delta\rho\,\Delta H\,\beta\,\Delta y}{f^2(y)\,\Delta x\,\Delta y)}\;,\;\text{or} \tag{3.10}$$

$$\frac{H}{t} = \frac{\beta\,g\,H}{f^2(y)}\,\frac{\Delta\rho}{\rho_0}\,\frac{H}{x}$$

The ratio $-(\,H/\,t)/(\,H/\,x)$ is the speed at which a line of constant H moves eastward. According to eqn (3.10) it has the constant value $-c_R(y)$ where

$$c_R(y) = \frac{\beta\,gH\,(\Delta\rho/\rho_0)}{f^2(y)} \tag{3.11}$$

is called the Rossby wave speed. The sign of eqn (3.10) says that geostrophic eddies move *westward* with this speed. Notice that $c_R(y)$ approaches infinity rapidly at the equator, where $f = 0$.

Taking typical values of $H = 300$ m, $\Delta\rho/\rho_0 = 3 \cdot 10^{-3}$, we find that $c_R(y)$ decreases from 1.27 m s^{-1} at 5°S or N to 0.08 m s^{-1} at 20°S or N and to 0.02 m s^{-1} at 40°S or N. At such speeds a Rossby wave would take about 6 months to cross the Pacific at 5° distance from the equator, but more like 20 years at 40°.

Rossby waves are a general phenomenon in planetary motion of fluids and gases and occur in the atmospheres of the earth and other planets as well. In the earth's atmosphere they are usually better known as atmospheric highs and lows and play a key role in determining the weather. They generally move eastward, carried by the fast-flowing Westerlies. Relative to the mean flow of air, however, their movement is westward, as it has to be according to our discussion. Current velocities in the ocean are much smaller than the Rossby wave speed at least near the equator, so oceanic Rossby wave movement in the tropics and subtropics is towards west.

If the ocean were purely geostrophic - i.e. if Rule 1 applied *exactly* outside the western boundary currents - then the depressions and bulges in thermocline depth seen for example in the 500 m map of Figure 2.5 or in Figure 2.8 would all migrate to the western boundary through Rossby wave propagation, and the ocean would come to a state of horizontally uniform stratification and no flow. Thus, there must be some process continually acting to replenish these bulges. What is this process, and how does it work? This is the subject of the next chapter.

CHAPTER 4

Ekman layer transports, Ekman pumping and the Sverdrup balance

Geostrophic flow, or the flow produced by a balance between the pressure gradient force and the Coriolis force, is frictionless flow. But momentum from the wind field is transferred into the ocean by friction; so frictional forces must be important in the surface boundary layer. The dynamics of this layer are best understood if it is assumed initially that the ocean interior is at rest, i.e. all isopycnal (and, as a consequence, all isobaric) surfaces are flat and steric height is constant everywhere. The balance of forces in the surface layer is then purely between the frictional force which transports momentum downwards, and the Coriolis force.

Ekman layer transports

The details of the flow in the surface layer, known as the Ekman drift after its first investigator, are complicated and depend on detailed knowledge of the coefficients of turbulent mixing. If fluctuations of periods shorter than a day are eliminated it is found that the current moves at an angle to the wind, turning further away from the wind direction and becoming weaker with depth. (Any of the text books mentioned in earlier chapters can be consulted for a description of the Ekman layer.) For the purpose of regional oceanography it is important to note that for a determination of total transport in the surface layer, knowledge about the details of the turbulent mixing process is not necessary. Theoretical analysis gives a relationship between the wind stress and the depth-integrated flow that develops in response to the wind which we formulate as

> *Rule 3: The wind-driven component of transport in the surface boundary or Ekman layer is directed perpendicular to the mean wind stress, to the left of the wind stress in the southern hemisphere and to the right in the northern hemisphere.*

The magnitude of the Ekman layer transport M_e is given by:

$$| M_e | = | \tau / f | \; .$$
(4.1)

Here, M_e is the wind-generated mass transport per unit width integrated over the depth of the Ekman layer, with dimensions kg m^{-1} s^{-1}; τ is the wind stress, and f the Coriolis parameter introduced earlier. The unit width is measured across the transport direction, i.e. in the direction of the wind stress. The difference between the wind stress direction and the direction of M_e develops because the Coriolis force acts perpendicular to the water movement; Rule 3 ensures that it opposes the wind stress, and eqn (4.1) ensures that it balances it in magnitude.

If we now relax the initial assumption of flat isopycnal surfaces, a pressure gradient force will be present throughout the water column, up to the surface, and geostrophic flow will develop. M_e is then the additional, non-geostrophic transport in the Ekman layer. As with our Rule 1 we note here again that our rules are not applicable near the equator, where according to eqn (4.1) Ekman layer transport grows beyond all bounds. In practical applications eqn (4.1) is valid to within about 2° of the equator, where quite large Ekman transports do occur in response to quite modest wind stresses.

Ekman pumping

Wind-induced currents are among the strongest currents in the oceanic surface layer. More importantly, the depth-integrated Ekman transports rank among the most important causes of flow in the deeper layers as well, throughout the upper one or two kilometres of the ocean. This comes about because the Ekman transports converge in some regions and diverge in others, and vertical flow develops at the bottom of the surface boundary layer to replace or remove the converging water masses. This process of generation of flow below the surface layer through vertical movement into and out of it is known as Ekman pumping. Because of its importance for the distribution of hydrographic properties below the surface layer it is worth looking at this process in some detail.

We see from eqn (4.1) that variations of Ekman layer transport can result from two factors, the change of the Coriolis parameter with latitude and the structure of the wind stress field. In most situations the variation of f is small and Ekman pumping is dominated by spatial variations of wind stress. Figure 4.1 shows some examples. Between the Trades and the Westerlies (boxes A and B) the wind generates a mean flow into the box, resulting in Ekman pumping or downwelling. Poleward of the maximum Westerlies (box C) the net flow is directed outward, giving negative Ekman pumping ("Ekman suction") or upwelling. In all situations (except near the equator) the vertical velocity w at the base of the Ekman layer (positive for downwelling, negative for upwelling) is given by the divergence of the Ekman transport: $-\rho_0 w = \partial M_e^x/\partial x + \partial M_e^y/\partial y$ where M_e^x and M_e^y are the components of M_e towards east and north respectively. From Rule 3 these components can be expressed, away from the equator, in tems of wind stress:

$$-\rho_0 w = \text{curl }(\tau/f) = \partial\,(\tau_y/f)\,/\,\partial x - \partial\,(\tau_x/f)\,/\,\partial y \qquad (4.2)$$

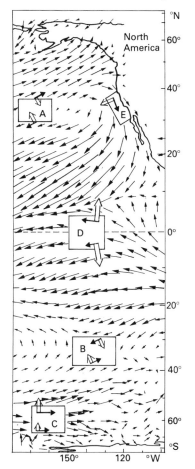

FIG. 4.1. Illustration of Ekman transport and Ekman pumping, based on the eastern Pacific Ocean. Full arrows indicate the wind, open arrows the Ekman transport resulting from the corresponding wind stress. Note the change of direction of the Ekman transport relative to the wind across the equator and the decrease of the Ekman transport, for equivalent wind stresses, with distance from the equator. Box *A* is between the Trade Winds and the Westerlies in the northern hemisphere; the Ekman transports at the northern and southern edges point into the box. The mean wind stresses along the east and west sides of the box do not add or subtract much to the transport budget. The result is net inflow into the box (convergence), or Ekman pumping and downwelling. Box *B* is the mirror image of *A* in the southern hemisphere; the Ekman transports are to the left of the wind, and there is again net inflow (convergence), or Ekman pumping and downwelling. Box *C* is located south of the maximum Westerlies; the Ekman transport is stronger at the northern edge than at the southern because the wind stress decreases towards the south. There is small net outflow (divergence), or Ekman suction and upwelling. Box *D* is on the equator. The winds stress does not vary much across the box, but the Coriolis parameter is small and changes sign across the equator, giving opposite directions for the Ekman transport north and south of the equator. The result is very large net outflow (divergence), or Ekman suction and upwelling. Box *E* is near an eastern coastline in the Trade Wind region. The wind stress is equatorward along the coast; the Ekman transport is directed offshore at the western edge but zero at the coast. The result is net outflow (divergence), or Ekman suction and upwelling.

where τ_x and τ_y are the x- and y-components of τ. (The curl of a field of vectors is a vector which measures the tendency of the vectors to induce rotation. It has three components (curl$_x$, curl$_y$, and curl$_z$); but in oceanography only the third component which relates to rotation around the vertical is of interest, and the index z is never written. A visualization of the relationship between Ekman pumping and wind stress curl is shown in Figure 4.2). Fig 4.3 gives a map of the annual mean of curl(τ/f) for the world ocean. Most of the ocean is characterized by Ekman transport convergence or downwelling, indicated by an extensive region of negative curl(τ/f) from the tropics to the temperate zone. Positive curl(τ/f) or upwelling is found on the poleward side of the core of the Westerlies.

The basis of Figure 4.3 is eqn (4.2), which is not defined near the equator and along boundaries. What happens there can be understood if we return to the box models of Figure 4.1. The equatorial region (box *D*) is often characterized by upwelling: In the equatorial Atlantic and eastern Pacific Oceans winds are eastward and fairly uniform; the change of sign of the Coriolis parameter across the equator produces Ekman transports of opposite signs at about 2°N and 2°S, i.e. Ekman suction and upwelling. Discussion of the situation in the equatorial Indian Ocean is left to Chapter 11 since it varies with the monsoon. Along most eastern boundaries (box *E*) the wind stress is directed equatorward; Ekman transport is directed offshore but cannot come out of the coast - again, upwelling must occur. Near western boundaries the dynamics are governed by the western boundary currents; Ekman layer dynamics are often strongly modified, or negligible, in these regions.

The Sverdrup balance

We now return to the question raised at the end of the last chapter: What replenishes the bulges in the subtropical thermocline, seen (for example) in the temperature distribution of Fig 2.5 at 500 m depth? The answer is found by combining the dynamics of the ocean interior and western boundary currents with the dynamics of the Ekman layer.

It was shown in the last chapter that in the absence of forcing, thermocline bulges in a $1^1/_2$ layer ocean move westward at Rossby wave speed. We obtained an equation for the bulge movement, or local depth variation of the thermocline:

$$\partial H/\partial t = \ (\beta \ g/f^2 \ (y) \ ) \ (\Delta\rho/\rho_0) \ H \ \partial H/\partial x \qquad (4.3)$$

Bulge movement is a consequence of geostrophic transport convergence of anticyclonic eddies in the west and divergence in the east as shown in Figure 3.4, so eqn (4.3) can be seen as a balance equation for the convergence of the flow field. Now, the bulges seen in Figure 2.5c represent the annual mean, i.e. a steady state; so we must have $\partial H/\partial t = 0$. We achieve this by adding the

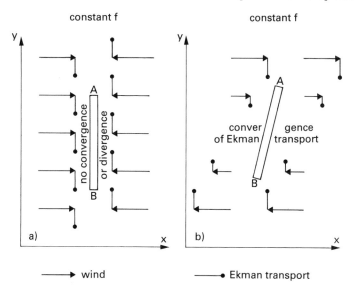

constant f   constant f

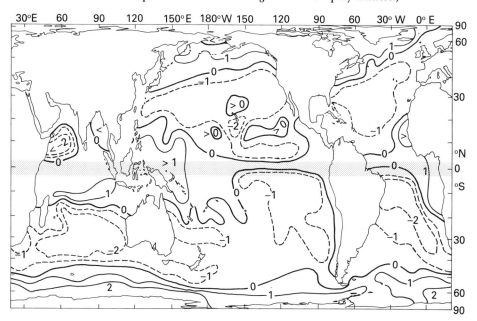

——→ wind   ——• Ekman transport

FIG. 4.2. A sketch of two wind fields and related Ekman transport, (a) with zero wind stress curl, (b) with non-zero wind stress curl.   The tendency of the wind field to induce rotation can be visualized by the drifting balloon *AB*; it does not rotate in (a) but rotates clockwise in (b). It is seen that Ekman pumping in the ocean occurs only where rotation is present in the atmosphere. (The variation of the Coriolis parameter has been neglected to simplify matters.)

FIG. 4.3. Annual mean distribution of curl(τ/f), or Ekman pumping, calculated from the distribution of Fig. 1.4 (10^3 kg m^2 s^{-1}). Positive numbers indicate upwelling. In the equatorial region (2°N - 2°S, shaded) curl(t/f) is not defined; the distribution in this region is inferred from the dynamical arguments of Fig. 4.1 and is not quantitative.

convergence of the Ekman transport (Ekman pumping). The change in thermocline depth produced by Ekman pumping is the vertical velocity at the base of the Ekman layer. Taking this quantity from eqn (4.2) and adding it to eqn (4.3) gives the steady state balance as

$$(\beta g / f^2(y)) (\Delta\rho/\rho_0) H \, \partial H/\partial x = \text{curl } (\tau /\rho_0 f) \qquad (4.4)$$

We can consider the right-hand side as continuously generating new bulges or depressions, which then migrate westwards as Rossby waves. The net effect is that in the steady state, an east-west gradient of H is set up which is in geostrophic balance, and that the mass divergence associated with this gradient matches the Ekman pumping convergence. The fact that energy propagates westward is no longer obvious in the steady state; but it does occur - the intense western boundary currents in each ocean are in fact dissipating the energy that has been fed into the ocean interior from the wind field by Ekman pumping. Because the energy which is fed into these narrow currents is the result of accumulation of energy from the wind field over the entire ocean surface, the western boundary currents are very intense, jet-like features. They achieve energy dissipation through bottom friction on the shelf and slope, and by spawning eddies and ejecting them into the open ocean.

Equation (4.4) is one form of an equation known as the *Sverdrup relation* for the $1^1/_2$ layer ocean. It was derived by Hans Ulrik Sverdrup and allows us to determine H everywhere for any ocean basin from a knowledge of H at its eastern boundary. (Knowledge of H at the western boundary is not of much use, since the Sverdrup relation includes only geostrophy and Ekman layer dynamics and does not apply to the western boundary currents.) We note that the right-hand side of eqn (4.4) is directly obtainable from the wind stress data. In other words, the depth variation of the thermocline, one of the key features of the oceans, can be determined from atmospheric observations alone! The left hand side of eqn (4.4), on the other hand, contains the essence of regional oceanography, namely the observations of temperature and salinity. In the $1^1/_2$ layer model which we used to simplify our description of ocean dynamics, the role of the observations is reduced to a determination of thermocline depth. In the more realistic case of a continuously stratified ocean the Sverdrup relation takes on the form (on multiplying both sides by f^2/β)

$$g \, \partial P/\partial x = f^2 / \beta \ \text{curl } (\tau/\rho_0 f) \quad , \qquad (4.5)$$

where

$$P(x,y) = \int_{z_0}^{0} h(x,y,z) \ dz \qquad (4.6)$$

has dimension m^2 and is the depth integral of the steric height h from an assumed depth of no motion z_0 to the surface: $\partial P/\partial x$ is the generalization of $H (\Delta\rho/\rho_0)\partial H/\partial x$ in the $1^1/_2$ layer model. Note that (4.5) is well-defined at the equator. Evaluation of the right-hand side of eqn (4.5) gives, using $\beta = df/dy$,

$$g \, \partial P/\partial x \; = \; f \, / \, (\beta \, \rho_0) \; \text{curl} \; (\tau) \; + \; \tau_x \, / \, \rho_0 \qquad (4.7)$$

All three terms in eqn (4.7) are well-defined everywhere, including the equator. The spurious "infinite" speed of Rossby waves at the equator does not affect the steady state which those waves bring about.

In the form shown in eqn (4.5), both sides of the Sverdrup relation are completely well-defined by available data - the right-hand side comes from the annual mean wind stress shown in Figure 1.4, while the left-hand side comes from the annual mean density field derived from the data shown in Figure 2.5. This provides us with a powerful means for testing the validity of our concept of ocean dynamics. How well, then, do the two sides of the Sverdrup relation match one another?

Figure 4.4 shows a global map of depth-integrated steric height calculated from atmospheric data, i.e. using the right-hand side of the Sverdrup relation (eqn (4.5)) in combination with appropriate boundary conditions to ensure that no mass flux occurs through the eastern boundary (Godfrey, 1989). The same quantity is also shown in Figures 4.5 and 4.6, this time calculated from oceanographic information, i.e. based on the left hand side of eqn (4.5) for two different depths of no motion, 1500 m and 2500 m. General agreement between the results from both data sets is observed in the subtropics. All three diagrams show a maximum value of P near the western boundary of each ocean at 30° - 40° N or S. The position of this maximum is well represented by Figure 4.4 in each basin. The number of contours from east to west in each ocean basin is fairly well represented, though the wind-based calculation produces a larger transport than the calculation based on ocean temperatures and salinities. This could be due, as has been suggested by researchers, to an inappropriately large choice of the constant C_d in the wind stress calculation of Hellerman and Rosenstein (1983).

It is also remarkable that over most of the oceans the choice of reference level for the integration of the oceanographic data fields does not influence the results very much. This indicates that flow between 1500 m and 2500 m depth is quite small, making both depths probably reasonably valid choices for a depth of no motion. Substantial differences between the strengths of the flows predicted by Figures 4.5 and 4.6 occur in the Antarctic region, indicating that at these latitudes substantial flow occurs below the 1500 m level. This aspect of the circulation will be discussed in detail in Chapter 6.

A major failing of the calculation from atmospheric data occurs west of the southern end of Africa, where it shows very strong north-south gradients of P.

These gradients imply an intense geostrophic flow across the South Atlantic Ocean to South America, which then flows about 500 km southward and returns to the Indian Ocean. In reality, such intense flows which pass through an open ocean region are unstable, particularly when they "double back" on themselves as in Figure 4.4 southwest of South Africa; the real Agulhas Current outflow turns eastward to join the Antarctic Circumpolar Current, as seen in Figures 4.5 and 4.6. The chapters on the Indian Ocean will take this up further.

The outflows from the western boundary currents of the North Atlantic and North Pacific Oceans are also not well represented by the model derived from the wind data. Both are underestimated in magnitude; and the Gulf Stream separation occurs near Labrador, rather than at its observed location near Cape Hatteras in the northern USA. However, considering the simplicity of the concept of Ekman pumping, Rossby wave propagation, and the Sverdrup balance as the result of an equilibrium between these two phenomena, it is pleasing to see to what large degree the atmospheric and oceanographic data sets available today confirm its validity. The model certainly accounts for a considerable fraction of what we see in the depth-integrated flow of the world ocean. The plan for the World Ocean Circulation Experiment (Figure 2.4) is therefore based on the concept that a systematic survey of temperature and salinity, augmented by direct current measurements through moorings in regions where the Sverdrup relation does not hold, will result in a reliable description of the oceanic circulation.

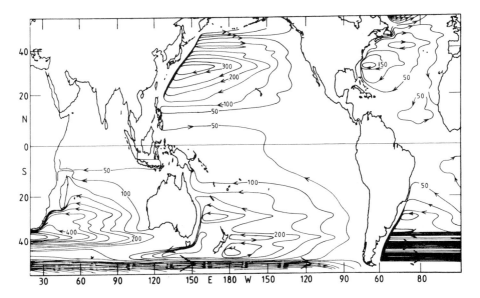

Fig. 4.4. Depth-integrated steric height P, calculated from the right-hand side of the Sverdrup relation (eqn (4.5)), using the data from Hellerman and Rosenstein (1983). Units are 10^1 m^2. For details of the integration procedure see Godfrey (1989).

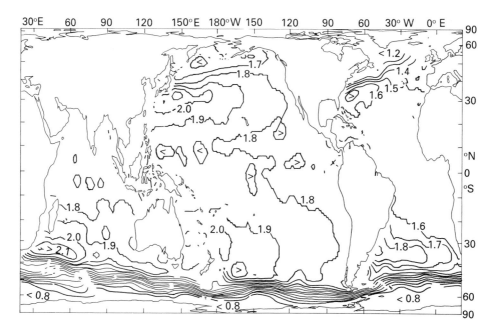

Fig. 4.5. Depth-integrated steric height *P*, from the left side of the Sverdrup relation (eqn (4.5)), using the data from Levitus (1982), for a depth of no motion of 1500 m. Units are 10^3 m^2.

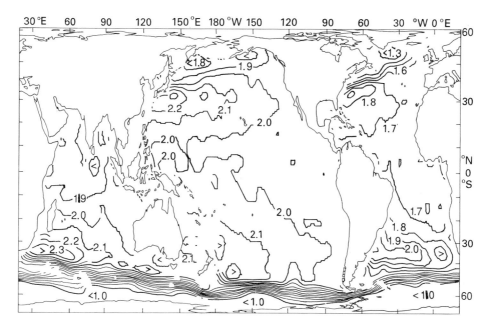

Fig. 4.6. Depth-integrated steric height *P*, from the left side of the Sverdrup relation (eqn (4.5)), using the data from Levitus (1982) for a depth of no motion of 2500 m. Units are 10^3 m^2.

The stream function and depth-integrated transports

Without doubt, depth-integrated steric height P is the best quantity to test the Sverdrup relation. Unfortunately, it does not provide the most convenient visualization of the flow field on oceanic scales, since its distribution combines variations of depth-integrated flow with the latitude-dependence of the Coriolis parameter. A better representation is achieved by introduction of a stream function which eliminates f from the picture. The direction of the total net transport between the assumed depth of no motion and the surface is then indicated by the direction of the streamlines, and its magnitude is proportional to the distance between them. Because there cannot be any flow through the coast, coastlines are streamlines by definition. If the coast is continuous it coincides with a single streamline, i.e. the value of the stream function along the coast is the same. It is also possible to estimate the flow around major islands such as Australia and New Zealand and allocate stream function values to their coasts (Godfrey, 1989).

If we neglect the small flow from the Arctic into the north Pacific Ocean and consider Bering Strait closed, the world ocean is surrounded by a single continuous coastline (formed by the Asian, African and American continents) and contains only few large islands (Antarctica, Greenland, Iceland, Australia, New Zealand, Madagascar). Figure 4.7 shows the volume transport stream function for this continuous domain. Since the stream function calculation is based on Sverdrup dynamics it can only proceed starting from eastern boundaries and extend to the outer edges of the western boundary currents. The differences in stream function values across the gap near the western coastlines give the

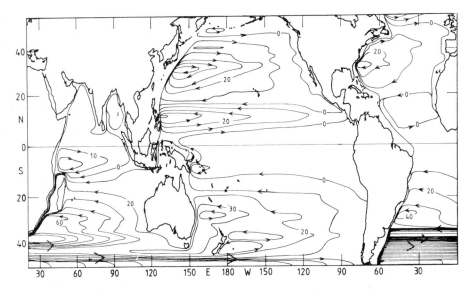

FIG. 4.7. Stream function for the world ocean, calculated from the wind stress data of Hellerman and Rosenstein (1983). Contour interval is 10 Sv.

transport of the western boundary currents. Details of the depth-integrated flow near the equator that are masked in Figure 4.4 by the variation of the Coriolis parameter (in particular the sign change of f across the equator) come out clearly in the stream function map. The North Equatorial Countercurrent, flowing eastward between 5° and 15°N in the North Pacific Ocean, is seen as a major feature. The corresponding feature in the South Pacific Ocean is extremely weak and restricted to an area south-east of Papua New Guinea. A strong Countercurrent is seen in the southern Indian Ocean. This asymmetrical distribution of currents reflects similar asymmetries of the annual mean wind stress field of Figure 1.4.

An important thing to note for later reference is the magnitude of the western boundary currents generated by the Sverdrup mechanism: 50 Sv for the Kuroshio, 30 Sv for the Gulf Stream and the Brazil Current, 25 Sv for the East Australian Current, and 70 Sv for the Agulhas Current; independent estimates of the flow between Australia and Indonesia of 16 Sv were used for obtaining the latter two transports. To give some feeling what these numbers mean we note that 50 Sv correspond to about 4000 cubic kilometres per day! When it comes to the discussion of individual oceans it will become clear that none of these transport estimates is very accurate, because regional effects modify the boundary currents. Nevertheless, there can be no doubt that the currents are wind-driven, in the sense that they provide continuity of transport for the Sverdrup regime of the ocean interior.

Another feature of Figure 4.7 worth noting is the existence, in each ocean basin, of circulations bounded by contours of zero stream function value. The circulation cells bounded by zero stream function contours are known as "gyres"; those lying between the maximum westerlies and Trades in Figure 4.7 are known as the *subtropical gyres*. Within each subtropical gyre, the wind stress curl generates equatorward depth-integrated flow in the ocean interior, which returns polewards in the western boundary currents. Poleward of the subtropical gyres in the Northern Hemisphere are the *subpolar gyres*. Flow within these is poleward in the interior and equatorward at the western boundary. The convergence of the western boundary currents of the subtropical and subpolar gyres produces the *polar fronts*, regions of strong horizontal temperature and salinity gradients where warm subtropical water meets cold subpolar water along the zero streamline contour (compare Figure 2.5). Further gyres are evident between 15°N and 10°S; they form part of the complex equatorial current system which will be discussed in later chapters and are not usually recognized as gyres by name.

It can be shown that if the winds were purely zonal and did not vary with longitude, the zero stream function contours would coincide with zero contours of wind stress curl. In other words, subject to the same assumptions, the boundaries of the gyres would coincide with the maximum westerly wind stresses, and with the maximum Trades. Purely zonal winds with no longitudinal variation are of course a somewhat poor description of the real mean winds; but the assumption is useful as a first approximation. The lines of zero stream function in Figure 4.7 do indeed either lie close to the maximum westerlies, or the maximum Trades, as can be seen by comparison with Figure 1.2.

Since the depth-integrated flow cannot cross streamlines, there is no *net* transport of mass between adjacent gyres (provided, that is, that the conditions of the Sverdrup circulation are met). This does not mean, however, that no water crosses from one gyre into another. In fact, transport in the surface layer across zero streamlines is maximum because the zonal wind stresses along these boundaries - and as a consequence the Ekman transport - are at a maximum. The surface layer transport is balanced by geostrophic transport below, so the *net* transport is zero. At the equatorward boundary of the subtropical gyres the Ekman transport is directed poleward and moves warm water away from the tropics, while the compensating deep geostrophic flow brings cold water into the tropics. Thus, although there is no net mass transport across the zero stream function contour, there is substantial poleward transport of heat. To give an example, a simple calculation with the winds of Figure 1.2 shows that in the north Pacific Ocean the northward Ekman transport across the southern edge of the subtropical gyre at about 16°N (the position of the strongest Trades) amounts to some 28 Sv, with equal and opposite (and colder) geostrophic flow below. Similarly, there is a net poleward Ekman flux of 29 Sv across the northern edge of the south Pacific subtropical gyre at about 14°S, again compensated by equal and opposite geostrophic flow of lower mean temperature, so that heat is transported polewards here also. Thus the simple wind-driven circulation of Figure 4.7 which gives such a good illustration of the use of Sverdrup dynamics can be very misleading when it comes to a discussion of the amount of heat carried by the ocean.

This concludes our discussion of ocean transport dynamics. A legitimate question at this point could be: If we go out to sea on a research vessel and make some observations, to which degree can we expect to see our picture of the large oceanic gyres and associated currents confirmed by the new data? A first answer is of course that Figures 4.5 and 4.6 are based on observations; they should therefore reflect the real world. A second answer is that they are based on annual mean data distributions. The question is then, can we expect to find the annual mean situation if we go out to sea and take a snap shot of the temperature and salinity distribution?

The answer is - yes, if we make our study area large enough. A synoptic survey of the subtropical north Atlantic Ocean would show the subtropical gyre. However, superimposed we would find a wealth of other structures, mainly eddies of all shapes and sizes. If we restricted our survey to a square of some 200 km base length we could find it very difficult to recognize the mean flow. The water in the ocean is constantly mixed, a major mixing mechanism being eddy mixing. Figure 4.8 (in the colour section) shows mean eddy energy in the ocean. The western boundary currents (Gulf Stream, Brazil Current, Kuroshio, East Australian Current, Agulhas Current) and the Circumpolar Current stand out as regions of particularly high eddy energy. Eddies in these regions are so energetic that the associated currents can reverse the direction of flow. In the western boundary

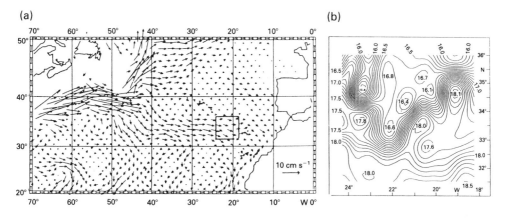

FIG. 4.9. Eddies in the Azores Current. (a) Mean surface flow based on historical
data. (b) The box of panel (a) enlarged, showing the Azores Current and eddies
as indicated by the temperature of the surface mixed layer, during April 1982.
(Horizontal temperature gradients indicate a sloping interface and associated eddies
in the manner of Fig. 3.3.) From Käse *et al.* (1985).

currents they turn the water movement from poleward to equatorward; poleward
mean flow is only observed if the study area is larger than the typical eddy size,
which is of the order of 200 km diameter. Away from the boundary currents eddy
energy goes down rapidly; but annual mean velocities decrease as well, and eddies
can still reverse the observed direction of mean flow if the study area is too small.
Figure 4.9 demonstrates this for the area of the Azores Current in the eastern
north Atlantic Ocean. In the example the cruise covered an area of roughly
550 · 550 km with 86 stations. At least three eddies are seen, but the Azores
Current is still visible meandering through the region between the eddies. A
smaller study area would not show much resemblance between the annual mean
and the actual situation.

CHAPTER 5

Water mass formation, subduction, and the oceanic heat budget

In the first four chapters we developed the concept of Ekman pumping, Rossby wave propagation, and the Sverdrup circulation as the steady-state balance between these two processes. We then showed how the *depth-integrated* flow in most ocean regions is well described by this concept. However, this is evidently not the full story of the oceanic circulation. The oceans carry heat from the tropics to polar latitudes, and they carry cold water from the poles towards the equator. The details of these transport processes, which are restricted to certain depth ranges, are hidden in the Sverdrup circulation. They have to be resolved if we want to understand the ocean's role in climate variability and climate change.

One thing we do know is that the ocean carries about as much heat towards the poles as the atmosphere does; but since its time scales are so much larger the ocean has a large capacity to act as a damping mechanism for rapid fluctuations in our climate. Conversely, much of the long-term variability of the climate may be related to the ocean as it slowly releases heat stored from earlier rapid climate changes. Topographic details play a large part in determining patterns of ocean heat transport. This is the point of interaction between regional oceanography and geophysical fluid dynamics. It is essential for theoretical studies to know to what degree regional oceanic features have to be part of the modelling process.

Generally speaking, most of the heat and freshweter exchange with the atmosphere occurs at the ocean surface in a layer which during most of the year is less than 150 m deep. Once a water parcel is removed from the surface layer its temperature and salinity do not change until it rises back up to the surface again, usually many years later. Evidence for this can be found in the temperature and salinity maps of Figures 2.5e and 2.5f for 2000 m depth. The temperature everywhere is less than 4.2°C, even in the tropics; such cold waters can only have acquired their low temperatures in the polar regions and moved into the tropics without much mixing with the warmer waters above. Since independent evidence obtained, for example, from ^{14}C dating sets the residence times of these deeper waters at hundreds of years, this lack of mixing is remarkable. Yet mixing is not entirely absent: Note that the highest temperatures at 2000 m occur just west of the Strait of Gibraltar, where they coincide with very high salinities. The pattern suggests that a "water mass" of relatively warm, salty water enters the Atlantic Ocean from the Mediterranean Sea and moves westward across the ocean, mixing as it goes.

This example illustrates a common feature found throughout the ocean. Water masses with well-defined temperature and salinity characteristics are created by surface processes in specific locations, which then sink and mix slowly with other water masses as they move along. Since these movements are so slow, it is usually unrealistic (and because of the presence of eddy motion impractical in any case) to measure them directly. It is easier to deduce these movements and the strength of mixing from the distribution of the water properties themselves.

The analysis of water mass movements and mixing can assist in understanding the deep oceanic circulation where the estimation of steric height relative to an assumed depth of no motion finds its limits. Steric height estimation provides valuable insights into the workings of the vigorously moving top kilometre of the ocean where accurate knowledge of the depth of no motion is not critical. It cannot resolve the slow flows at great depth. The best hope to achieve a complete description of the circulation in the ocean at all depths is to combine water mass analysis with our knowledge of the constraints on possible flows posed by geostrophy and Ekman dynamics. This approach, which infers water movement from the effect it has on the distribution of oceanic parameters, is known as "inverse modelling". This is a field of active research, mathematically quite complex and beyond the scope of this book. An understanding of the processes involved in water mass formation and of the life history of water masses is, however, required in regional oceanography.

Water masses, water types, and T-S diagrams

We begin with a brief review of concepts and definitions. In the past, oceanographers have used terms such as water mass and water type rather loosely to describe waters with common or outstanding properties. For the purpose of a quantitative description of the transport of water properties in the ocean it is necessary to introduce unambiguous definitions, even though they may not always correspond with earlier uses of the terms. We define a *water mass* as a body of water with a common formation history. An example of water mass formation is the cooling of surface water near the Antarctic continent, particularly in the Weddell Sea, which increases the density and causes the water to sink to great depth. All water which originates from this process shares the same formation history and is called Antarctic Bottom Water. It is found in all oceans well beyond its formation region, extending even into the northern hemisphere.

Common names of known water masses usually relate to their major area of residence. Unfortunately, this can give rise to ambiguity since the same name may be used for a well defined water mass or simply for water found in a certain region. To avoid this confusion we adopt the convention that water masses are always identified by capitals. For example, "Bottom Water" can stand for Antarctic, Arctic, or other Bottom Water but always refers to a water mass, while water found at the bottom of an oceanic region may be referred to as "bottom water" without implying that it is a known and well defined water mass. Likewise, we use the term "intermediate water" occasionally for intrusions of water at intermediate depth; in contrast, "Intermediate Water" is used to indicate well-defined water masses.

It is important to note that exclusive occupation of an oceanic region by a single water mass occurs only in the formation regions. As the water masses spread across the ocean they mix, and several water masses are usually present at an oceanic location. However, water masses occupy a measurable volume, which is the sum of the volumes occupied by all its elements regardless of their present whereabouts. It is possible to determine the percentage contribution of all water masses to a given water sample, because the water mass elements retain their properties, in particular their potential temperature and their salinity, when leaving the formation region. Water masses can therefore be identified by plotting temperature against salinity in a so-called T-S diagram. An elementary description of T-S diagrams and their use can be found, for example, in Dietrich *et al.* (1980). Figure 5.1 is an example of a T-S diagram from a tropical ocean region and shows how observational data can be used to identify water masses from their T-S combinations. It is seen that the properties of Central Water in the Coral Sea correspond closely to those in its formation region, indicating that little mixing with other water masses occurred along its way. In contrast, the intermediate and deep water masses are not present with their original T-S values; their properties are modified by mixing with water above and below, and their presence is indicated by salinity or temperature extrema in the vicinity of the T-S combinations found in their formation regions.

From Figure 5.1 it is seen that T-S relationships alone are insufficient to describe a water mass. Particularly in the upper ocean water masses undergo property changes in response to atmospheric conditions, as indicated in Figure 5.1 by the increase in the standard deviation as the surface is approached. For a full description of a water mass it is necessary to include information about the degree of spatial and long-term variability during its formation, as expressed through its standard deviation. Some water masses, such as the Intermediate Water in Figure 5.1, require only a single T-S combination (T-S point) and a standard deviation; others, such as the Central Water, require a set of T-S combinations, or a T-S relationship, together with a standard deviation envelope. Since data sets to determine standard deviations require observations over several years and are not always available, most books identify water masses by one or more T-S points without standard deviations. Points in the T-S diagram are called *water types*, and water mass definition points are known as *source water types*. In reality, very little - if any - of the water belonging to a water mass has exactly the T-S properties of the corresponding source water types. But most T-S values are very close, within the (often unknown) standard deviation. Numerical T-S values for water masses given in later chapters have to be understood in this way. In general, the standard deviation is very small for abyssal water masses but increases rapidly as the upper layers are approached.

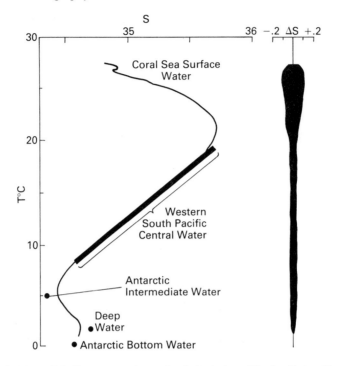

F$_{IG}$. 5.1. Mean T-S diagram and standard deviation ΔS of salinity (for given temperatures) in the eastern Coral Sea, in comparison to water mass definitions in the south Pacific Ocean. Heavy dots and the heavy line indicate water mass properties in the formation regions, which for all but Surface Water are located far outside the Coral Sea. The standard deviation was determined by comparing stations in the region with a space average and does not include variability in time. Similar standard deviations can be derived for temperature and other properties.

The seasonal, tropical, and permanent thermoclines

Most water masses are formed at the ocean surface. This is a region of strong mixing, which produces uniformity of properties above a layer of rapid property change. The term thermocline was occasionally used for this layer in the last two chapters. In the context of water mass formation it is necessary to sharpen our definition of a thermocline.

Oceanographers refer to the surface layer with uniform hydrographic properties as the *surface mixed layer.* This layer is an essential element of heat and freshwater transfer between the atmosphere and the ocean. It usually occupies the uppermost 50 - 150 m or so but can reach much deeper in winter when cooling at the sea surface produces convective overturning of water, releasing heat stored in the ocean to the atmosphere. During spring and summer the mixed layer absorbs heat (moderating the earth's seasonal temperature extremes by storing heat until the following autumn and winter; this aspect is discussed in detail in Chapter 18), and the deep mixed layer from the previous winter is covered by a shallow layer of warm, light water. During this time mixing does not reach very deep, being achieved only by

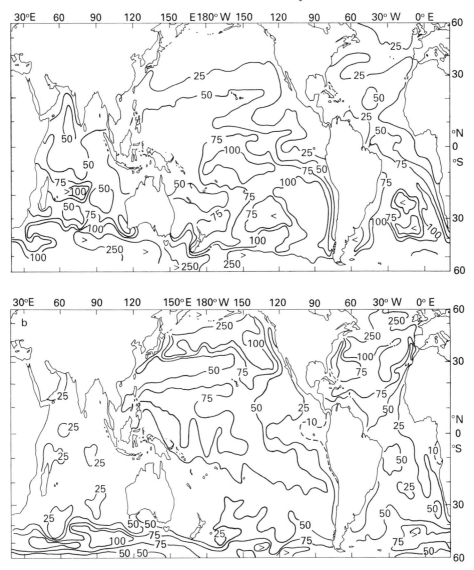

FIG. 5.2. Mean depth of the surface isothermal layer (m). (a) August - October, (b) February - April. Contouring levels are 10 m (dashed line), 25 m, 50 m, 75 m, 100 m, 250 m. Adapted from Sprintall and Tomczak (1990).

the action of wind waves. Below the layer of active mixing is a zone of rapid transition, where (in most situations) temperature decreases rapidly with depth. This transition layer is called the *seasonal thermocline*. Being the bottom of the surface mixed layer, it is shallow in spring and summer, deep in autumn, and disappears in winter (An example can be seen in Figure 16.27). In the tropics, winter cooling is not strong enough to destroy the seasonal thermocline, and a shallow feature sometimes called the *tropical thermocline* is maintained throughout the year. Figure 5.2 shows the thickness of the surface isothermal layer. It was obtained using the data from Levitus (1982) by extracting the depth where the temperature differed from the temperature at the surface by more than 0.5°C and is representative of the depth of the seasonal thermocline.

Mixed layer dynamics are quite complex, and we refer the reader to other text books (for example Pickard and Emery, 1990) for details. The point of interest here is that turbulent energy levels in the winter mixed layer drop drastically, by a factor of 1000 or more, after it is covered by lighter water during spring - mixing then becomes so slight that the water characteristics established before the turbulence falls off become "frozen". These layers of "fossilized mixing", which retain their signatures for long time spans, are the source layers of most water masses, and their generation is the essence of water mass formation.

The depth range from below the seasonal thermocline to about 1000 m is known as the *permanent* or *oceanic thermocline*. It is the transition zone from the warm waters of the surface layer to the cold waters of great oceanic depth and provided the model for the interface of our $1^1/_2$ layer representation of the ocean. The temperature at the upper limit of the permanent thermocline depends on latitude, reaching from well above 20°C in the tropics to just above 15°C in temperate regions; at the lower limit temperatures are rather uniform around 4 - 6°C depending on the particular ocean. Wherever the word thermocline is used without further specification in this book, the term refers to the permanent thermocline. Again, a detailed description of the various thermoclines and their seasonal life cycle can be found in other textbooks (Dietrich *et al.*,1980; Pickard and Emery, 1990).

Subduction

What maintains the permanent thermocline and prevents its erosion from mixing with the waters below and above? The principal factors are the combination of water mass formation and Ekman pumping. Figure 4.3 tells us that the subtropics are a region of negative $curl(\tau/f)$, which means that water is pumped downwards. As this water is not denser than the underlying water, it gets injected into intermediate depths, following the isopycnal surface of its own density. This process, which is known as *subduction* and illustrated in Figure 5.3, is responsible for the formation of the water masses in the permanent thermocline. Its intensity varies with the seasons, partly in response to variations in the strength of the Ekman pumping but mainly because of the seasonal development of the seasonal thermocline: The summer mixed layer depth is shallower than the depth of the winter mixed layer; the water trapped between (the fossilized mixing region) is available for subduction. If, for example, in late autumn and winter the bottom of the mixed layer (z_1 in Figure 5.3) progresses downward faster than the surface water moves downward as a result of Ekman pumping, water pumped from the surface during summer is caught by the expanding surface mixed layer before it can escape into the permanent thermocline, and the properties of the water subducted into the permanent thermocline are determined by the surface water properties during late winter only. In other words, although subduction is a permanent process, water mass formation occurs only in late autumn and winter. This can be verified by comparing the properties of the surface mixed layer in the Subtropical Convergence with those of the permanent thermocline in the tropics, i.e. by comparing T-S diagrams along the lines *ABCD* and *A'B'C'D'* in Figure 5.3. An example of such a comparison from the Indian Ocean is shown in Figure 5.4; it demonstrates that over the T-S range of the thermocline the two T-S diagrams are nearly identical in late winter (August - October) but differ during all other seasons.

Water masses subducted into the thermocline are commonly known as Central Water. The term was introduced sixty years ago to differentiate between thermocline water of the central north Atlantic Ocean (now known as North Atlantic Central Water) and water from the shelf area to the west. It is now used to identify thermocline water masses in all three oceans.

Studies of T-S diagrams and of property distributions on isopycnal surfaces led to the conclusion that mixing across isopycnal surfaces is generally much weaker than mixing on isopycnal surfaces. The principle does not apply universally; western boundary currents, the Equatorial Undercurrent, and frontal regions are among the regions where mixing across density surfaces, or *diapycnal mixing*, contributes significantly to the exchange of properties. But in general, and over vast ocean regions, it is safe to neglect diapycnal mixing as a first guess for the oceanic circulation.

Figure 5.5 summarizes the discussion of thermoclines and subduction. Intermediate Water, which spreads just below the permanent thermocline, is also produced by subduction. Although the driving agent is not Ekman pumping but mixing and convection the mechanism is the same, movement along isopycnal surfaces towards the equator.

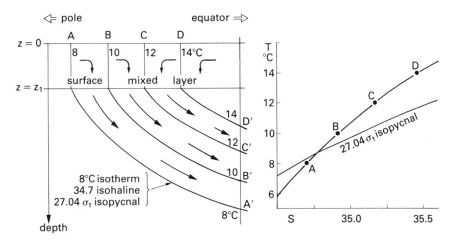

FIG. 5.3. Sketch of water mass formation by subduction in the Subtropical Convergence. The T-S diagram shows both the meridional variation of temperature and salinity between stations *A* and *D*, and the vertical variation equatorward of station *D* from the surface down along the line *A'B'C'D'*. For more detail, see text.

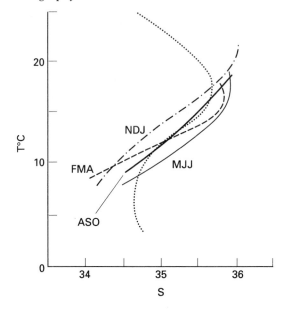

FIG. 5.4. A comparison of T-S diagrams across the Subtropical Convergence (STC), along 102.5°E between 30°S and 45°S, and across the permanent thermocline at 20°S. T-S diagrams across the STC are shown for August - October (ASO), November - January (NDJ), February - April (FMA), May - July (MJJ). The remaining   curve is the vertical T-S diagram at 20°S. Adapted from Sprintall and Tomczak (1993).

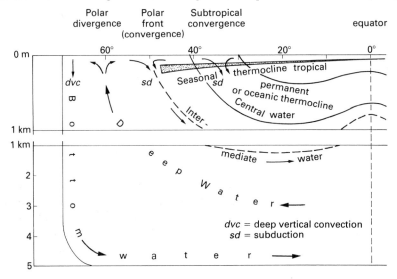

FIG. 5.5. Meridional section through a hypothetical ocean, showing the permanent thermocline, the seasonal and tropical thermoclines, the various surface convergences and divergences, and the major water masses. Note the scale change at 1000 m depth, which underestimates the volume of Deep and Bottom Water. On the other hand, the importance of the tropics does not come out in this graph since   it   does not show the convergence of meridians towards the poles.

The barrier layer

Traditionally, it has been assumed that the depth over which the temperature is uniform can be used as an indication of the depth of the mixed layer, i.e. the layer affected by surface mixing processes. This assumption probably developed more out of necessity than physical argument, since many upper ocean observations lack information on salinity and some way had to be found to determine the depth of the surface mixed layer from temperature information alone. The better data base of today allows us to check that assumption, thereby gaining a better understanding of the processes of water mass formation. Figure 5.6 shows the thickness of the surface isopycnal layer, i.e. the layer over which density does not change. This map was obtained by determining the depth where density is larger than the density at the surface by an amount which corresponds to the temperature change of 0.5°C used in the construction of Figure 5.2. This amount depends strongly on temperature and also on salinity. To make sure that the distributions of Figures 5.2 and 5.6 are comparable, the density increment is not constant across the map but was evaluated locally from the surface temperature and salinity. Figure 5.7 shows a map of the differences between the calculated isothermal and isopycnal layer thicknesses.

If the classical assumption were correct, the differences shown in Figure 5.7 should be close to zero everywhere. There are indeed large ocean areas which display very small differences. But we also notice regions where the difference is clearly not zero. In the tropics, the difference between isothermal and isopycnal layer depth is often positive, indicating a density change within the isothermal layer. The density change is caused by salinity stratification. In these regions, the halocline is the true indicator of mixed layer depth. The tropical surface water has low salinities and high temperatures and therefore very low densities; it spreads in a thin layer over the top of the water column. Subtropical surface water, on the other hand, has high salinities; when this water is subducted in regions of Ekman pumping it cannot be pushed very deep before it spreads sideways, forming a salinity maximum below the surface layer. If the temperature gradient is insignificant, the result of both processes is a halocline within the isothermal layer.

The layer between the halocline and the thermocline is now referred to as the *barrier layer*, because of its effect on the mixed layer heat budget. The mixed layer receives large amounts of heat from solar radiation. In a steady state situation this source of heat has to be balanced by one or more sinks. In the absence of a barrier layer, i.e. where the surface mixed layer reaches down into the thermocline, an important sink is located at the bottom of the mixed layer where cold water is entrained from below. The presence of a barrier layer means that the water entrained into the mixed layer has the same temperature as the water above. There is therefore no heat flux through the bottom of the mixed layer, and other sinks have to come into play to prevent a permanent rise of mixed layer temperature. The dynamics of the

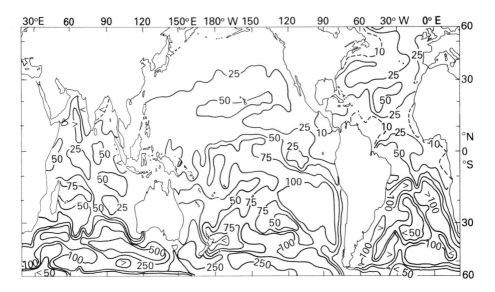

FIG. 5.6a. Mean depth of the surface isopycnal layer (m), August - October. Contouring levels are 10 m (dashed), 25 m, 50 m, 75 m, 100 m, 250 m, 500 m. Adapted from Sprintall and Tomczak (1990).

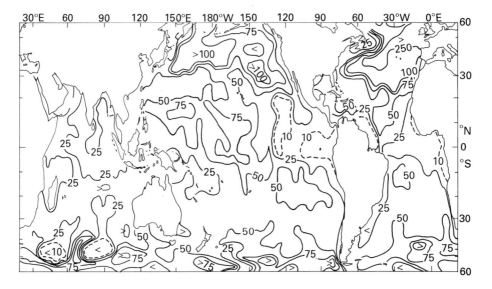

FIG. 5.6b. Mean depth of the surface isopycnal layer (m), February - April. Contouring levels are 10 m (dashed), 25 m, 50 m, 75 m, 100 m, 250 m, 500 m. Adapted from Sprintall and Tomczak (1990).

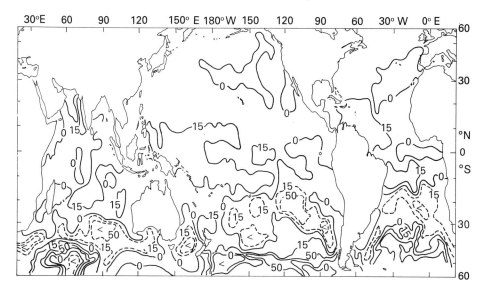

FIG. 5.7(a). Mean depth difference (m) between isothermal and isopycnal layer (barrier layer thickness), August - October. Adapted from Sprintall and Tomczak (1990).

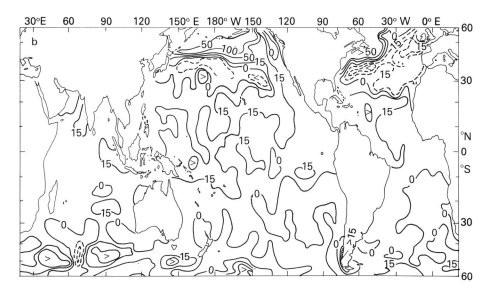

FIG. 5.7(b). Mean depth difference (m) between isothermal and isopycnal layer (barrier layer thickness, February - April. Broken contours indicate negative differences. Adapted from Sprintall and Tomczak (1990).

prevent a permanent rise of mixed layer temperature. The dynamics of the three regions which show a barrier layer through most of the year, the western Pacific, the equatorial Atlantic, and the Bay of Bengal in the Indian Ocean, differ distinctly from each other; they will be discussed in more detail in the chapters which deal with the individual oceans.

Another region of positive difference in isothermal vs isohaline layer depth is in the polar latitudes where a thermocline does not exist: The surface water, constantly cooled by the atmosphere, is so dense that it sinks virtually to the bottom of the ocean (the refinements of this process will be discussed in Chapters 6 and 7), leaving the water column isothermal to great depth. The salinity, on the other hand, is seasonally influenced by freshwater release from the ice shelf and from icebergs. The resulting reduction in density can inhibit the sinking of water, and a halocline can develop. This process, seasonal development of a pycnocline as a result of freshwater flux from the ice regions, is indicated by the large differences in Figure 5.7 south of 50°S and during February to April near Newfoundland and Labrador and south of Alaska and the Aleutian Islands.

The third region of non-zero differences in Figure 5.7 is located in the subtropics and displays negative values, indicating that the first density change below the surface is found at a depth *greater* than the isothermal layer thickness. This can only occur if the density change produced by the change in temperature across the thermocline is compensated by an appropriate salinity change. To explain this feature we note from Figure 2.5 that in the subtropics where the water of the permanent thermocline is subducted, sea surface temperature and salinity both decrease rapidly towards the poles. The T-S diagram which describes the meridional variation of temperature and salinity across the subduction zone thus nearly follows an isopycnal (compare Figure 5.3). As a consequence, the isothermal layer thickness, say z_1 at station C, is evaluated correctly by the 0.5°C criterion of Figure 5.2. The equivalent density criterion, however, is not exceeded until a greater depth z_2 is reached, because of the compensating salinity effect. It is seen that the difference in layer thickness shown in Figure 5.7 for the subtropics does not, as in the case of the tropics, result from different isothermal and isohaline layer thickness. We might even regard it as an artefact produced by the subduction process. On the other hand, the discussion in Chapters 9, 12, and 15 will show that the negative differences in the subtropics are a reliable indicator for subduction of thermocline water.

The picture that emerges particularly in the subtropical gyres is a set of independent flows on isopycnal surfaces which, when depth-integrated, collectively satisfy the Sverdrup relationship. In the following chapters, we shall use the Sverdrup circulation as a guide for the discussion and fill in the depth-dependence by looking at water mass properties where necessary. One instant where this need will arise is whenever we want to evaluate oceanic transports of heat or salt, because both involve integrals over products of velocity and temperature or salinity which require knowledge of the distribution of currents with depth. To see this, consider the example of warm water moving poleward with a velocity of 0.4 m s^{-1} in the upper 1000 m and cold water moving equatorward with 0.1 m s^{-1} between 1000 m and 5000 m. The net transport of heat is obviously poleward, but the net mass transport is zero. Clearly, the vertically integrated flow patterns of Chapter 4 are inadequate for estimating heat or salt transports.

CHAPTER 6

Antarctic oceanography

The region of the world ocean bordering on Antarctica is unique in many respects. First of all, it is the only region where the flow of water can continue all around the globe nearly unhindered and the circulation therefore comes closest to the situation in the atmosphere. Secondly, the permanent thermocline (the interface $z = H(x,y)$ of Figure 3.1) reaches the surface in the Subtropical Convergence (Figure 5.5) and does not extend into the polar regions; temperature differences between the sea surface and the ocean floor are below 1°C close to the continent and generally do not exceed 5°C, *i.e.* 20% of the difference found in the tropics (Figure 6.1). What this means is that our $1^1/_2$ layer ocean model cannot be applied to the seas around Antarctica.

It may seem strange that having spent five chapters on a discussion of temperate and tropical ocean dynamics, we now begin our regional discussion with a region that does not fit the earlier picture. However, our earlier discussion is not entirely irrelevant; it taught us how to get an idea of a region's dynamics by identifying

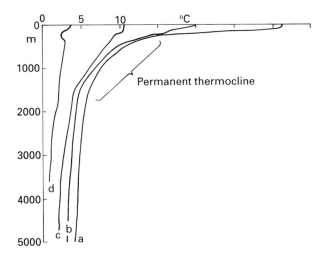

FIG. 6.1. Temperature profiles for different climatic regions near 150°W (Pacific Ocean). (a) tropical (5°S), (b) subtropical (35°S), (c) subpolar (50°S), (d) polar (55°S). The temperature scale is correct for the polar profile; other profiles are shifted successively by 1°C. Note the shallowness of the warm surface layer and the absence of the permanent thermocline in the polar region. Data from Osborne *et al.* (1991).

67

the important forces and looking at their balance. The dynamics relevant for Antarctic waters are those of the "ocean interior" of Figure 3.1, i.e. geostrophy. In the tropics and subtropics, where density varies rapidly across the permanent thermocline, a small tilt of the thermocline produces a large horizontal pressure gradient. It is thus possible to balance all flow geostrophically across the thermocline and reduce velocities to virtually nothing below (this is the essence of the $1^1/_2$ layer model). In Antarctic waters density variations with depth are small and the pressure gradient force is more evenly distributed over the water column. As a result, currents are not restricted to the upper few hundred metres of the ocean but extend to great depth. Observations in Drake Passage show mean current speeds of 0.01 - 0.04 m s^{-1} at 2500 m depth, or 10-30% of the speeds observed at the 500 m level. It is therefore easy to see why the Circumpolar Current has the largest mass transport of all ocean currents: It moves a slab of water more than 2000 metres thick with speeds comparable to other surface currents.

Another aspect which makes the Antarctic region unique is the unlimited communication with all other oceans. The fact that the hydrology of all ocean basins cannot be understood without insight into what goes on in Antarctic waters, provides us with the reason why we are looking at this region first.

Many oceanographers refer to the region around the continent of Antarctica as the *Southern Ocean*. The International Hydrographic Bureau, which is the authority responsible for the naming of oceanic features, does not recognize a subregion of the world ocean of that name but includes its various parts in the other three oceans. Thus, the area between 146°55'E and 67°16'W (about 40%) is considered part of the Pacific Ocean, the area from 20°E to 146°55'E (about 35%) part of the Indian Ocean, and the area between 67°16'W and 20°E (the remaining 25%) part of the Atlantic Ocean. These definitions were developed and agreed upon before the central features of ocean dynamics discussed in the previous chapters were established. From an oceanographic point of view, subdivisions of the world ocean should reflect regional differences in its dynamics. The Southern Ocean certainly deserves its own name on that ground. While we took care not to use the term Southern Ocean for the chapter heading, we shall adopt it in this book from now on and define it as the region south of a line where the tropical/temperate dynamics of Figure 3.1 break down. This occurs where the permanent thermocline reaches the surface, i.e. in the Subtropical Convergence. Realistically, this is not a well defined line but a broad zone of transition, between tropical/temperate and polar ocean dynamics. Its southern limit is marked by a frontal region of limited width known as the Subtropical Front, which will be discussed in detail following the sections on the topograpy and on the wind regime. Within the limitations set by the time variability of the Subtropical Front, the surface area encompassed by the Southern Ocean represents roughly 77·10^6 km^2, or 22% of the surface of the world ocean.

Bottom topography

Since we expect the Circumpolar Current to be present at all depths we can anticipate that in the Southern Ocean the topography of the ocean floor has a much larger impact on the currents, and on the hydrology in general, than in any other ocean. Figure 6.2 shows that the Southern Ocean consists of three major basins, where the depth exceeds 4000 m, and three major ridges. The *Amundsen, Bellingshausen,* and *Mornington Abyssal Plains,* sometimes collectively called the Pacific-Antarctic Basin, extend eastward from the Ross Sea towards South America and belong fully to the Pacific sector of the Southern Ocean. They are separated from the basins of the temperate and tropical Pacific Ocean by the Pacific-Antarctic Ridge and the East Pacific Rise in the west and the Chile Rise in the east. The *Australian-Antarctic Basin,* which is located in the Indian Ocean sector, stretches westward from the longitude of Tasmania to the Kerguelen Plateau. The South-East Indian Ridge separates it from the Indian Ocean to the north, but communication with the basins of the eastern Indian Ocean below 4000 m depth is possible via a gap at 117°E. The *Enderby* and *Weddell Abyssal Plains,* also known as the Atlantic-Indian Basin, form part of the Atlantic and Indian Ocean sectors and reach westward from the Kerguelen Plateau to the Weddell Sea. They are bounded by the Mid-Atlantic and South-West Indian Ridges but at the 4000 m level well connected with the Argentine Basin in the western Atlantic and the basins of the western Indian Ocean. It is worth specific mention that at the 4000 m depth level the eastern Atlantic Ocean and the Pacific Ocean in general have no direct connection with the Southern Ocean.

More important for the dynamics than the basins are the ridges that separate them. The *Scotia Ridge,* which connects Antarctica with South America and contains numerous islands, is located about 2000 km east of Drake Passage, a narrow constriction where the southern tip of South America reaches 56°S while the Antarctic Peninsula extends to 63°S. At the 500 m depth level, the width of Drake Passage is about 780 km. The Scotia Ridge is generally less than 2000 m deep, but some openings exist at the 3000 m level. The combined effect of Drake Passage and the Scotia Ridge on the Circumpolar Current is quite dramatic: The current accelerates to squeeze through the gap and hits the obstacle at increased speed. It emerges highly turbulent and shifts sharply northward. The shift is a result of several factors, including deflection by the Coriolis force and changes in bottom depth along its path; however, the dynamics are too complex to be considered here in detail.

The *Kerguelen Plateau,* which carries a few isolated islands, reaches and nearly blocks the 2000 m level, although most of its broad plateau is between 2000 m and 3000 m deep. It leaves a narrow gap between itself and Antarctica through which flow can occur below the 3000 m level. No significant departure of the current direction across the plateau is observed.

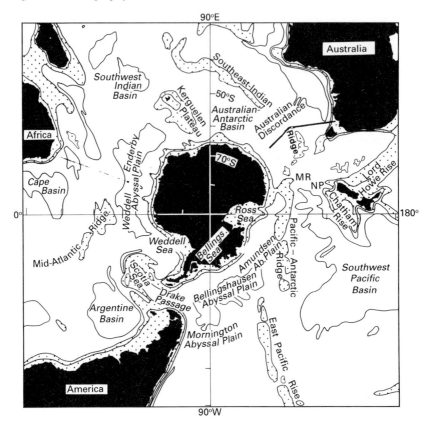

FIG. 6.2. Bottom topography of the Southern Ocean. The 1000, 3000, and 5000 m isobaths are shown, and regions less than 3000 m deep are stippled. Heavy lines near 20°E and 140°E indicate the location of the sections shown in Fig. 6.8. MR: Macquarie Ridge, NP: New Zealand Plateau.

Finally, south of eastern Australia and New Zealand the *Macquarie Ridge*, the *Pacific-Antarctic Ridge*, and the *South-East Indian Ridge* combine to form the third obstacle to the Circumpolar Current that reaches the 3000 m level. The only gap at that depth is located just south of the Macquarie Ridge, at 59°S. On the other hand, much of the Macquarie Ridge reaches above 2000 m, and the ridge carries three islands. The *Campbell Plateau*, a large expanse of water less than 1000 m deep, reaches 54°S just east of the ridge. As a result of the complicated topography and Coriolis force influence, the Circumpolar Current shows a clear northward deflection in this region.

The wind regime

Most of the information required in this section is already included in Figures 1.2 - 1.4 and 4.3; but for the discussion of the Southern Ocean, projection on polar coordinates gives a better representation of the continuity around Antarctica. Figures 6.3 - 6.5 show the relevant maps. The surface pressure map

(Figure 6.3) shows, in both summer and winter, a ridge of high pressure at about 25°-35°S, with highest pressure over each ocean basin, and a trough at about 65°S, just north of the Antarctic continent. The mean geostrophic wind is evidently westerly between the trough and the ridge; but the mean pressure distribution gives only a weak impression of the mean wind stress, which is proportional to the mean value of the square of the wind velocity. The circumpolar belt of westerly winds is dominated by frequent storms, which start in the north and angle southeastwards to die near 65°S where the winds turn into easterlies. It is these storms which determine the mean wind stress. The wind stress figures of Figure 6.4, which are based on observations made by merchant vessels, include the transient storms but not necessarily the effect of wind bursts associated with squalls; they have to be seen as reliable but low estimates. The distribution of curl(τ/f) (Figure 6.5) has to be taken with similar caution.

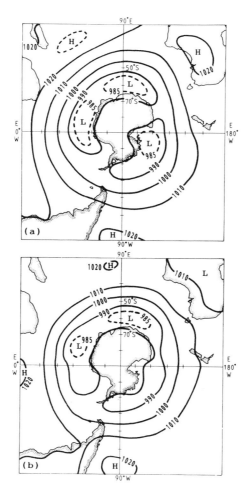

FIG. 6.3. Sea level air pressure (hPa) over the Southern Ocean. (a) July mean; (b) January mean. From Taljaard et al. (1969)

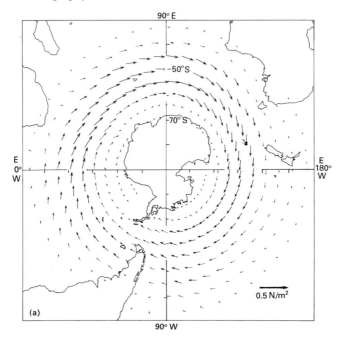

FIG. 6.4. Mean wind stress over the Southern Ocean (see Figure 1.4 for sata sources). (a) annual.

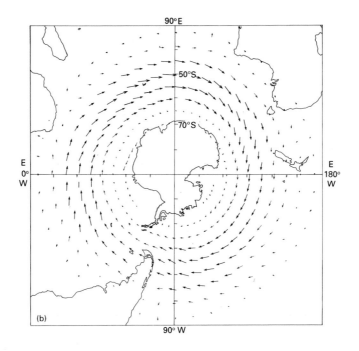

FIG. 6.4. Mean wind stress over the Southern Ocean (see Figure 1.4 for sata sources). (b) summer (December - February). See (a) for units and scale.

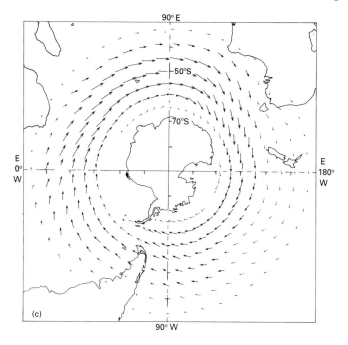

Fɪɢ. 6.4. Mean wind stress over the Southern Ocean (see Figure 1.4 for sata sources). (c) winter (June - August) mean. See (a) for units and scale.

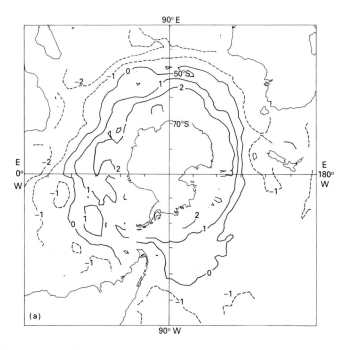

Fɪɢ. 6.5. Mean curl(τ/f) over the Southern Ocean (10^{-3} kg m^2 s^{-1}, from Figure 4.3). (a) annual. Continued page 74.

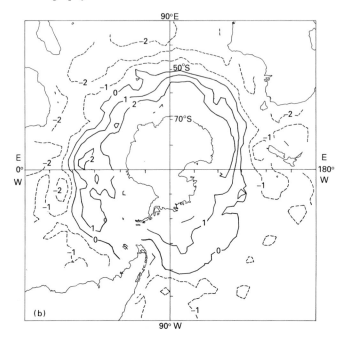

FIG. 6.5. Mean curl(τ/f) over the Southern Ocean (10^{-3} kg m^2 s^{-1}, from Figure 4.3). (b) summer (February - April).

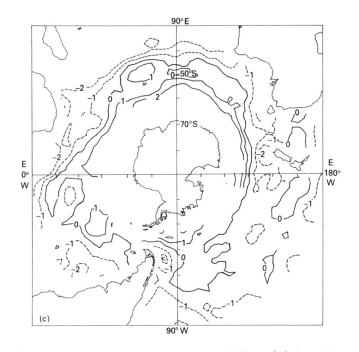

FIG. 6.5. Mean curl(τ/f) over the Southern Ocean (10^{-3} kg m^2 s^{-1}, from Figure 4.3). (c) winter (August - October) mean. Note the uniformity of conditions over all seasons despite variations in wind speed (Figure 6.4).

Close to the Antarctic continent the wind stress shows a reversal from eastward to westward, indicating the presence of Polar Easterlies along the coast. In the northern hemisphere the combination of West Wind Drift, Polar Easterlies, and meridional coastlines produces ocean currents known as the subpolar gyres (which will be discussed in later chapters). The Southern Ocean is devoid of meridional barriers, so the Polar Easterlies drive the *East Wind Drift*, a narrow coastal current which flows westward against the dominant eastward flowing Circumpolar Current. Both currents together are the southern hemisphere equivalent of the subpolar gyres.

One observation which stands out clearly in satellite data on the oceanic wave climate is the combined effect of consistently large wind speeds with little variation in wind direction, and no land barriers which could impede the build-up of a fully developed sea. The combination of infinite fetch around Antarctica and high average wind stress makes the Southern Ocean the region with the largest average wave height. Figure 6.6 shows results from GEOSAT, a satellite launched in March 1985 and active until October 1989. Wave data from GEOSAT were not available to non-military applications during the first eigtheen months of its mission; the data used for Figure 6.6 are therefore restricted to a three-year period between November 1986 and October 1989. This represents the best available estimate of annual mean conditions at present. The figure shows the Southern Ocean as a region of extreme average wave height.

Convergences and divergences

The Subtropical Convergence (STC) was introduced in Chapter 5 as the subduction region for Central Water (Figures 5.3 and 5.5). It is of some 1000 km meridional extent and corresponds to the region of negative $curl(\tau/f)$ in Figure 6.5. The geographic definition of the Southern Ocean is tied to the southern limit of the STC, so it is desirable to define some line across the ocean surface as this limit. Observations show that in the southern STC temperature and salinity do not vary uniformly from north to south; there exists a narrow band around Antarctica where the salinity changes rapidly between 35.0 and 34.5 from north to south and temperatures drop rapidly as well. The feature, which runs parallel to the contour of zero wind stress curl some 5 - 10 degrees north of it, is called the *Subtropical Front*. Figure 6.7 shows three such features: the Subtropical Front, the Antarctic Polar Front indicative of the Antarctic Convergence, and the Antarctic Divergence. It has become accepted terminology to call the region between the continent and the Antarctic Polar Front the Antarctic Zone and the region between the Antarctic Polar and Subtropical Fronts the Subantarctic Zone. The positions of the fronts were constructed from data collected during passages of oceanographic vessels to Antarctica and back. No objective method was used in establishing the lines; rather, they represent an attempt of classical oceanography to interpret a patchy and noisy data set in the framework of a steady state. The observations indicate that at the surface the

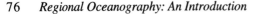

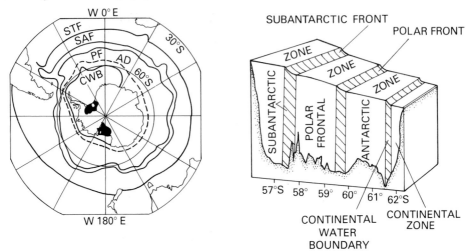

FIG. 6.7. Convergences and divergences of the Southern Ocean and a schematic representation of the zonation in the Southern Ocean. STF: Subtropical Front, SAF: Subantarctic Front, PF: (Antarctic) Polar Front, AD: Antarctic Divergence (dashed line), CWB: Continental Water Boundary. The vertical section is derived from data in Drake Passage where the zonation can extend to the bottom; it generally extends down to the level of Circumpolar Water. The dark regions indicate the Weddell and Ross Sea ice shelves.

transition from the Subantarctic Zone to the Antarctic Zone occurs in two distinct steps rather than one, the so-called *Subantarctic Front* and the *Polar Front* proper (Figure 6.8). A complete zonation of the Southern Ocean therefore includes a Polar Frontal Zone between the Subantarctic and Antarctic Zones, bounded by the two fronts (Figure 6.7). Close to the continent a separate water mass of uniform temperature and low salinity is found in the upper 500 m, separated from water of the Antarctic Zone by another frontal region, the continental water boundary.

In reality, the positions of the fronts and divergences vary greatly in time, and the intensity of sinking and rising motion is variable as well. There can be no doubt that the largest variability occurs in the Subtropical Front. Observations like those shown in Figure 6.9 usually indicate that the band of strong horizontal temperature and salinity gradients does not simply extend in a zonal direction but includes meanders, convolutions and eddies of various sizes. They also indicate large shifts in the meridional position of the front, probably in response to variations in the wind stress field. An idea of these variations can be gained if it is recalled that the wind is geostrophic, too, and meridional shifts of the boundary between the Trades and the Westerlies are coupled with similar shifts in atmospheric isobars. Figure 6.10 shows, for the five years 1972 - 1977, the southernmost position of the 1015 hPa isobar, which on average (Figure 6.3) coincides with the position of the Subtropical Front (Figure 6.7). Seasonal variability appears small in the Indian, Atlantic, and eastern Pacific sectors; but in the western and central Pacific sector the difference between summer and

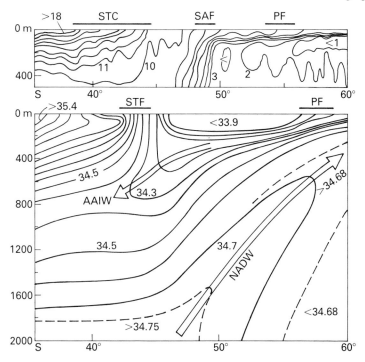

Fɪɢ 6.8. Hydrological sections through the Southern Ocean (summer conditions). (Top) Temperature section in the eastern Indian sector. The Polar Frontal Zone is indicated by the 3 - 9°C isotherms, and the split into the Subantarctic (SAF) and Polar (PF) Fronts by the crowding of isotherms at the surface around 7 - 9°C and 5 - 6°C. The Subtropical Front is indicated by the crowding of isotherms near 13 - 15°C within the Subtropical Convergence (STC). From Edwards and Emery (1982). (Bottom) Salinity section in the eastern Atlantic sector. Crowding of isohalines near 34.5 indicates the Subtropical Front. The Antarctic Divergence is located poleward of the section; near 65°S its salinity maximum is found just below 150 m. Upwelling of North Atlantic Deep Water (NADW) towards the divergence is indicated by the rise of the salinity maximum and sinking of Antarctic Intermediate Water (AAIW) from the Polar Front by the associated salinity minimum. Based on Bainbridge (1980). The different degree of detail between the two sections is the result of very different station density. See Fig. 6.2 for locations of sections.

winter can exceed 10 degrees in latitude. This is the same region where interannual variability is highest (Figure 6.10b), although the Indian sector also displays large differences, particularly in summer.

How this variability in the atmospheric conditions translates into variability of the oceanic conditions is not known, but it can reasonably be argued that variations in the position of the Subtropical Front might be larger in regions of strong meridional shifts of the boundary between the Trades and the Westerlies than elsewhere. Comparison of synoptic surveys of the Subtropical Front (Figures 6.8a and 6.9) with the long-term mean (Figure 2.5a) reveals that at any particular time, property gradients across the front are much stronger than the maximum meridional gradient indicated by the mean property distributions. The fact that the front does not show up in the mean is most likely the result of averaging

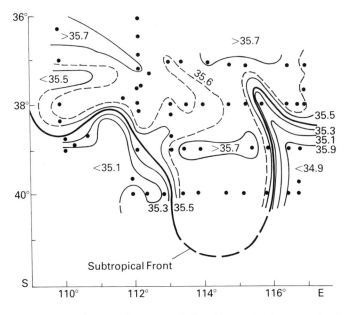

F<small>IG</small>. 6.9. Sea surface salinity in the eastern Indian Ocean showing meander formation in the Subtropical Convergence. The meander formed by the Subtropical Front was also seen in the track of a drifting buoy a few days before the cruise. Dots are station positions. From Cresswell *et al.* (1978).

over a relatively narrow frontal region which changes its position over time. The Antarctic Polar Front and the Divergence appear to be less mobile than the Subtropical Front and therefore show up stronger in the long-term mean, but they, too, display high variability at least in time. This is seen in the paths of buoys tracked by satellite which in the region between the Polar Front and the Divergence typically show rapid movement for several hundred kilometres, followed by longer periods of quite slow movement.

The Antarctic Divergence, on the other hand, is linked with a meridional salinity maximum. At the surface the maximum is masked by low salinity from high precipitation and additional melting of ice, but it is clearly discernible below 150 m (Figure 6.8b). It is produced by upwelling of water with high salinity. The upwelling is unique in that the water reaching the surface comes from great depths; in the Atlantic Ocean, it is lifted from between 2500 m and 4000 m. The deep upwelling occurs for two reasons. Firstly, there is equatorward movement in the Intermediate Water and above, and again in the Bottom Water below 4000 m depth. Poleward movement must therefore occur in the intermediate depth range for reasons of mass conservation, and this water must be lifted to the surface somewhere, to replace the water which sinks to form the Intermediate and Bottom Waters. Secondly, the fact that the Southern Ocean is continuous around Antarctica above the level of the Scotia Ridge precludes net southward movement of water above 2500 m.

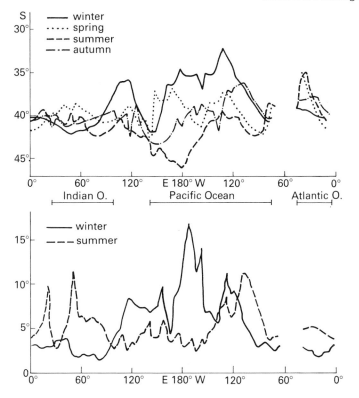

FIG. 6.10. Southernmost position of the 1015 hPa isobar during 1972-1977: (a) seasonal mean, (b) interannual variability, expressed as range (difference between largest northward and southward departure from the mean). Seasons are defined as: summer, December - March; autumn, April - May; winter, June - September; spring, October - November. Note the extreme seasonal and interannual variability in the Western Pacific Ocean. From Streten (1980).

To see this, consider the pressure distribution sketched in Figure 6.11. Along a circle of latitude through Drake Passage pressure must be continuous above the sill depth. Expressed in other words, a net zonal pressure gradient cannot exist, and there cannot be any *net* poleward geostrophic flow in the layer above the sill depth. Only below the depth of the sill can a zonal pressure gradient be supported (in the form of a pressure difference across Drake Passage, with the higher pressure in the west as sketched in Figure 6.11). If at that latitude the Westerlies produce northward Ekman transport in the surface layer, the water moving away from the Divergence can only be supplied be geostrophic southward flow below the sill depth of Drake Passage. The southward motion occurs principally just behind the sill, i.e. in the Atlantic Ocean. It is therefore mainly North Atlantic Deep Water which rises in the Antarctic Divergence. Integrated around Antarctica along 55°S, the northward Ekman transport - calculated from the wind stress data - is about 15 Sv, which is close to the southward flow of North Atlantic Deep Water.

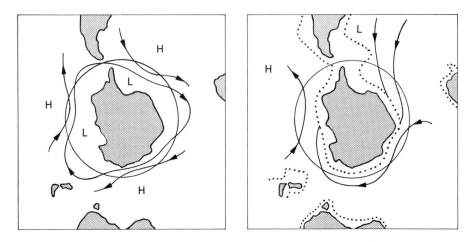

FIG. 6.11. Sketch of a pressure map in the Southern Ocean: (a) above, (b) below the Scotia Ridge sill depth. Above the sill depth, isobars have to be continuous around Antarctica. Geostrophic flow that enters the region encircled by a given latitude through Drake Passage has to leave it again. Below the sill depth, a pressure differential is supported across the sill; inflow into the region is possible.

The importance of the Scotia Ridge for the world ocean circulation was demonstrated in a numerical model (Gill and Bryan, 1971) who showed that the location and intensity of the Antarctic Divergence depends strongly on the sill depth in Drake Passage: a deepening of the passage would weaken the Divergence; closing the passage completely would suppress the Divergence entirely. Topography apparently also plays an important role in determining the width and location of the various frontal zones. In the Indian Ocean sector, the Subantarctic and Subtropical Fronts merge near 95°E in the vicinity of the mid-ocean ridge (Edwards and Emery, 1982), and the Antarctic Polar Front merges with both above the Kerguelen Plateau near 65°E, eliminating the Subantarctic Zone completely (Gamberoni *et al.*, 1982).

Precipitation and ice

If collecting wind data in the Southern Ocean is difficult, collecting rain and snowfall data on the deck of a ship in gale force seas is positively unpleasant. It is therefore not surprising that our information on the mass exchange between the atmosphere and the ocean is particularly scarce in that region. A broad band of relatively high precipitation surrounds Antarctica, centred at about 50°S, the regions of the strongest winds. Since evaporation in these high latitudes is very low, the mass budget between ocean and atmosphere is dominated by fresh water gain for the ocean.

The effect of precipitation on sea surface salinity is augmented by the loss of salt from the surface in winter, as brine rejected by sea ice sinks to great depth, and the melting of ice in summer; this explains the low salinity of near-surface water in the Antarctic region (Figure 2.5b). The sea ice does not extend far past

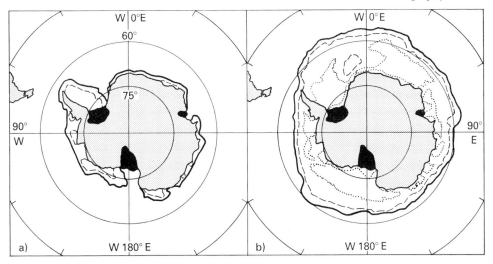

FIG. 6.12. Mean seasonal ice conditions in Antarctica, based on satellite data for 1973 - 1976. (a) Late summer (February), (b) late winter (October). Black areas indicate ice shelf (regions where the ice sits on the ocean floor): the ice shelves of the Weddell Sea at 50°W, the Ross Ice Shelf at the date line, and the Amery Ice Shelf at 70°E. The solid, broken, and dotted lines indicate ice coverage of 15%, 50%, and 85%.

Antarctica in summer, but it covers an area the size of the continent in late winter (Figure 6.12). Estimates of ice coverage from satellite data (Gloersen and Campbell, 1988) give the average ice extent for 1978-1987 as $3.5 \cdot 10^6$ km^2 in summer and $18 \cdot 10^6$ km^2 in winter, of which 18% was open water. The extent of open water in otherwise ice-covered regions (polynya) varied greatly; large polynya were seen in the central Weddell Sea during three of the first four winters but not in later years. Icebergs can be found further north than sea ice, as far as 50°S at any season, their great mass preventing melting within a season.

Hydrology and water masses

Having established the main features of Southern Ocean dynamics we start the discussion of its hydrology by looking at a meridional section. Figure 6.8b shows salinity along a section in the eastern south Atlantic Ocean. The general southward and upward movement of high salinity North Atlantic Deep Water from depths below 2000 m is reflected in the shape of the isohalines. A substantial portion of this water comes to within 200 m of the surface at the Antarctic Divergence where it warms the surface water, melting the sea ice and the snow that falls on it, and sinks again at the Antarctic Polar Front. By the time it is subducted it can no longer be recognized as Deep Water, having been warmed and diluted by rain and snow on its northward passage, and is then known as the low salinity Antarctic Intermediate Water.

Modification of properties in the vicinity of the fronts is particularly strong during winter when convection creates a deep surface layer with water of uniform temperature and salinity in a region of usually strong horizontal and vertical gradients. Water in such layers is often called Mode Water and the winter water in the Subantarctic zone referred to as *Subantarctic Mode Water*. This water is not a water mass but contributes to the Central Water of the southern hemisphere. In the extreme east of the south Pacific Ocean it is responsible for the the the formation of Antarctic Intermediate Water (McCartney, 1977; England *et al.*, in press).

The intense mixing processes which form the water masses of the Southern Ocean come out clearly if a T-S diagram of surface observations along a meridional line of stations is compared with T-S diagrams from stations in the Antarctic and Subantarctic zones (Figure 6.13). Both profiles start at the T-S point of Antarctic Bottom Water and end on the surface T-S curve but do this in a distinctly different way. In the Antarctic zone, water at the surface has very low temperatures, ranging down to the freezing point of -1.9°C, and low salinities as a result of ice melting in summer. In a hydrographic station in this zone (the dotted line in Figure 6.13) the influence of this low surface salinity is felt in the upper 100 - 250 m; this water is called *Antarctic Surface Water*. In the Subantarctic zone, surface water has a larger temperature and salinity range since seasonal variations of solar heating, rainfall, and evaporation become more important. The temperature range of this *Subantarctic Upper Water* spans 4 - 10°C in winter and 4 - 14°C in summer, with a salinity varying between 33.9 and 34.9 and reaching as low as 33.0 in summer as the ice melts. This produces a shallow surface layer of low salinity and an intermediate salinity maximum between 150 m and 450 m depth, as seen in the T-S data of the station from the Subantarctic zone (the

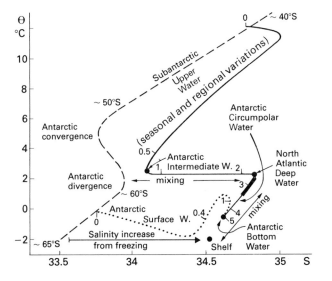

FIG. 6.13. A T-S diagram for a station in the Subantarctic (full line) and in the Antarctic zone (dotted line), and a T-S diagram from surface observations along crossings of the Antarctic Polar Front and Divergence (dashed line). Depth on the vertical T-S profiles is indicated in km.

full line in Figure 6.13). The difference between the full line and the dashed line in the T-S range of the Upper Water indicates that the figure compares data from different seasons. There are also variations between the various sectors of the zone, with lowest salinities in the Pacific and highest in the Atlantic sector.

The transformation of North Atlantic Deep Water into Antarctic Intermediate Water is seen in the T-S diagram as a mixing process between the deeper waters and surface water at the Antarctic Divergence. South of Australia Intermediate Water consists of some 60% Subantarctic Upper Water and 40% Circumpolar Water; only in the extreme eastern south Pacific Ocean and in the Scotia Sea of the Atlantic sector, where the T-S properties of Subantarctic Mode Water resemble those of Antarctic Intermediate Water, is Intermediate Water apparently formed in direct contact with the atmosphere. Formation of *Antarctic Circumpolar Water* through mixing of North Atlantic Deep Water and Antarctic Bottom Water

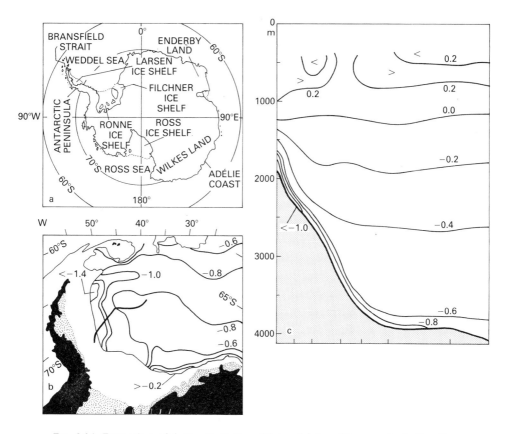

FIG. 6.14. Formation of Antarctic Bottom Water. (a) Locality map, including the regions where deep convection occurs, (b) bottom potential temperature (°C) in the Weddell Sea - the stippled area indicates ice shelf, and the thin line is the approximate 3000 m contour, (c) a vertical section of potential temperature (°C) in the Weddell Sea. The position of the section is shown by the heavy line in (b). From Warren (1981a)

is indicated by the straight line formed by T-S data between the Deep Water and the water on the Antarctic shelf.

It has of course to be remembered that the deep vertical and meridional circulation occurs on the background of intense zonal flow. This is particularly evident when the formation of Antarctic Bottom Water is investigated in detail. Its origin lies in deep convection at the continental shelf driven by the freezing of sea ice (Figure 6.13), but its final properties are shaped during intense mixing with the water of the Circumpolar Current (Circumpolar Water) while sinking to the bottom. It is therefore incorrect to say that the formation process for Antarctic Bottom Water is convection alone; rather, it is a combination of convection and subsurface mixing. It is seen that the properties of Circumpolar and Bottom Water are defined in a process of mutual interaction which draws on the properties of North Atlantic Deep Water as well.

The areas where convective sinking occurs (Figure 6.14a) are believed to be relatively limited in size. The only location where sinking to the ocean floor by convective overturning has been identified from data is Bransfield Strait; but the water which sinks there is collected in an isolated trough of between 1100 m and 2800 m depth and of little consequence for Bottom Water formation. In all other areas (the Weddell Sea, the Ross Sea, and probably also along the Adélie Coast and Enderby Land) the sinking occurs underneath the ice and is difficult to verify directly. There are, however, sufficient data which show the effect of the sinking. In the Weddell Sea, which probably contributes most to Bottom Water formation, the water flows westward under the influence of the Coriolis force as it sinks, forming a thin layer of extremely cold water above the continental slope (Figure 6.14c). It mixes with the overlying water, which is recirculated with the large cyclonic eddy in the central Weddell Sea. This water, known as *Weddell Deep Water*, has very stable properties; its potential temperature usually is above 0.4°C and below 0.7°C. It is renewed by surface cooling and subsequent convection in the ice-free central part (polynya) of the Weddell Sea (Gordon, 1982). The opportunity for the water on the slope to mix with Weddell Deep Water is enhanced by the fact that sinking does not occur along the shortest possible path but in nearly horizontal motion along the slope. On reaching 65°S some of the water gets injected into the Circumpolar Current, where it continues to mix with the Circumpolar Water. The properties of the sinking water are somewhat known from ships that have been trapped in the ice over winter. They measured temperatures around the freezing point (about -1.9°C) and salinities of 34.7 - 34.9. By the time the water leaves the Weddell Sea its temperature has risen to -0.8°C. The further path of Antarctic Bottom Water can be followed by looking at the potential bottom temperature map (Figure 6.15); it indicates the Adélie Shelf and the Ross Sea as other important regions where cold - and saline - water is injected from the surface. Eventually, the water spreads from the Circumpolar Current (i.e. with the properties of Circumpolar Water) into all three oceans. At that stage its properties are best described as 0.3°C and 34.7 salinity.

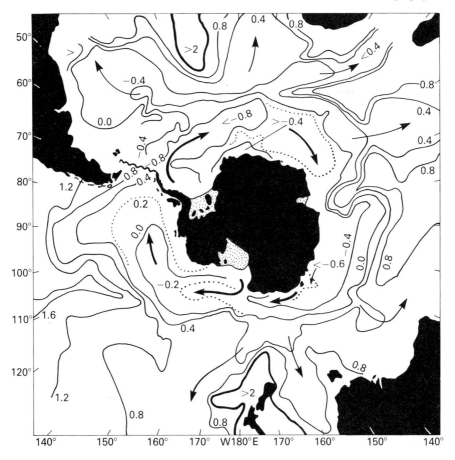

FIG. 6.15 Bottom potential temperature in the Southern Ocean. Isotherms are drawn every 0.4°C, with the exception of the New Zealand Plateau and the Mid-Atlantic Ridge where the 2°C isotherm follows the 0.8°C isotherm. The arrows show inferred movement of Antarctic Bottom Water. Southward intrusions of high potential temperatures reflect ridges less than 4000 m deep. Northward extensions of low potential temperatures indicate movement of Antarctic Bottom Water over sills; the deeper the sill, the lower the temperature. Adapted from Gordon (1986b).

Although much of the above discussion and all of the figures are based on modern data, the best way of summarizing the hydrography of the Southern Ocean is to reproduce a block diagram designed half a century ago. Figure 6.16 shows the interplay of strong zonal currents, meridional flow caused by deep convection, convergences and divergences, and water mass formation and spreading.

Estimation of zonal and meridional flow

As mentioned earlier, the southward flow of North Atlantic Deep Water is estimated at 15 Sv from oceanographic data. It is opposed by northward flow of 2.5 - 5 Sv from the formation of Antarctic Bottom Water. Northward Ekman

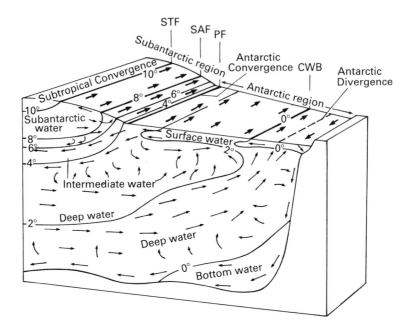

Fig 6.16. Block diagram of the circulation in the Southern Ocean. From Sverdrup
et al. (1942), with the addition of frontal locations.(STF: Subtropical Front, SAF:
Subantarctic Front, PF: Polar Front, CWB: Coastal Water Boundary).

transport in the West Wind belt must make up the balance. Although most of
the zonal transport occurs in the Atlantic sector, closure of the transport budget
involves all three oceans. This is mainly caused by that part of the Ekman layer
flow that contributes to the formation of Antarctic Intermediate Water; some
of the Ekman transport therefore leaves the Antarctic sector and is recirculated
through the Indian and Pacific Oceans before it can ultimately contribute to
the replacement of the water sinking in the north Atlantic Ocean.

When it comes to estimating the zonal transport, the difficulty is in the
determination of an acceptable depth of no motion. Again, the sill depth of the
Scotia Ridge plays an important role. The observations of currents in Drake
Passage mentioned earlier came from moorings which were deployed for a one-
year period in 1978. They showed that below 2500 m depth there was much short
term current fluctuation but little annual mean current. From these and later
observations (Whitworth and Peterson, 1985) spanning a total period of about
four years, the mass transport for the Circumpolar Current through Drake Passage
was determined as 128 ± 15 Sv, with maximum variations of over 50 Sv within
two months and an indication of a winter minimum in July. Most of the mean
transport could be accounted for by geostrophic flow above an assumed depth
of no motion of 2500 m. The 50 Sv fluctuations were associated mostly with
changes in sea level gradient across the passage with very little density change;
the corresponding current variations were therefore uniform in depth. Earlier

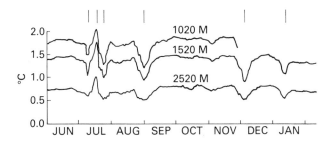

FIG. 6.17. A time series of temperature (°C) obtained at a mooring in central Drake Passage, showing the passage of five cold-core rings and one warm-core ring. From Pillsbury and Bottero (1984).

observations, for a one-year period and again in Drake Passage (Bryden and Pillsbury, 1977), gave an average of 139 Sv but a total range between 28 and 290 Sv. More definite estimates will become available with the completion of the World Ocean Circulation Experiment (WOCE). Current speeds are generally low, between 0.05 and 0.15 m s^{-1}, because of the large width and depth of the current, although 0.5 m s^{-1} and even 1 m s^{-1} have been observed in jets associated with frontal regions on occasions. Because of their enhanced horizontal density gradients and associated geostrophic currents, the frontal regions carry most of the transport; observations from Drake Passage (Nowlin and Clifford, 1982) indicate that above 2500 m, 75% of the total flow occurs in the frontal zones, which occupy only 19% of the cross-sectional area.

Estimating the meridional heat flux, a key element in the global heat budget, is even more difficult. The transfer of heat from the Circumpolar Current to the atmosphere has been estimated (Gordon and Owens, 1987) as $3 \cdot 10^{14}$ W. This heat loss must be balanced against poleward oceanic heat flux across the current. The primary movers of heat appear to be the large eddies generated by the current in interaction with topography. Little is known about the frequency of eddy formation and the life expectancy of individual eddies. Satellite altimeter observations (Figure 4.8) imply that they are not uniformly distributed along the path of the Circumpolar Current but are more frequent east of the Scotia Ridge and in the region of the Macquarie Ridge (between Tasmania and New Zealand). These regions therefore are likely to play a major role in the poleward transport of heat. Temperature records from Drake Passage (Figure 6.17) indicate the passage of five cyclonic (i.e. cold core) eddies and one anticyclonic (warm core) eddy over a period of eight months. The eddies were of 30 - 130 km diameter, extended to at least 2500 m depth and were moving northward across the Circumpolar Current at about 0.04 m s^{-1}. This northward movement of (on average) cold water has to be compensated by poleward movement of comparatively warmer water and therefore represents a poleward flux of heat. Provisional estimates (Keffer and Holloway, 1988) give values of $1.3 - 5.4 \cdot 10^{14}$ W, enough to balance the estimated heat loss to the atmosphere.

CHAPTER 7

Arctic oceanography; the path of North Atlantic Deep Water

The importance of the Southern Ocean for the formation of the water masses of the world ocean poses the question whether similar conditions are found in the Arctic. We therefore postpone the discussion of the temperate and tropical oceans again and have a look at the oceanography of the Arctic Seas.

It does not take much to realize that the impact of the Arctic region on the circulation and water masses of the World Ocean differs substantially from that of the Southern Ocean. The major reason is found in the topography. The Arctic Seas belong to a class of ocean basins known as mediterranean seas (Dietrich *et al.*, 1980). A mediterranean sea is defined as a part of the world ocean which has only limited communication with the major ocean basins (these being the Pacific, Atlantic, and Indian Oceans) and where the circulation is dominated by thermohaline forcing. What this means is that, in contrast to the dynamics of the major ocean basins where most currents are driven by the wind and modified by thermohaline effects, currents in mediterranean seas are driven by temperature and salinity differences (the salinity effect usually dominates) and modified by wind action. The reason for the dominance of thermohaline forcing is the topography: Mediterranean Seas are separated from the major ocean basins by sills, which limit the exchange of deeper waters.

The circulation in mediterranean seas can be divided into two classes, depending on the freshwater budget at the surface (Figure 7.1). If evaporation over the mediterranean sea exceeds precipitation, freshwater loss in the upper layers increases the density of the surface waters, resulting in deep vertical convection and frequent renewal of the water below the sill depth. The circulation at the connection between the mediterranean sea and the ocean basin consists of inflow of oceanic water in the upper layer and outflow of mediterranean water in the lower layer. The inflow is driven by the freshwater loss in the mediterranean sea; in addition, the density difference between the salty mediterranean water and the fresher oceanic water causes outflow of mediterranean water in the deeper part of the connecting channel with a compensating inflow above it. The outflowing water sinks until it reaches

89

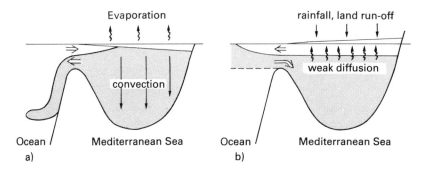

Fɪɢ. 7.1. Schematic illustration of the circulation in mediterranean seas; (a) with negative precipitation - evaporation balance, (b) with positive precipitation - evaporation balance.

the depth where the density of the oceanic water matches its own; it then spreads on this density surface, where it can be traced through the ocean basin by its high salinity. (A prominent example is the Eurafrican Mediterranean Sea; its impact on the salinity of the Atlantic Ocean has already been noted in the discussion of Figure 2.5.) Because the salinity of the oceanic water is increased as it passes through the mediterranean sea, this type of mediterranean sea is also known as a concentration basin.

If, on the other hand, precipitation over the mediterranean sea exceeds evaporation, the freshwater gain drives an outflow into the ocean basin through the upper layer. It also decreases the surface density, and the resulting density difference at the sill causes inflow of oceanic water through the lower layer and additional compensating outflow in the upper layer. A strong pycnocline is established, and renewal of the deeper waters is strongly inhibited. Inflow of oceanic water is usually the only renewal process of any significance, and if the connection across the sill is narrow or the deep water volume large, deep water renewal is not always sufficient to prevent the depletion of oxygen in the deep basins. In these cases the sea is devoid of life (apart from sulfur-reducing bacteria) below the pycnocline. This type of mediterranean sea is also known as a dilution basin.

Mediterranean seas are a special class of marginal seas, which are defined as those parts of the World Ocean that are separated from the major deep ocean basins by topographic features such as islands or bay-like coastline configurations. Examples of marginal seas are some of the major shelf regions, e.g. the North Sea or the East China Sea, and topographically semi-enclosed ocean regions, e.g. the Tasman Sea or the Bay of Bengal. While the circulation and stratification in these marginal seas may be strongly modified by thermohaline or tidal forcing it is still dominated by the wind. Mediterranean seas are the only marginal seas where thermohaline forcing dominates. (For readers familiar with estuarine dynamics, mediterranean seas can be defined as those marginal seas which display a circulation of the estuarine type.)

Bottom topography

A look at the topography of the Arctic Seas (Figure 7.2) clearly establishes their mediterranean character. The major connection with the three oceans is to the Atlantic Ocean where a 1700 km wide opening exists along a large oceanic sill running from Greenland across to Iceland, the Faroe Islands and Scotland. Approximate sill depths are 600 m in Denmark Strait (between Greenland and Iceland), 400 m between Iceland and the Faroe Islands, and 800 m in the Faroe Bank Channel (between the Faroe Islands and Scotland). Minor openings to the Atlantic Ocean exist through the Canadian Archipelago, mainly through Nares Strait and Smith Sound with a sill depth of less than 250 m and Barrow Strait and Lancaster Sound with about 130 m sill depth. The connection with the Pacific Ocean through Bering Strait is only 45 m deep and 85 km wide and of little consequence for the Arctic

Fɪɢ. 7.2. Bottom topography of the Arctic Mediterranean Sea. The 1000, 3000, and 5000 m isobaths are shown, and regions less than 3000 m deep are stippled.

circulation (It is important for the global freshwater balance; see Chapter 18).

Within the confines of the Arctic Mediterranean Sea are the Greenland, Iceland, and Norwegian Seas, and the Arctic or North Polar Sea proper which includes the various regions of the large Siberian shelf area, i.e. (beginning at Bering Strait and moving westward) the Chukchi, East Siberian, Laptev, Kara, Barents, and White Seas, and the Lincoln and Beaufort Seas on the Greenland-Canadian-Alaskan shelf. The Greenland, Iceland, and Norwegian Seas communicate with the North Polar Sea through Fram Strait, which between Greenland and West Spitsbergen (the westernmost island of the Svalbard group) is 450 km wide and generally deeper than 3000 m, with a sill depth somewhat less than 2500 m. The North Polar Sea itself is structured through three ridges into a series of four basins. The nomenclature for these features is not uniform; the GEBCO charts identify the Canada Basin with a depth of 3600 - 3800 m, the Makarov Basin with approximately 3900 m depth, the Amundsen Basin with depths of 4300 - 4500 m, and the Nansen Basin with depths between 3800 m and 4000 m. The ridges between these basins are the Alpha and Mendeleyev Ridge system which rises to between 1200 m and 1500 m, the Lomonossov Ridge with depths between 850 m and 1600 m, and the Arctic Mid-Ocean Ridge, sometimes referred to as the Nansen Ridge, which reaches 2500 m depth. The Amundsen and Nansen Basins and the Arctic Mid-Ocean Ridge are often combined into what is then called the Eurasian Basin. Herman (1974) lists other names.

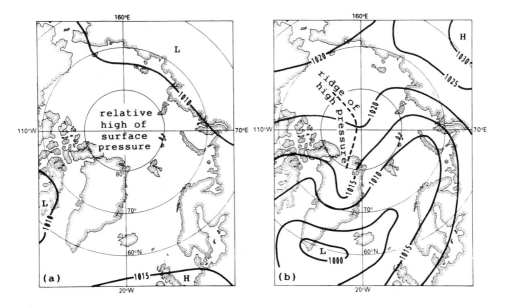

Fig. 7.3. Air pressure (hPa) at sea level over the Arctic Mediterranean Sea. (a) July mean, (b) January mean, both for the period 1950 - 1980. Data from University of East Anglia (1992).

The Arctic Mediterranean Sea has a total volume of $17 \cdot 10^6 \, \text{km}^3$, which is 1.3% of the World Ocean. If the seas south of Fram Strait are excluded, the Arctic Sea proper covers an area of about $12 \cdot 10^6 \, \text{km}^2$ and contains $13 \cdot 10^6 \, \text{km}^3$ of water. This represents about 3% of the World Ocean area but only 1% of its volume. The reason is found in the large expanse of shelf area. Along the American coast the shelf is only 50 - 90 km wide, but on the Siberian side its width exceeds 800 km in most places. The shelf is also rather shallow, 20 - 60 m in the Chukchi Sea, and probably similar in the East Siberian Sea, 10 - 40 m in the Laptew Sea, an average depth of 100 m in the Kara Sea, and 100 - 350 m in the Barents Sea. It represents nearly 70% of the surface area of the Arctic Sea. Numerous large rivers empty into the Arctic shelf seas, reducing their salinity. These shallow shelf areas therefore greatly influence surface water conditions in the Arctic Mediterranean Sea.

The wind regime

Again, Figures 1.2 - 1.4 contain the relevant information but projection on polar coordinates gives the better representation. Figures 7.3 - 7.4 show air pressure and surface winds over the Arctic region. The comments made about the validity of wind stress estimates over the Southern Ocean are even more relevant here, since the Arctic ocean includes ocean basins at very high latitudes where wind observations are extremely sparse. Nevertheless, while the winds of Figure 7.4 may not give the magnitudes of all wind stresses right they give some indication of their directions.

High pressure in the vicinity of the north pole determines the wind system over the Arctic Mediterranean Sea through the year. It is more prominent during winter when it takes the form of a ridge from the Canada Basin towards northern Greenland. Pressure gradients are reduced during summer, but the pressure near the pole is still higher than over the continents. Most of the Arctic seas are therefore under the influence of the Polar Easterlies and display anticyclonic (westward) surface circulation, in contrast to the Southern Ocean where the effect of the Polar Easterlies is only noticeable in a weak westward current along the Antarctic continent. Winds are much stronger in winter, and the annual mean resembles the January distribution (Figure 7.4). Over the Greenland and Norwegian Seas the wind system is dominated by the Icelandic atmospheric low, which generates cyclonic water movement.

The estimated surface circulation, shown in Figure 7.5, is derived from drift tracks of research stations on ice islands and ships, trapped in the ice either accidentally or deliberately. The most famous of these crossings of the polar seas was the drift of the research vessel Fram, built by the Norwegian explorer Fritjof Nansen to withstand the pressure of the ice, during 1893 - 1896. Earlier, the American vessel Jeanette had been caught in the ice of the Chukchi Sea near Bering Strait in November 1879 and crushed by the ice in June 1881;

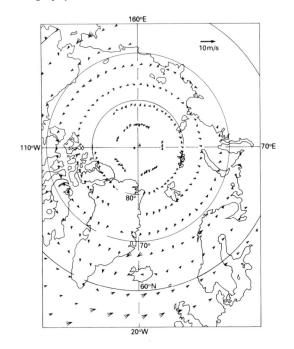

FIG. 7.4. Surface winds over the Arctic Mediterranean Sea. (a) annual mean. See Figure 1.2 for data sources.

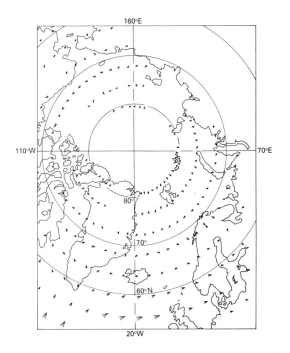

FIG. 7.4. Surface winds over the Arctic Mediterranean Sea. (b) July mean. See Figure 1.2 for data sources and (a) for units and scale.

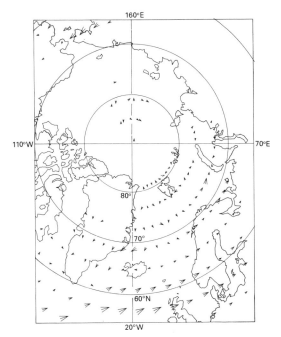

Fig. 7.4. Surface winds over the Arctic Mediterranean Sea. (c) January mean. See
Figure 1.2 for data sources and (a) for units and scale.

wreckage from that vessel was recovered in 1884 on the southwest coast of
Greenland. Other early observations include the drifts of the Maud from
Bering Strait to the New Siberian Islands in 1918 - 1925 and of the Sedov
which took $2^1/_2$ years to drift from the New Siberian Islands to Spitsbergen
during 1937 - 1940.

These and the numerous drift tracks from ice stations (such as NP1, NP4
and T3 in Figure 7.5) of the last forty years establish a picture of mean
westward circulation around a centre of motion close to the centre of the
atmospheric high, with outflow from the Arctic Mediterranean Sea along the
coast of Greenland and inflow along the Norwegian coast. The fact that inflow
and outflow both occur near the surface is due to a combination of factors.
The width of the Greenland-Iceland-Faroe-Scotland Ridge allows the Coriolis
force to exert an influence on the currents; it concentrates the outflow of
low salinity water in the East Greenland Current on the western side, leaving
room for inflow in the Norwegian Current on the eastern side. The inflowing
water has its origin in the temperate and subtropical gyres of the North
Atlantic Ocean. Its low density is due to the high temperatures in the Gulf
Stream extension. Current speeds in the East Greenland and Norwegian
Currents are usually in the vicinity of 0.2 m s^{-1} but can reach 0.5 m s^{-1} on
occasions. Data from current meter moorings deployed in the East Greenland
Current for one year (Foldvik *et al.*, 1988) show a decrease of average current

speed with depth to about 0.05 m s^{-1} at 600 m; the passage of eddies can increase these values more than threefold or reverse the flow. Generation of strong currents in the polar anticyclonic gyre by the wind is inhibited by permanent ice coverage; average speeds are close to 0.02 m s^{-1} (2 km per day). Some indication of eastward flow produced by the West Wind Drift is seen on the Siberian shelf. However, measurements in these regions are sparse, and the currents are likely to be influenced by coastline topography and river inflow; so most of the estimates of flow on the shelf are hypothetical. Eastward flow on the Alaskan shelf is well documented as a wind-driven extension of the inflow through Bering Strait.

Precipitation and ice

Compared to conditions for data collection in the Southern Ocean, collection of rainfall and snowfall data on the stable platforms of drifting ice stations is no serious challenge. Reasonable information on rainfall and snowfall is therefore available. Precipitation is low in the region of the Polar Easterlies but significant in the subpolar regions which are dominated by the West Wind Drift and its associated large variability and high storm frequency. Since most of the precipitation occurs over ice which does not melt until it is exported from the Arctic region, local snowfall does not play a major role in the oceanic mass budget. The major contribution comes from precipitation over Siberia and the resulting river run-off, estimated in total as 0.2 Sv. With evaporation over ice being comparatively low, the Arctic Mediterranean Sea is a dilution basin, i.e. its outflow is fresher on average than its inflow.

Precipitation over Greenland feeds the glaciers which produce the several thousand icebergs annually found in the East Greenland, West Greenland, and Labrador Currents. A few of them - and in winter quite a few, between 50 and 100 - reach the area south of the Newfoundland Bank and enter the main shipping route between North America and Europe. When the cruise liner Titanic hit one of these on her maiden voyage in 1912 and sank, taking 1490 lives, the International Ice Patrol Service was established. Between March and July it monitors and reports the positions of all icebergs that may become a danger to transatlantic shipping.

A remarkable feature of the ice distribution is the extreme southward extent of the ice-affected region along the American continent and the extreme northward extent of the region which is permanently ice-free along the Norwegian coast (Figure 7.6). Nowhere else in the world ocean can ports at 70° latitude be reached by sea during the entire year, as is the case with the Norwegian cities of Tromsö and Hammarfest; and nowhere else do icebergs reach 40° (the latitude of southern Italy) as they do south of the Newfoundland Bank. This is of course the result of the temperature difference between the outgoing and incoming water which is so marked that it can

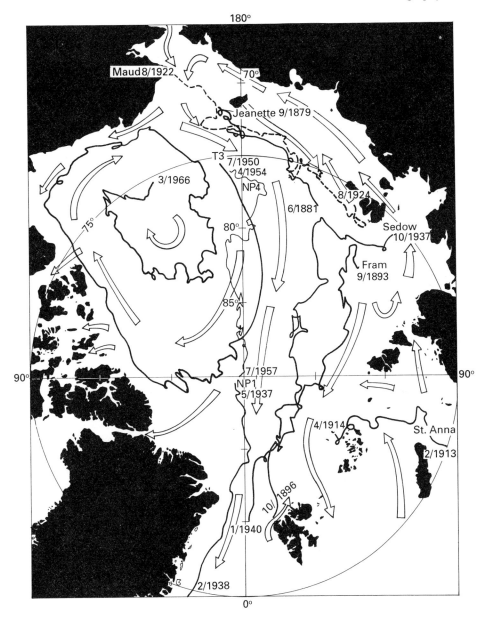

FIG. 7.5. Mean circulation of surface waters, with some tracks of vessels and ice stations.

even be seen in severely space and time averaged data (for example in Figure 2.5). The average temperature of the Norwegian Current is 6 - 8°C; the average temperature of the East Greenland Current is below -1°C. This temperature difference easily compensates the effect of salinity on density, which is higher (34 - 35) in the Norwegian Current and along the Norwegian coast than in the East Greenland Current (30 - 33); so the water is denser

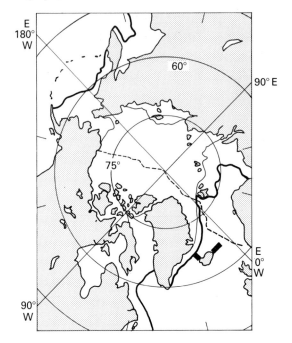

FIG. 7.6. Ice coverage in the Arctic ocean, based on satellite data from 1974 - 1976. The thick line shows the extent of sea ice in late winter (March), the thin line the ice extent in late summer (September). The heavy bars near Iceland give the southern ice limit during March 1968; they are included to indicate the degree of interannual variability. The broken line shows the location of the section shown in Fig. 7.8.

on the western side and isopycnals slope downward to the east, in accordance with the thermal wind relation (Rule 2a). The boundary between the two currents, which coincides more or less with the ice boundary, is characterized by shear-generated eddies, of 10 - 20 km diameter and with a life span of 20 - 30 days, which affect the movement of water to a depth of several hundred metres.

Estimates of variations in ice extent from satellite observations (Gloersen and Campbell, 1988) indicate an area of $8 \cdot 10^6$ km^2 for the summer season and $15 \cdot 10^6$ km^2 for winter, of which 14% is open water. The average life span of large polynya was 75 days, half the life span of polynya in the Southern Ocean where ice movement is less constrained by coastlines, favouring the development of large regions of open water.

Hydrology and water masses

The mediterranean character of the Arctic Seas comes out clearly when we now look at the circulation and water masses at and below the sill depth. Since precipitation over the Arctic Mediterranean Sea exceeds evaporation, the region acts as a dilution basin for the Atlantic Ocean. Water exchange across the sill follows the scheme of Figure 7.1b, with three modifications.

The first modification occurs in the surface layer where - as mentioned before - the Coriolis force restricts outflow into the Atlantic Ocean to the western side of the sill. The second modification concerns the renewal of the water in the deep basins. The water below the sill depth is several degrees colder than the water entering the Arctic Seas from the Atlantic Ocean and is thus denser. The inflowing water therefore does not sink to the bottom but spreads through an intermediate layer, and a hydrographic station from anywhere in the Arctic Mediterranean Sea (the Norwegian Sea being the only exception) displays a layering of three water masses (Figure 7.7). Finally, as a third modification it will be seen that outflow of water from the Arctic Seas is not restricted to the surface layer.

We begin the discussion of the water masses by looking at *Arctic Bottom Water*. For a long time it was believed that its formation region is in the Greenland or Norwegian Seas. It is now known that the formation process involves the interplay of two sources, Greenland Sea Deep Water and water from the Arctic shelf regions. *Greenland Sea Deep Water* is formed during winter in the central Greenland Sea, where the cooling of surface water causes intense vertical convection. Sinking of water to the bottom occurs in events, clearly related to the passage of storm systems; the events last less than a week and are limited to regions a few kilometres across. In each event, individual cooling cycles occur on even smaller time and space scales. At the beginning of each cycle the surface layer is quite fresh (see Figure 7.7), and concentration of

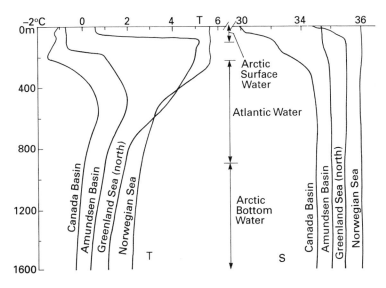

FIG 7.7. Temperature (T °C) and salinity (S) profiles in the Arctic Mediterranean Sea. The scale is correct for the Canada Basin; other profiles are offset by 1 unit from each other. Arctic Surface Water is not present in the Norwegian Sea, where inflow of Atlantic Water extends to the surface. From Coachman and Aagaard (1974) and Osborne *et al.* (1991).

salt is required to initiate sinking. This is achieved by ice formation. Eventually the density increase is sufficient to overcome the barrier posed by the warm but saline water below. Sinking sets in and is compensated by upwelling of the warm water which melts the ice, bringing the cycle to an end. The Greenland Sea is therefore never completely ice-covered, and formation of Greenland Sea Deep Water occurs in mainly open water (Rudels and Quadfasel, 1991).

The other source of Arctic Bottom Water is found in the Arctic shelf regions (Aagaard et al., 1985b). The low salinity of shelf water, which reflects input of freshwater from rivers, facilitates ice formation. As salt is rejected from the ice, the salinity of the water below the ice is increased. The shelf regions therefore produce a large variety of water bodies through the seasons, some fresh, some salty, some very salty, but all very cold. When the salinity exceeds 35 - which has been observed on occasions in the Chukchi Sea; water in 300 m deep depressions of the Barents Sea has also been found to have salinities above 35 and temperatures below -1.8°C - the water is dense enough to sink to the bottom of the basins of the Arctic Sea, thus contributing to the formation of Arctic Bottom Water. Fjords of the Svalbard island group have also been found to contribute to this input of high salinity water.

Salinities in the Arctic Bottom Water are generally close to 34.95 but highest in the Canada Basin (Figure 7.8). The contribution of the shelf regions to Bottom Water formation is evident from the fact that salinities in the Canada Basin are higher than could be explained from surface salinities in the Norwegian or Greenland Seas and therefore cannot be the result of mixing with the inflowing high-salinity water alone.

Figure 7.9 is a schematic representation of Arctic Bottom Water formation. The Norwegian Sea, which apparently does not cool enough to experience deep winter convection, nevertheless plays an important role in the formation process as a mixing basin where Arctic Bottom Water obtains its final characteristics. Greenland Sea Deep Water, the densest component, is confined to the centre of a cyclonic gyre in the centre of the Greenland Sea. Its temperature is consistently below -1.1°C, lower than the temperature of Norwegian Sea Deep Water (-0.95°C), and the lowest of all bottom water in the Arctic Mediterranean Sea. It entrains from its flanks water from the depths of the Amundsen and Nansen Basins which contains the contribution from the Arctic shelf. A salinity maximum in the East Greenland Current at 1500 m depth is interpreted as evidence for the presence of the shelf contribution; Smethie *et al.* (1988) estimate the amount of Arctic Bottom Water produced on the shelf at up to 0.1 Sv and the sinking of surface water at 0.5 Sv. Further downstream, about 1 Sv of Greenland Sea Deep Water enters the southern Norwegian Sea, mixes with Norwegian Sea Deep Water, and recirculates into the basins of the Arctic Sea. Eastward flow in a 2000 m deep channel north of Jan Mayen has been verified through direct current meter measurements

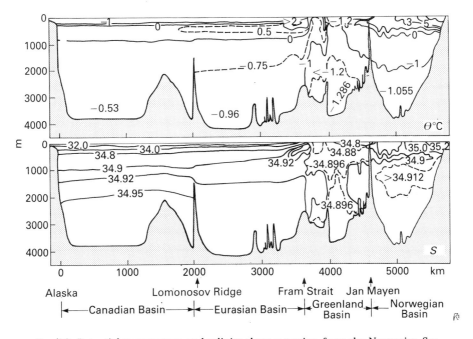

FIG. 7.8. Potential temperature and salinity along a section from the Norwegian Sea into the Canada Basin. The penetration of Atlantic Water - from the surface in the Norwegian Sea to the depth range between Surface and Bottom Water - is indicated by the temperature and salinity maxima near 500 m depth. Note that the temperature and salinity increments between isotherms and isohalines are not constant over the respective ranges. See Fig. 7.6 for location of the section. From Aagaard *et al.* (1985b).

which for the depth range 1700 - 2000 m gave average current speeds of 0.05 m s^{-1} for a seven month period in 1981 and 0.08 m s^{-1} for a ten month period in 1983 and 1984 (Sælen, 1988).

From the point of view of world ocean water masses, the product of the complex mixing process indicated in Figure 7.9 can justifiably be called Arctic Bottom Water, since it not only fills all deep Arctic basins but also plays an important role in the deep circulation of the world ocean (as will be discussed below). The temperature in the Arctic Bottom Water varies slightly between basins and in each basin shows an apparent increase with depth due to the increase in pressure (potential temperature, which is shown in Figure 7.8, is uniform with depth). In the Norwegian Sea its potential temperature is close to -0.95°C. Further north its temperature is determined by sill depths; in most basins its temperature is between -0.8°C and -0.9°C, but the Lomonossov Ridge prevents Arctic Bottom Water colder than -0.4°C from entering the Canada Basin (Figure 7.8). The effect of the Lomonossov Ridge sill depth, which is near 1500 m, comes out clearly in observations of tritium and radiocarbon concentration (Östlund *et al.*, 1987). They indicate a water renewal time of 30 years for Arctic Bottom Water in the Amundsen and

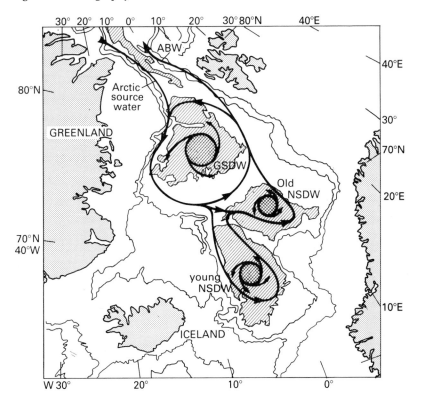

FIG. 7.9. Diagram of Arctic Bottom Water formation showing the circulation at and below 2000 m depth. Approximate 1000 m, 2000 m, and 3000 m contours are shown, with basins deeper than 3000 m shaded. "Arctic source water" is the water from the Amundsen and Nansen Basins which contain contributions from the Arctic shelf. GSDW: Greenland Sea Deep Water, NSDW: Norwegian Sea Deep Water. Deep convection occurs in the cross-hatched areas. ABW: Arctic Bottom Water, the product of the mixing process. After Smethie *et al.* (1988)

Nansen Basins but 700 years in the Canada and Makarov Basins; in the upper 1500 m, water in all basins is renewed at a rate of 30 years.

Above Arctic Bottom Water is the Atlantic Water, which occupies the depth range between about 150 m and 900 m. Its classification as a water mass is justified since it enters the Arctic Mediterranean Sea from the Atlantic Ocean with distinct properties and can be regarded as formed outside the region under consideration. Atlantic Water has the same salinity as Bottom Water but is much warmer (up to 3°C near Spitsbergen), being effectively the summer version of Norwegian Sea Deep Water (and a source for Atlantic Bottom Water when winter cooling sets in). It is also warmer than Arctic Surface Water; but its high salinity makes it denser than the Surface Water. In a hydrographic station Atlantic Water is therefore seen as a temperature maximum (Figure 7.7), at a depth of 150 m near Spitsbergen and progressively deeper to 500 m in the Canada Basin. The maximum is gradually eroded by mixing, and a plot of temperature at the depth of the temperature

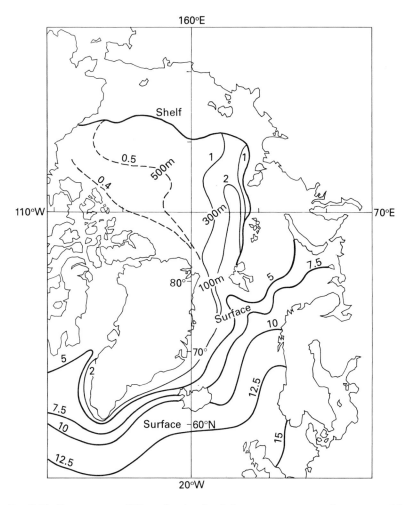

Fig. 7.10. Temperature (°C) at the depth of the temperature maximum caused by inflow of Atlantic Water during summer. Approximate depths are indicated.

maximum (Figure 7.10) indicates the path of Atlantic Water through the Arctic Mediterranean Sea rather well. It is seen that the water enters the North Polar Sea with the West Spitsbergen Current and flows in cyclonic movement, opposite to the circulation of the surface layer above. Movement is again slow, of the order of 0.02 m s^{-1}, except in the West Spitsbergen Current which displays speeds similar to the other two major surface currents.

Arctic Surface Water, which occupies the depth range from the surface to 150 - 200 m, has temperatures close to the freezing point (-1.5°C to -1.9°C), with little variation over depth. Salinity, on the other hand, varies strongly and is sometimes used to distinguish a surface layer of 25 - 50 m thickness and a sub-surface layer below. In the surface layer, Arctic Surface Water shows very much the same characteristics throughout the Arctic Mediterranean Sea, with salinity depending strongly on the degree of ice melting or freezing and varying accordingly between 28 and 33.5. The sub-surface layer is characterized by a strong salinity gradient but uniform temperature. Water in this layer is the product of intense mixing on the Siberian shelf which the Atlantic Water enters through a series of canyons. Production of shelf water with salinities not high enough to contribute to Arctic Bottom Water formation but sufficiently high to bring the water into contact with Atlantic Water has been estimated at 2.5 Sv (Aagaard *et al.*, 1981), not much less than the flow of Atlantic Water into the Arctic Sea. With its very low temperature (near the freezing point) the sub-surface layer of sunken shelf water acts as a heat shield for the surface layer: By entraining Atlantic Water in the canyons and reducing its temperature sufficiently, the water of the sub-surface layer prevents the Atlantic Water from melting the ice layer above.

A remarkable feature of the sub-surface layer is that it contains the swiftest currents of the Arctic Seas. They are usually of the order of $0.3 - 0.6 \text{ m s}^{-1}$, last only for up to a fortnight, and appear to be linked with the movement of subsurface eddies, or lenses. The lenses are an indication for the intensity of the mixing; they are a few tens of kilometres in size and carry water with distinct property characteristics into a region with different water properties, increasing the area of contact between the different water masses where mixing can occur. The residence time for water in this layer is correspondingly low; tracer observations put it at about ten years.

The sub-surface layer of Arctic Surface Water is also the depth where inflow of water from the Pacific Ocean through Bering Strait has some impact. In summer, water from the Bering Sea is warmer than Arctic Surface Water (2 - 6°C, occasionally up to 10°C) but saltier (31 - 33) and spreads at 50 - 100 m depth, producing a temperature maximum (just as Atlantic Water does between 150 m and 500 m). In winter, it is of comparable temperature (around -1.6°C) but again saltier (32 - 34) and spreads at about 150 m depth, producing a temperature minimum. Using these temperature inversions, the presence of water from the Bering Sea can be verified in the sub-surface layer of most of the Canada Basin.

Figure 7.11 summarizes the character of the Arctic Mediterranean Sea as a dilution basin in a T-S diagram. Although the volume of the surface water masses is small (and the T-S diagram grossly misleading in this respect) their importance for the modification of water properties is evident. The dilution effect can be seen along the path of the Labrador Current into the region of the Polar Front in the Atlantic Ocean.

Mass and heat budget

Estimation of mass and heat transport in the Arctic Seas is easier than in the Antarctic region, since water movement is mainly restricted to the layers above the sill depths and can be determined reasonably well through

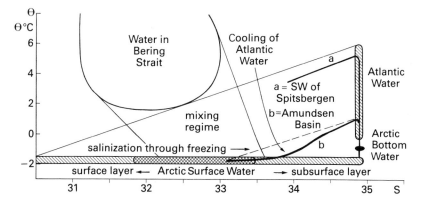

FIG. 7.11. T-S diagram of the Arctic Mediterranean Sea, showing the T-S properties of the water masses and two examples of station data (curves *a* and *b*). The hatched areas give T-S properties of source water masses; the thin lines limit the regions of all possible and observed T-S combinations produced by mixing. T-S curves from individual stations such as curves *a* and *b* generally follow straight mixing paths within the delineated range. Cooling of Atlantic Water through mixing with Surface Water of the subsurface layer is indicated by the departure of the T-S curve from a straight mixing line in the Amundsen Basin.

geostrophy. However, transport variations can be quite large, and knowledge of long-term mean transports in the major contributors to the mass budget is still unsatisfactory.

The Arctic Mediterranean Sea receives water from the Norwegian Current, through Bering Strait, and from river run-off. On average, the same amount of water leaves the Arctic Mediterranean Sea through the East Greenland Current, the Canadian Archipelago, as meltwater and ice, and in the overflow of Arctic Bottom Water discussed below. The Norwegian Current transports about 10 Sv of Atlantic Water northward. Some 4 Sv leave the Norwegian Sea towards the Atlantic Ocean, as outflow of Arctic Bottom Water across the Greenland-Iceland-Scotland Ridge (see below). Of the remaining 5 - 6 Sv, the West Spitsbergen Current carries 3 - 5 Sv into the Amundsen and Nansen Basins, while 1 Sv flows through the Barents Sea and enters the Arctic region between Franz Josef Land and Novaya Semlya. Transport through Bering Strait is well documented (Coachman and Aagaard, 1988) as ranging between 0.6 Sv in winter and 1.1 Sv in summer, while river run-off is estimated at 0.2 Sv. Outflow in the East Greenland Current is estimated at about 3 - 5 Sv, which includes 0.1 - 0.2 Sv (4000 - 5000 km^3 per year) of meltwater and ice. Transport through the Canadian Archipelago into the Atlantic Ocean via Baffin Bay is estimated at 1 - 2 Sv.

A summary of the mass budget is presented in Figure 7.12. Obviously the figures are adjusted to give zero total balance. However, the various contributions differ by an order of magnitude, and errors in the estimates for the major components exceed most of the minor contributions to the budget. All figures, particularly those for the major components, have to be

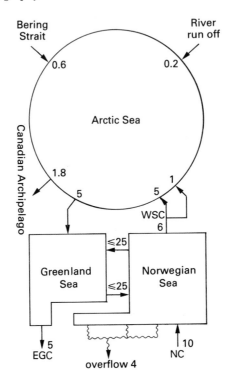

FIG. 7.12. Mass budget for the Arctic Mediterranean Sea and its three major sub-divisions. EGC: East Greenland Current, NC: Norwegian Current, WSC: West Spitsbergen Current. Volume transports are in Sv.

seen as representing our best knowledge to date but are not final.

An unresolved feature of the budget is the volume flow in the recirculation of the Norwegian and Greenland Seas. Recent estimates based on tracer measurements give figures for the recirculation of Arctic Bottom Water in the region as low as 1 Sv, while direct current measurements, which include the recirculation of Atlantic Water (Coachman and Aagaard, 1974), lead to figures as high as 25 Sv. The huge differences are of little consequence for estimates of the mass exchange between the Arctic Mediterranean Sea and the Atlantic Ocean but are crucial for a correct assessment of the mixing processes in the seas south of Fram Strait.

The Arctic region plays a major role in the world's climate, and changes in heat content of the Arctic Mediterranean Sea are likely to influence future climate trends in most parts of the world. Establishing an Arctic heat budget is therefore an important task. Among its prerequisites is a correct mass budget, since heat advection by currents is responsible for the net heat gain of the region. The above estimates of volume flow require some refinement for that purpose; for example, the 5 Sv transported in the East Greenland Current consist of about 1 - 2 Sv of Arctic Surface Water and 3 - 4 Sv of Atlantic Water. Using this kind of split and the average temperature and salinity of both water masses, the heat and freshwater transport of the East Greenland Current can be established. It is found (Aagaard and Greisman, 1975) that on average, the Arctic Sea gains $108 \cdot 10^{12}$ W (most of it through the import of warm Atlantic Water in the West Spitsbergen Current but up to 30% through the export of ice and another 15% through the export of Arctic Surface Water in the East Greenland Current, through the Canadian Archipelago, and through Bering Strait). This heat flux must compensate for the heat loss experienced at the surface. The total salt budget is of course balanced; but the split into components gives an idea of the degree of mixing experienced in the Arctic Sea: The Atlantic Water which enters the region with the West Spitsbergen Current imports about $250 \cdot 10^6$ kg of salt per year; the Atlantic Water which leaves with the East Greenland Current exports only $185 \cdot 10^6$ kg. The remainder is exported with the Arctic Surface Water through the Canadian Archipelago and again in the East Greenland Current.

Fate of Arctic Bottom Water, and path of North Atlantic Deep Water

The mass budget of the Arctic Mediterranean Sea includes an outflow of some 4 Sv of Arctic Bottom Water from the Norwegian Sea into the Atlantic Ocean. Arctic Bottom Water is the densest water of the world ocean, but it is not found anywhere outside the Arctic region. The question arises what happens to the 4 Sv that cross the Greenland-Iceland-Faroe-Scotland Ridge.

The answer to this question is well known today and leads us to the formation of North Atlantic Deep Water, which we met during our discussion of the Southern Ocean. Figure 7.13 shows the path of what is called the Arctic Bottom Water overflow. The deepest passage is in the Faroe Bank Channel and has an average transport of 1 Sv. Overflow over the Iceland-Faroe sill also amounts to 1 Sv. A third path is through Denmark Strait, which contributes 2 Sv to the overflow. During the passage across the sill the Coriolis force keeps the overflowing water to the right (Figure 7.14); so the overflow produces intense currents of Arctic Bottom Water along the continental shelves of Iceland and Greenland which cross the depth contours at a very small angle toward greater depth. The two eastern transports combine to form a uniform flow south-east of Iceland, entraining some 3 Sv of water from the side. This flow continues along the Mid-Atlantic Ridge until it encounters the Gibbs Fracture Zone, a break in the ridge deep enough to allow passage of water below 3000 m depth. The western overflow increases its transport by entraining at least another 3 Sv of water along its path. It follows the continental slope around southern Greenland where it is joined by the other overflow component. This component is slightly warmer and more saline (1.8 - 3.0°C and 34.98 - 35.03 salinity) than the western component (0.0 - 2.0°C

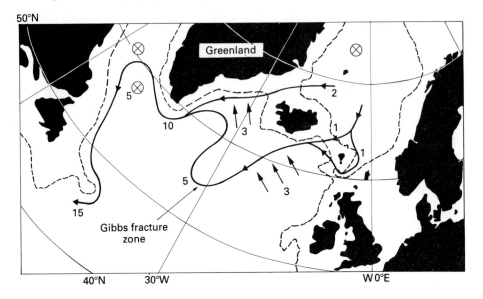

FIG. 7.13. The path of Arctic Bottom Water overflow. Numbers indicate volume transport in Sv. Deep convection is indicated by ⊗. The broken line is the 1000 m contour.

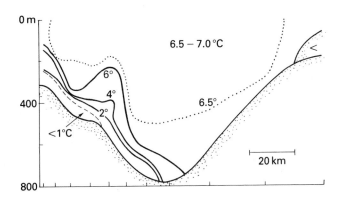

FIG. 7.14. Temperature section across Denmark Strait at about 65°N, demonstrating the concentration of Arctic Bottom Water overflow on the right of the channel. Ticks along the horizontal scale indicate station positions. From Worthington (1969).

and 34.88 - 34.93), a result of the higher temperatures and salinities of the water entrained in the east.

Observations of tritium concentrations in the overflow have shown (Peterson and Rooth, 1976) that the source of the overflow is not water from the bottom of the Norwegian Sea but from depths close to 1000 m. This is of course to be expected since the water has to pass over shallow sills and does that only intermittently when water residing behind the sill is lifted up by a hundred metres or so. The triggering mechanism for this uplift are atmospheric disturbances. Intense storm systems locally generate cyclonic winds. The corresponding Ekman transports are set up within a matter of hours, and they point outward from the centre of the low pressure region, causing upwelling. This is accompanied by a depression of the sea surface which, by our Rule 1a of Chapter 3, is reflected in a rise of the thermocline. The Greenland-Iceland-Faroe-Scotland Ridge is in the West Wind region and therefore sees plenty of storm systems passing by. Each storm lifts the water behind the sill above the sill depth and produces an overflow event. The combined effect of all these events is a flow of 4 Sv of Arctic Bottom Water into the eastern Atlantic Ocean.

Like the Greenland Sea, the Labrador Sea is a region of intense surface cooling and deep winter convection. On arriving in the Labrador Sea from the southern (Atlantic) entrance, Arctic Bottom Water is therefore mixed particularly in winter with water which sank from the surface. This water has similar salinities (near 34.9) but higher temperatures (about 3.5°C). The resulting mixture is known as North Atlantic Deep Water. Arctic Bottom Water is absorbed entirely by this new water mass and cannot be traced beyond the Labrador Sea.

Estimates of the rate of sinking in the Labrador Sea vary and are mostly based on indirect arguments because of the lack of direct winter observations. The total flow of North Atlantic Deep Water from the Labrador Sea is believed to be somewhere around 15 Sv; these figures were derived from estimates of mean heat loss to the atmosphere in the north Atlantic Ocean. They indicate sinking of 5 Sv of water in the Labrador Sea. The Deep Water which leaves the Labrador Sea is more saline but distinctly warmer than Antarctic Bottom Water. It therefore does not spread along the bottom of the ocean but above the Antarctic Bottom Water (which explains its name Deep Water).

Although North Atlantic Deep Water forms in the far north of the Atlantic Ocean it has a large impact on the water properties of the world ocean as a whole. We already met this water mass near Antarctica, where it upwells from between 2000 m and 4000 m into the upper layer and leaves the Atlantic Ocean with the Circumpolar Current. What this means is that the Atlantic Ocean loses surface water to the deeper layers near Labrador at the rate of 15 Sv, while the other oceans gain 15 Sv of water through the Circumpolar Current. There has to be a return flow somewhere. This problem was investigated by Gordon (1986a) who, on the basis of transport estimates and exchanges of heat and salt, developed the picture sketched in Figure 7.15: North Atlantic Deep Water spreads below the permanent thermocline into the three oceans, where it is slowly lost to the surface layer through weak but continuous upwelling. It returns in the surface layer, from the Pacific Ocean through the passages of the Indonesian seas and from the Indian Ocean with the Agulhas Current extension.

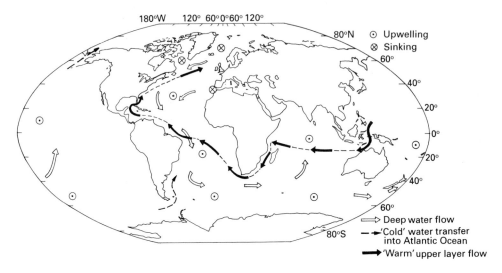

FIG. 7.15. The recirculation path of North Atlantic Deep Water (NADW) through the world ocean. Open arrows indicate flow of NADW and Antarctic Intermediate Water produced in the western south Atlantic Ocean from upwelling of NADW at the Antarctic Divergence. Dots and crosses indicate movement into and out of the thermocline. Full arrows indicate flow of Central Water. Broken arrows indicate deep flow of NADW which is considered insignificant in Gordon's model. Adapted from Gordon (1986a).

The general concept of Gordon's circulation scheme — with some modifications, discussed in the appropriate chapters — has become an accepted part of oceanography today, and there is no doubt that it describes a valid recirculation path for North Atlantic Deep Water. Whether it is the major path is not certain, and the transport estimates involved have yet to be firmly tested. According to Figure 7.15 the rate of North Atlantic Deep Water formation should be equal to the rate of water loss of the Agulhas Current to the Atlantic Ocean which, as we shall see in the discussion of the Indian Ocean, carries most of its water back to the east. It also assumes that all North Atlantic Deep Water upwells in the Southern Ocean and no fraction of it enters the Atlantic Ocean again from the Pacific Ocean after one or more complete cycles in the Circumpolar Current, to recirculate northward. However, in Chapter 9 it will be seen that North Atlantic Deep Water can be identified in the abyssal layers of the Pacific Ocean. How much of it passes through Drake Passage into the Atlantic Ocean is unknown at present. There is also evidence that large amounts of NADW make their way through Drake Passage in the form of Antarctic Intermediate Water (Rintoul, 1991).

The second aspect concerns changes of temperature and salinity along the recirculation. From Figure 7.16 it can be seen that temperature along the path mainly reflects warming during the upwelling and cooling during the sinking. Salinity, on the other hand, undergoes repeated changes, indicating various episodes of mixing with surrounding waters, dilution from rainfall,

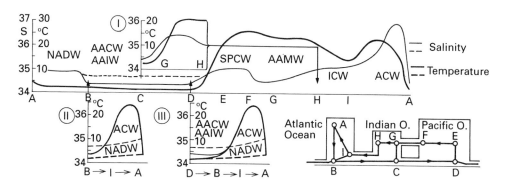

FIG. 7.16. Temperature and salinity along recirculation paths of North Atlantic Deep Water. Pathways are indicated in the lower right. The main diagram follows A-B-C-D-E-F-G-H-I-A, which is Gordon's (1986a) path. Diagram I takes the shorter route A-B-C-G-H-I-A through the Indian subtropical gyre (thus avoiding the reduction of salinity in the AAMW), diagram II recirculates NADW along A-B-I-A within the Atlantic Ocean. Lastly, diagram III gives the path A-B-C-D-B-I-A through Drake Passage. Full lines indicate paths which involve upwelling into the thermocline, broken lines indicate recirculation at depth; heavy lines are temperatures, thin lines salinities. Water masses, described in detail in the chapters on the individual oceans, are abbreviated as NADW: North Atlantic Deep Water, AACW: Antarctic Circumpolar Water (broken lines), AAIW: Antarctic Intermediate Water (full lines), ICW: Indian Central Water, SPCW: South Pacific Central Water, AAMW: Australasian Mediterranean Water, ACW: Atlantic Central Water.

and evaporation. Occasionally such episodes occur in well defined regions and influence the water properties to such a degree that the water can be identified as a new water mass. As an example, the low salinity water which leaves the Indonesian Archipelago has become known as Australasian Mediterranean Water. From the point of view of the recirculation path of Figure 7.15 it could be said that it is just another transformation of North Atlantic Deep Water. A detailed heat and mass balance between ocean and atmosphere along the surface layer part of the proposed path can assist to establish the relative roles of the various circulation paths.

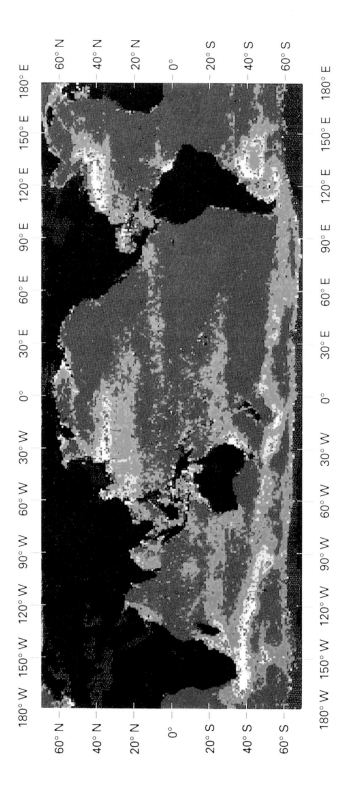

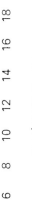

centimeters rms

2   4   6   8   10   12   14   16   18   20   22

Fig. 4.8. Annual mean of eddy energy in the ocean as observed by satellite altimeter during December 1986 - November 1987. The eddies are detected by measuring the shape of the sea surface, which bulges downward in cyclonic and upwards in anticyclonic eddies as explained in Figs 2.7 and 3.3. The quantity shown is the standard deviation of observed sea level (cm) from the mean sea level over the observation period. From Fu *et al.* (1988).

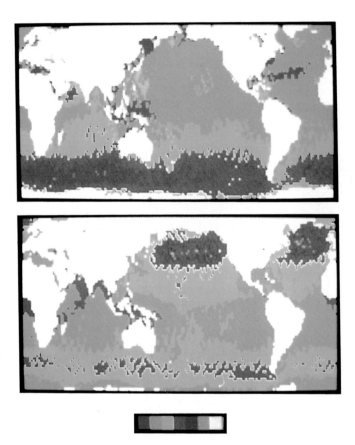

FIG. 6.6. Average significant wave height $H_{1/3}$ for the world ocean during the three-year period November 1986 - October 1989, determined from GEOSAT data. (a) July, (b) January. ($H_{1/3}$ is approximately the mean height of the 1/3 highest waves.) Dark blue indicates $H_{1/3} < 1$ m; each colour step indicates an increase of 1 m. Note the prevalence of large waves ($H_{1/3} > 5$ m) in the Southern Ocean throughout the year, the increase in wave height during winter in both hemispheres, and the large waves in the western Arabian Sea during July (see Chapter 11).

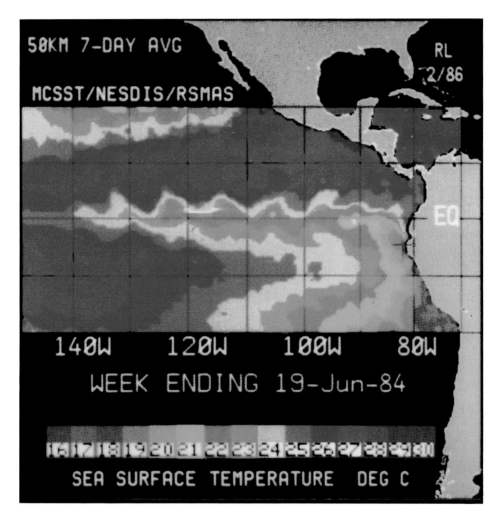

FIG. 8.14a. Satellite image of sea surface temperature in the eastern equatorial Pacific Ocean showing cool water along the equator resulting from upwelling and waves of about 1000 km wavelength in the region of the largest gradient. From Legeckis (1986).

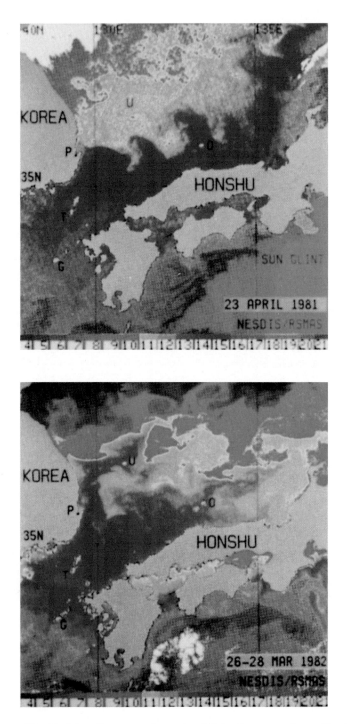

FIG. 10.8. The Tsushima Current and associated Polar Front seen in satellite images of sea surface temperature. (a) In April 1981, (b) in March 1982. From Kim and Legeckis (1986).

CHAPTER 8

The Pacific Ocean

After these two lengthy excursions into polar oceanography we are now ready to test our understanding of ocean dynamics by looking at one of the three major ocean basins. The Pacific Ocean is not everyone's first choice for such an undertaking, mainly because the traditional industrialized nations border the Atlantic Ocean; and as science always follows economics and politics (Tomczak, 1980), the Atlantic Ocean has been investigated in far more detail than any other. However, if we want to take the summary of ocean dynamics and water mass structure developed in our first five chapters as a starting point, the Pacific Ocean is a much more logical candidate, since it comes closest to our hypothetical ocean which formed the basis of Figures 3.1 and 5.5. We therefore accept the lack of observational knowledge, particularly in the South Pacific Ocean, and see how our ideas of ocean dynamics can help us in interpreting what we know.

Bottom topography

The Pacific Ocean is the largest of all oceans. In the tropics it spans a zonal distance of 20,000 km from Malacca Strait to Panama. Its meridional extent between Bering Strait and Antarctica is over 15,000 km. With all its adjacent seas it covers an area of $178 \cdot 10^6$ km^2 and represents 40% of the surface area of the world ocean, equivalent to the area of all continents. Without its Southern Ocean part the Pacific Ocean still covers $147 \cdot 10^6$ km^2, about twice the area of the Indian Ocean.

All adjacent seas of the Pacific Ocean are grouped along its west coast. Some of them (such as the Arafura and East China Seas) are large shelf seas, others (e.g. the Solomon Sea) deep basins. In contrast to the situation in the Indian and Atlantic Oceans, adjacent seas of the Pacific Ocean exert little influence on the hydrology of the main ocean basins. The Australasian Mediterranean Sea, the only mediterranean sea of the Pacific Ocean, is a major region of water mass formation and an important element in the mass and heat budgets of the world ocean; but its influence on Pacific hydrology is of only minor importance, too, far less than its effect on the hydrology of the Indian Ocean.

Before considering the Pacific topography in detail it is worth looking at the world ocean as a whole. Figure 8.1 shows that a system of interoceanic ridges, the result of tectonic movement in the earth's crust, structures the world ocean into a series of deep basins. The major feature of this system is a continuous

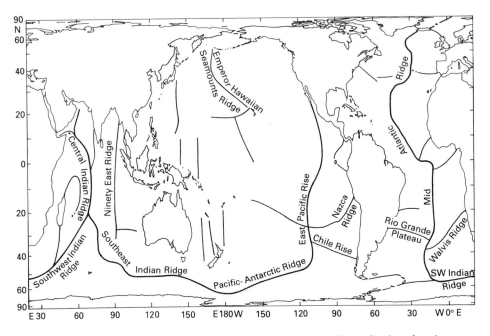

FIG. 8.1. The interoceanic ridge system of the world ocean (heavy line) and major
secondary ridges. Structures with significant impact on ocean currents and properties
are labelled.

mountain chain that stretches from the Arctic Mediterranean Sea through the
Atlantic and Indian Oceans into the Pacific Ocean and ends in the peninsula
of Baja California. Numerous fracture zones cut deep into the slopes of this chain.
To map them all in reliable detail will require an enormous amount of ship time
and remains a task of the future. In the large-scale maps of this book the details
cannot be shown anyway, but they are important where they connect deep basins
which would otherwise be isolated. To give an idea of the real topography,
Figure 8.2 gives an example of such a fracture zone from the Atlantic Ocean,
on the original scale of the GEBCO charts. It is obvious that the world ocean
has not been surveyed to that amount of detail and many passages for the flow
of bottom water are not accurately known.

The interoceanic ridge system divides the Atlantic and Indian Oceans into
compartments of roughly equal size. In the Pacific Ocean it runs close to the
eastern boundary, producing divisions of the southeastern Pacific Ocean similar
in size to the Atlantic and Indian basins. The vast expanse of deep ocean in the
central and northern Pacific Ocean, on the other hand, is subdivided more by
convention than topography into the *Northeast Pacific, Northwest Pacific, Central
Pacific,* and *Southwest Pacific Basins* (Figure 8.3). Further west, New Zealand and
the Melanesian islands provide a natural boundary for two adjacent seas of the
Pacific Ocean, the Tasman and Coral Seas, while in the north the *West* and *East
Mariana Ridges* and the *Sitito-Iozima Ridge* offer a natural subdivision.

Communication between the Southern Ocean and the Pacific basins is much

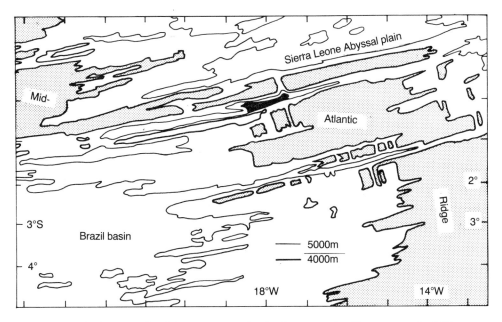

FIG. 8.2. An example of fracture zones in the interoceanic ridge system: the Mid-Atlantic Ridge at the equator. The ridge stretches from northwest to southeast as a series of depths <4000 m(shaded). It is cut by the Romanche Fracture Zone at the equator (identifiable by the Romanche Deep with depths >6000 m, shown in black) and the Chain Fracture Zone at 2° - 3°S. Two other fracture zones can be seen north of the equator. The figure is a simplified reproduction of part of a GEBCO chart, on the same scale.

more restricted by the topography than in the other oceans. Flow of water from the Australian-Antarctic into the Southwest Pacific Basin and the Tasman Sea is blocked below the 3500 m level. Flow from the Amundsen Abyssal Plain into the Southwest Pacific Basin is possible to somewhat greater depth but not below 4000 m. The *Peru* and *Chile Basins* are closed to the north and west at the 3500 m level but connected with the *Mornington Abyssal Plain* and with each other at slightly greater depth, probably somewhere around 3600 - 3800 m.

A unique feature of the Pacific Ocean is its large number of seamounts, particularly in the Northwest and Central Pacific Basins. Seamounts are found in all oceans, but the vulcanism of the northwestern Pacific Ocean produces them in such numbers that in some regions they cover a fair percentage of the ocean floor (Figure 8.4). This may have an impact on the dissipation of tidal energy. Their effect on mean water movement is probably negligible.

The wind regime

The atmospheric circulation over the Pacific Ocean is shown in Figures 1.2 - 1.4. The northern Trades are the dominant feature in the annual mean. They are comparable in strength to the Trades in the Atlantic and southern Indian Ocean and make it difficult to see why the ocean received its reputation as the "pacific",

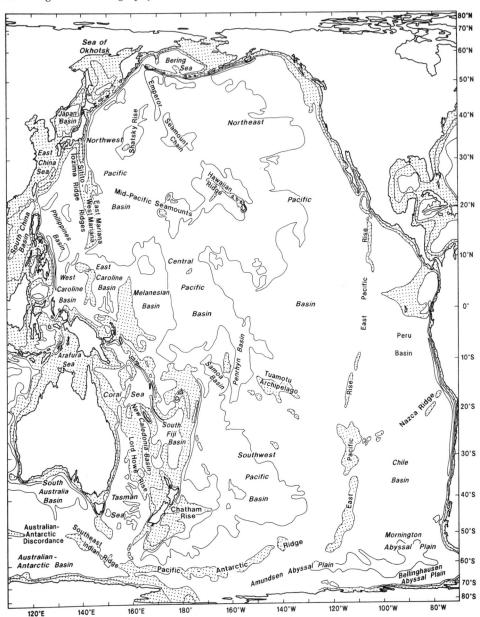

Fig. 8.3. Topography of the Pacific Ocean. The 1000, 3000, and 5000 m isobaths are shown, and regions less than 3000 m deep are stippled.

or peaceful, ocean. The justification for the name is found in the southern hemisphere where east of 170°W the Trades are moderate or weak but extremely steady. Seasonal variations are also smaller south of the equator, since the belt of high pressure located at 28°S during winter is maintained during summer (January), pushed southward to 35°S by the heat low over the Australian continent, Papua New Guinea, and the Coral Sea. East of 170°W the distribution of air

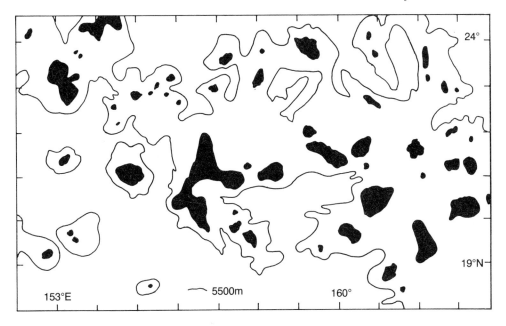

Fɪɢ. 8.4. Seamounts in the Northwest Pacific Basin. The 5500 m contour is shown, depths less than 3500 m are indicated in black. The peaks of most shallow structures are less than 2000 m below the sea surface; some of the larger structures may carry more than one peak. Simplified from a GEBCO chart and on the same scale.

pressure changes little, and the Trades and Westerlies display correspondingly little seasonality there. The effect of the Australian summer low is felt west of 170°W; in the northern Coral Sea across to Vanuatu it produces a monsoonal wind pattern: during winter (June - September) the Trades provide southeasterly air flow, during summer (December - March) the Northwest Monsoon blows from Papua New Guinea and Cape York.

The Trades and the Westerlies of both hemispheres are stronger in winter (July in the south, January in the north) than in summer. North of 55°N this is also true for the polar Easterlies which are barely noticeable in July but very strong in January when the Aleutian low and the Asian high are fully developed; the cyclonic winter circulation associated with the Aleutian low is so strong that it determines the annual mean. The Asian winter high extends a fair way over the ocean and produces a wind reversal over the East and South China Seas and the region east of the Philippines; these regions thus experience monsoonal climate, with Northeast Monsoon during winter (December - March) and Southwest Monsoon during summer (June - September). The monsoon seasons and winds are the same as in the Indian Ocean, both monsoon systems being in fact elements of the same large seasonal wind system produced by the seasonal heating and cooling of the Asian land mass.

The Intertropical Convergence Zone (ITCZ) is located at 5°N, indicated by a minimum in wind speed (the Doldrums). A second atmospheric convergence known as the South Pacific Convergence Zone (SPCZ) extends from east of Papua

New Guinea in a southeastward direction towards 120°E, 30°S. In the annual mean it is not so much seen as a wind speed minimum but more as a convergence in wind direction. Both convergences are regions of upward air movement and thus cloud formation. They are prominent features in satellite-derived maps of cloud cover (Figure 8.5) and will be addressed in more detail when their effect on rainfall and surface salinity is discussed in the following chapter. Contrary to widespread opinion, both the ITCZ and the SPCZ are not regions of no wind; though winds are generally weak, completely calm conditions are encountered during not more then 30% of the year.

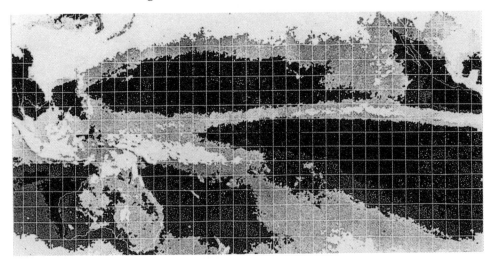

FIG. 8.5. The Intertropical Convergence Zone (ITCZ) and the South Pacific Convergence Zone (SPCZ) as seen in satellite cloud images. The figure is a composite of many months of observations, which makes the cloud bands come out more clearly. It covers the region 40°S - 40°N, 97°E - 87°W; the grid gives every 5° latitude and longitude.

The integrated flow

We saw in Chapter 4 how the depth-integrated flow can be derived independently either from atmospheric or oceanic data. We now return to the relevant figures for a more detailed look at the situation in the Pacific Ocean.

Generally speaking, the integrated flow field derived from atmospheric data (Figure 4.4) compares well with the fields derived from oceanic data with different assumed depths of no motion (Figures 4.5 and 4.6). The most prominent feature is the strong subtropical gyre in the northern hemisphere, consisting of (Figure 8.6) the North Equatorial Current with strongest flow near 15°N, the Philippines Current, the Kuroshio, the North Pacific Current, and the California Current. The circulation in the subtropics of the southern hemisphere is weaker; but the gyre is again well resolved from both data sets. The high degree of agreement in the region of weak flow east of 160°W is particularly remarkable: flow away from the Circumpolar Current is northeastward south of 30°S, where

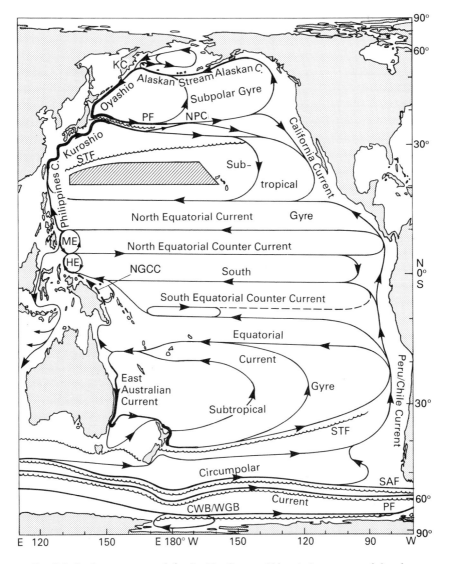

FIG. 8.6. Surface currents of the Pacific Ocean. Abbreviations are used for the Mindanao Eddy (ME), the Halmahera Eddy (HE), the New Guinea Coastal (NGCC), the North Pacific (NPC), and the Kamchatka Current (KC). Other abbreviations refer to fronts: NPC: North Pacific Current, STF: Subtropical Front, SAF: Subantarctic Front, PF: Polar Front, CWB/WGB: Continental Water Boundary / Weddell Gyre Boundary. The shaded region indicates banded structure (Subtropical Countercurrents). In the western South Pacific Ocean the currents are shown for April - November when the dominant winds are the Trades. During December - March the region is under the influence of the northwest monsoon, flow along the Australian coast north of 18°S and along New Guinea reverses, the Halmahera Eddy changes its sense of rotation and the South Equatorial Current joins the North Equatorial Countercurrent east of the eddy. Flow along the STF is now called the South Pacific Current (Stramma *et al.*, in press).

it turns northwestward for a while before joining the general westward flow of the South Equatorial Current. This is one of the remotest regions of the world ocean - no shipping lanes pass through it, the distances to ports are too long for most research vessels to reach it, no islands offer refuelling facilities. This makes exploration of this part of the southern subtropical gyre an expensive undertaking. Until more information is obtained from drifting buoys and satellite data, the integrated flow field will remain our best information on currents in the region. Sverdrup dynamics should work particularly well there, and the flow pattern seen in Figures 4.4 - 4.7 should find confirmation from field observations.

More details are revealed in the stream function map (Figure 4.7). It shows the South Equatorial Current, centred around 15°S, and the Peru/Chile Current as major components of the southern subtropical gyre and indicates the existence of western boundary currents along the coasts of Australia and New Zealand. The split near 18°S into the southward flowing East Australian Current and northward flow in the Coral Sea and northward transport across the equator east of Papua New Guinea have been confirmed by recent field observations. The stream function map also reveals the existence of an Equatorial Countercurrent near 5°N, fed from both subtropical gyres. The current's position coincides with that of the Doldrums, where it flows against the direction of the prevailing weak winds.

An indication of a subpolar gyre in the northern hemisphere is seen north of 50°N. Eastward transport in this gyre is again achieved by the North Pacific Current; the circulation is completed by the poleward and westward flowing Alaska Current, the Alaskan Stream, the southern part of the East Kamchatka Current, and the Oyashio.

In summary, the integrated flow indicates the presence of six western boundary currents: the southward flowing Oyashio between 60°N and about 45°N; the northward flowing Kuroshio between about 12°N and 45°N; the inshore edge of the Mindanao Eddy which flows southward from about 12°N to 5°N; a northward flowing unnamed current between 18°S and 5°N; the southward flowing East Australian Current between 18°S and Tasmania; and another southward current along the east coast of New Zealand. Compared to observations, the start and end latitudes for all boundary currents are quite accurate. An exception occurs in the case of New Zealand; observations show that the Tasman Current only travels to the south end of North Island, while a cold current flows northward along South Island. This may be the result of inadequacies in the wind data for the seldom-travelled region east of South Island.

A marked difference between the integrated flow fields derived from atmospheric and oceanic data is seen in the meridional gradients of integrated steric height just to the east of Japan (a similar phenomenon occurs in the north Atlantic Ocean). This difference does not reflect any inadequacy of the wind field; rather, it is a failure of Sverdrup dynamics which assumes broad, slow flow. The Oyashio and Kuroshio are neither broad nor slow, and they meet head-on off Japan. The Kuroshio advects warm water northward, causing steric height to be larger than it otherwise might be within the Kuroshio. Similarly, the Oyashio's advection of cold water reduces steric height below what we would expect from extending the Sverdrup relationship close to the western shore. Thus the gradient between the two is intensified, and the outflow from both boundary currents is narrower and stronger than we would expect from the Sverdrup model. Narrow and strong flow (though still much broader than in reality) is indicated in Figures 4.5 and 4.6 which are based on oceanic observations averaged over many years. In contrast, the wind-based flow fields (Figures 4.4 and 4.7) spread the outflow unrealistically over more than 10 degrees of latitude.

The equatorial current system

When the structure of the circulation is investigated in detail it is found that significant elements of the current field do not show up clearly in the vertically integrated flow. Details of the three-dimensional structure are revealed in field observations, which we shall now review. We divide the discussion into the three major components of the circulation, the equatorial, western boundary, and eastern boundary currents, and begin with the equatorial current system.

Figure 8.7 is a schematic summary of the various elements of the equatorial current system in the Pacific Ocean. It is seen that the system has a banded structure and contains more elements of eastward flow than could be anticipated from the integrated flow field, which indicated only the presence of the North Equatorial Countercurrent. The most prominent of all eastward flows is the *Equatorial Undercurrent* (EUC). It is a swift flowing ribbon of water extending over a distance of more than 14,000 km along the equator with a thickness of only 200 m and a width of at most 400 km. The current core is found at 200 m depth

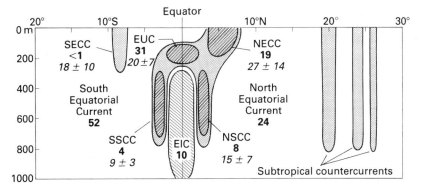

FIG. 8.7. A sketch of the structure of the equatorial current system in the central Pacific Ocean (170°W). Eastward flow is shaded, strong eastward flow hatched upwards to the right; strong westward flow (the EIC) is hatched upwards to the left. All westward flow north of 5°N constitutes the North Equatorial Current, westward flow south of 5°N outside the EIC represents the South Equatorial Current. EUC = Equatorial Undercurrent, EIC = Equatorial Intermediate Current, NECC and SECC = North and South Equatorial Countercurrents, NSCC and SSCC = North and South Subsurface Countercurrents. Transports in Sverdrups are given for 155°W (bold figures; based on observations from April 1979 - March 1980) and 165°E (italics, based on January 1984 - June 1986).

in the west, rises to 40 m or less in the east and shows typical speeds of up to 1.5 m s^{-1}. Surface flow above the EUC is usually to the west, and the EUC does not appear in reports of ship drift. Although it is the swiftest of all equatorial currents its existence remained unknown to oceanographers until 1952 when it was discovered by Townsend Cromwell and Ray Montgomery. None of the theories of ocean dynamics at the time predicted eastward subsurface flow at the equator. The discovery of the Atlantic Equatorial Undercurrent by Buchanan 80 years earlier (see Chapter 14) had been forgotten, and the discovery of the Pacific EUC was therefore a major event in oceanography; for a few years after Cromwell's death the Undercurrent was called the Cromwell Current.

In hydrographic sections the EUC is seen as a spreading of the isotherms in the thermocline (Figure 8.8). This weakening of the vertical temperature gradient occurs for two reasons. Firstly, it shows the "thermal wind" character of the Undercurrent (rule 2a in Chapter 3): above about 150 m, eastward current *increases* with depth, and isotherms slope downward on either side of the current; between 150 m and 250 m, eastward current *decreases* with depth, and isotherms slope upward. A second reason becomes apparent when we look at the processes that drive the Undercurrent. We noted in Chapter 3 that at the equator geostrophy works only for zonal flow. This is indeed shown by the fact that the thermal wind equation holds for the Undercurrent, i.e. the *meridional* component of the pressure gradient (indicated by the north-south slope of the isotherms) is in geostrophic balance. But the pressure gradient at the equator also has a *zonal* component, a result of the Trades which dominate the tropics and subtropics from 30°S to 30°N and produce an accumulation of warm water in the western Pacific Ocean. The accumulation of water is evident in any hydrographic section along the equator as a downward slope of the thermocline towards the west (Figure 8.9); according to our Rule 1a this indicates a westward rise of the sea surface. The sea level difference between the Philippines and Central America amounts to about 0.5 m and produces a zonal pressure gradient which is unopposed by a Coriolis force. As a result, the current below the wind-driven

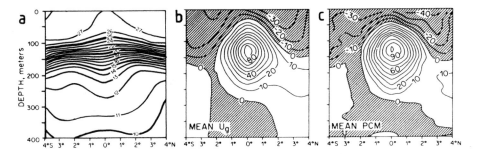

Fig. 8.8. The Equatorial Undercurrent during February 1979 - June 1980 near 155°W. (a) Mean temperature (°C), (b) mean geostrophic zonal velocity (10^{-2} m s^{-1}), (c) mean observed zonal velocity (10^{-2} m s^{-1}). Note the spreading of the isotherms at the equator. From Lukas and Firing (1984).

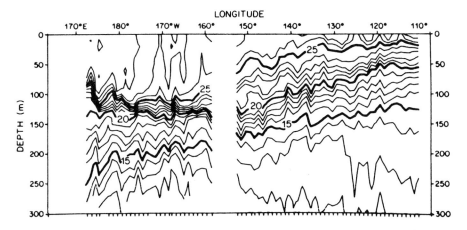

FIG. 8.9. A hydrographic section along the equator. Note the variation in the thickness of the nearly isothermal layer (temperatures above 26°C) from 100 m in the west to less than 20 m in the east, and the upward slope of the thermocline (temperatures between 15°C and 20°C) from west to east from 200 m to 70 m. From Halpern (1980).

surface layer accelerates down the pressure gradient (i.e. from west to east) until friction between the current and its surroundings prevents further acceleration of the flow and establishes a steady state. Friction is associated with mixing, and the Equatorial Undercurrent is therefore a region of unusually strong mixing. This leads to a weakening of the gradients normally found in the thermocline and contributes to the observed spreading of isotherms.

Recent observations indicate that in the depth range of the western Pacific thermocline, exchange between the northern and southern subtropical circulation systems is very limited, the separation between both being located at the southern flank of the North Equatorial Countercurrent. Evidence for strong separation is found in the T-S characteristics. Figure 8.10 shows T-S curves from the region north of Papua New Guinea. The change from high salinity water of southern origin to low salinity northern water occurs within 250 km between 1°S and 2°N. That this separation of the circulation is maintained towards the east is seen in the distribution of tritium introduced into the ocean from atmospheric bomb tests during the late 1950s and early 1960s. These tests were all performed in the northern hemisphere; tritium entered the thermocline through subduction at the northern Subtropical Convergence and quickly reached the equatorial current system. In 1973 - 1974 tritium levels surpassed 4 TU (1 TU = 1 tritium atom per 10^{18} hydrogen atoms) north of 3°N and had reached 9 TU near 12°N. In comparison, tritium values south of 3°N were close to 1.5 TU (Fine *et al.*, 1987).

From the location of the separation zone north of the equator it can be concluded that the Equatorial Undercurrent belongs entirely to the southern circulation system. Observations show that its source waters originate nearly exclusively from the southern hemisphere. Most of the 8 Sv transported by the EUC past 143°E can be traced back to the South Equatorial Current (Figure 8.11).

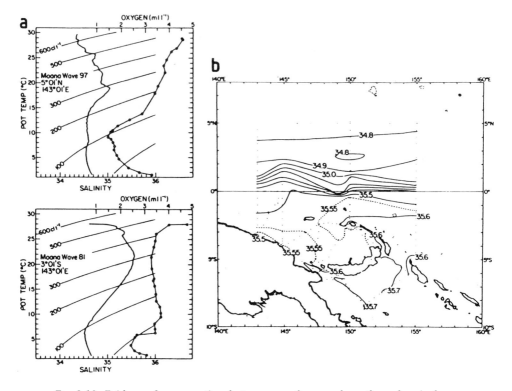

FIG. 8.10. Evidence for separation between northern and southern hemisphere circulation systems in the Pacific thermocline. (a) T-S diagrams and T-O$_2$ diagrams (identified by circles) from two stations north of Papua New Guinea, (b) salinity on an isopycnal surface located at approximately 180 m depth. Note the difference in maximum salinity and the crowding of the isohalines between the equator and 3°N. From Tsuchiya *et al.* (1989).

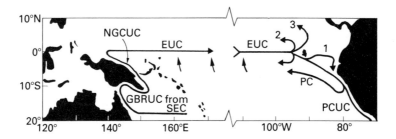

FIG. 8.11. The path of the Equatorial Undercurrent (EUC). SEC: South Equatorial Current. GBRUC: Great Barrier Reef Undercurrent. NGCUC: New Guinea Coastal Undercurrent. PCUC: Peru/Chile Undercurrent. PC: Peru Current (extends to the surface). Flow along path (1) does not occur during July-November. Path (2) is a contribution to the northern flank of the South Equatorial Current, path (3) a smaller contribution to the North Equatorial Current. Based on Tsuchiya *et al.* (1989) and Lukas (1986).

The transport of the EUC increases downstream, reaching 35 - 40 Sv in the east. Based on the tritium distribution it must be assumed that the water drawn into the EUC along its way also stems mainly from the south. The southern origin of its waters allows the EUC to be identified as a subsurface salinity maximum. Figure 8.12 shows seasonal mean T-S diagrams from the termination region of the Undercurrent. The seasonal variation of salinity at temperatures above 13°C indicates that the EUC flows strongest during January - June but is much weaker in July - December.

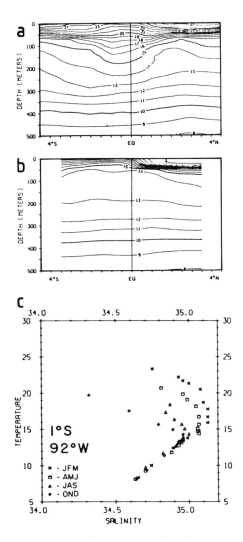

FIG. 8.12. Seasonal variability of the Equatorial Undercurrent in the termination region at 92°W. (a) Mean temperature (°C) for January - March, (b) mean temperature (°C) for October - December, (c) seasonal mean T-S curves. Low salinity and the absence of isotherm spreading in October - December indicate the absence of the Undercurrent. From Lukas (1986).

The second most important eastward flow in the equatorial current system is the *North Equatorial Countercurrent* (NECC). It is prominent in the integrated flow which shows it being fed by western boundary currents both from the south and the north (Figure 4.7). Its annual mean transport decreases uniformly with longitude, from 45 Sv west of 135°E to 10 Sv east of the Galapagos Islands. In its formation region the NECC participates in the Mindanao Eddy. At its other end it turns north on approaching the central American shelf, creating cyclonic flow close to the continent. According to rule 2 of Chapter 3 cyclonic motion is associated with a rise of the thermocline in its centre. In the termination region of the NECC this effect is known as the Costa Rica Dome, a minimum in thermocline depth near 9°N, 88°E (Figure 8.13).

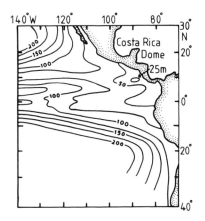

Fɪɢ. 8.13. Annual mean depth of the thermocline in the eastern Pacific Ocean, showing the Costa Rica Dome. After Voituriez (1981).

The NECC varies seasonally in strength and position. During February - April when the Northwest Monsoon prevents the South Equatorial Current from feeding the NECC (see below) the Countercurrent is fed only from the north. It is then restricted to 4 - 6°N with a volume transport of 15 Sv and maximum speeds below 0.2 m s^{-1}; east of 110°W it disappears altogether. During May - January the NECC flows between 5°N and 10°N with surface speeds of 0.4 - 0.6 m s^{-1}. It is then fed from both hemispheres, a fact somewhat at odds with the tritium data west of the dateline, which place the separation zone between the circulation of the hemispheres to the south of the NECC and indicate little NECC contact with the southern circulation. The likely answer is that the water that enters the NECC west of 140°E from the south is again lost to the south before reaching the dateline. Significant loss of water from the current is indicated by the strong eastward decrease of its transport. Historical data indicate that in the eastern Pacific Ocean most of this loss occurs to the south.

The major westward components of the equatorial current system are the *North Equatorial Current* (NEC) and the *South Equatorial Current* (SEC). Both are directly wind-driven and respond quickly to variations in the wind field. They are therefore strongly seasonal and reach their greatest strength during the winter of their respective hemispheres when the Trades are strongest. The NEC carries about 45 Sv with speeds of 0.3 m s^{-1} or less; it is strongest in February. The SEC is strongest in August when it reaches speeds of 0.6 m s^{-1}. Its transport at the longitude of Hawaii (155°W) is then about 27 Sv; this decreases to 7 Sv in February. In the eastern Pacific Ocean between 110°W and 140°W horizontal shear between the SEC and the NECC is so large that wave-shaped instabilities develop along the separation zone between the two currents. They are seen as fluctuations of the meridional velocity component and steric height with periods of 20 - 25 days and wavelengths of 1000 km; satellite observations of sea surface temperature show them as cusped waves along the temperature front between both currents (Figure 8.14). The instability disappears during March - May when the SEC and NECC flow with reduced strength.

On approaching Australia the South Equatorial Current bifurcates near 18°S; part of it feeds the East Australian Current, while its northern part continues northward along the Great Barrier Reef and through the Solomon Sea and passes through Vitiaz Strait to feed the North Equatorial Countercurrent and the Euqatorial Undercurrent. This northern path is suppressed near the surface during the Northwest Monsoon season (December - March) but continues below the

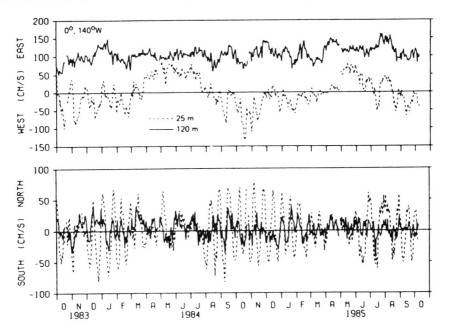

FIG. 8.14. Instabilities at the front between the South Equatorial Current and the North Equatorial Countercurrent in the eastern Pacific Ocean. (a) (see the colour section at the end of the book) Satellite image of sea surface temperature; (b) (above) daily means of current components at the equator, 140°W. Note that the 20 - 25 day oscillations do not occur in the Equatorial Undercurrent (120 m level) and are restricted to the meridional component. Note also the absence of oscillations during March - June. From Halpern *et al.* (1988).

then prevailing southward surface flow as the Great Barrier Reef Undercurrent (Figure 8.11). The SEC therefore continues to feed the Equatorial Undercurrent during the monsoon season but does not supply source waters for the North Equatorial Countercurrent during those months.

The *Equatorial Intermediate Current* (EIC) is an intensification of westward flow within the general westward movement of the SEC. Observations over 30 months at 165°E gave an average westward transport of 7.0 ± 4.8 Sv with speeds above 0.2 m s^{-1} near 300 m. At 150 - 160°W its core is consistently found with speeds above 0.1 m s^{-1} near 900 m. At the same latitudes the cores of the *South Subsurface Countercurrent* and the *North Subsurface Countercurrent* are usually located near 600 m. An explanation for the existence of these currents is still lacking. Recent observations indicate that the banded structure of currents at the equator continues to great depth (Figure 8.15). Below the permanent thermocline currents exceeding 0.2 m s^{-1} are quite rare in the open ocean, and the existence of such currents near the equator indicates that the dynamics of the equatorial region cannot be explained by our $1^1/_2$ layer model. It is possible that the EIC, NSCC, and SSCC are integral parts of a dynamic system that reaches much deeper than the thermocline. The fact that the Costa Rica Dome is a permanent feature despite strong seasonality of the North Equatorial Countercurrent indicates that

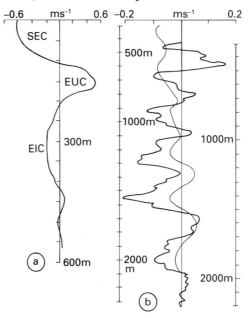

FIG. 8.15. Evidence for banded structure of currents at the equator. (a) The South Equatorial Current (SEC), Equatorial Undercurrent (EUC), and Equatorial Intermediate Current (EIC) at 165°E; (b) deep equatorial currents at 150 - 160°W during 1980 (solid line, right depth scale) and during March 1982 - June 1983 (thin line, left depth scale). The cores of all current bands coincide if the entire current system during 1982/83 is shifted upward some 130 m. Note the different depth and velocity scales. Adapted from Delcroix and Henin (1988) and Firing (1987).

the NSCC also plays a part in maintaining the thermocline structure in the Dome.

The *South Equatorial Countercurrent* is a weak eastward surface current not seen in current maps based on ship drifts but persistently found in results of geostrophic calculations. This may be due to lack of ship traffic, high variability in space and time, or both. Typical surface speeds are below 0.3 m s^{-1} at 170°E, giving a transport of about 10 Sv. Like its northern counterpart the stronger NECC, it is located at a minimum of annual mean wind stress and is therefore strongly seasonal. It is strongest during the Northwest Monsoon (the cause of the wind stress minimum, February - April) and barely seen during winter. In both seasons its strength decreases rapidly east of the dateline (see Figure 8.7), and it may be absent from the eastern Pacific Ocean during most months.

Superimposed on the zonal circulation of Figure 8.7 is weak but important meridional movement. The most important element is at the equator where the Ekman transport in the South Equatorial Current is to the right in the northern hemisphere and to the left in the southern hemisphere. This produces a surface divergence and *equatorial upwelling* in the upper 200 m of the water column. The resulting vertical movement can only be determined by indirect means. Using an array of current meter moorings and applying the principle of continuity of mass between diverging flow, it has been estimated as of the order of 0.02 m per day. This is about one order of magnitude smaller than vertical movement in coastal upwelling regions, but the effect is clearly seen in the sea surface temperature (Figures 2.5a and 8.14a). Observations of tritium near the equator are consistent with a vertical transport of 47 Sv, indicating that upwelling is an important element of the current system. The meridional motion associated with the upwelling is also essential for the heat balance of the tropical Pacific Ocean. The heat input received at the surface is balanced by advection of cold water, but zonal advection does not achieve much in that respect in the tropics where the east-west temperature gradients are small. It is therefore mainly meridional advection and upwelling of colder subsurface water that balances the heat input.

In concluding the discussion of the equatorial current system it has to be pointed out that all its elements can change dramatically from one year to the next and that speeds and transports given above are therefore not necessarily representative for particular years. The variations are linked with the ENSO phenomenon which is the topic of Chapter 19. To give an idea of the changes that occur we only mention here that the EUC has been observed to disappear entirely for several months during the mature phase of an ENSO year, while the transport of the NECC increased to 70 Sv - twelve months later it was reduced to 2 Sv. Further discussion of these changes is postponed to Chapter 19.

The *Subtropical Countercurrents* in the region 20 - 26°N are also permanent features of geostrophic current calculations. They extend to the bottom of the thermocline and often to the 1500 m level. At the surface they can be found in ship drift data with speeds reaching 0.15 m s^{-1}. These eastward flows are located in the centre of the subtropical gyre and therefore not strictly part of the equatorial current system; they are mentioned here for completeness. They do not exist east of the Hawaiian Islands and seem to be a modification of the Sverdrup circulation caused by the presence of that major barrier in the middle of the subtropical gyre - model calculations by White and Walker (1985) indicate that they would not exist if the Hawaiian archipelago were removed. However, banded current structure with alternating eastward and westward flow exceeding 0.5 m s^{-1} has also been reported from the region north of the Hawaiian Ridge (Talley and deSzoeke, 1986); so a final explanation remains to be developed. Similar subtropical countercurrents in the southern hemisphere can be expected from the Society Islands and the south Pacific islands further west. Evidence for a South Subtropical Countercurrent in the Coral Sea was presented by Donguy and Henin (1975).

Western boundary currents

We begin the discussion of western boundary currents with the *Kuroshio* or "black (i.e. unproductive) current". All western boundary currents have a number of features in common: They flow as swift narrow streams along the western continental rise of ocean basins; they extend to great depth well below the thermocline; and they separate at some point and continue into the open ocean as narrow jets, developing instabilities along their paths. These features result from general hydrodynamic principles and reflect the balance of forces in the western boundary regions of the subtropical and subpolar gyres (the closure of the Sverdrup regime). Additional characteristics are imposed by the topography and give each boundary current its own individuality. The characteristic feature of the Kuroshio is that it has several quasi-stable paths. A complete description of the Kuroshio system therefore includes a number of alternative pathways (Figure 8.16). The current begins where the North Equatorial Current approaches the Philippines and continues northward east of Taiwan. It crosses the ridge that connects Taiwan with the Okinawa Islands and Kyushu and continues along the continental rise east of the East China Sea. As the ridge is less than 1000 m deep the current is relatively shallow in this region. It responds to the ridge crossing by forming the East China Sea meander. The meander shows some seasonality in strength and position, increasing in amplitude and moving northeastward in winter. Oscillations with periods of 10 - 20 days and wavelengths of 300 - 350 km occur along the Kuroshio front but the path along the East China Sea is quite stable otherwise. The Tsushima Current branches off from the Kuroshio near 30°N (see Chapter 10).

South of Kyushu the Kuroshio passes through Tokara Strait, a passage also not deeper than 1000 m, and bends sharply to the left. Downstream from Tokara Strait it has been observed near the 1000 - 1500 m isobaths to be only 600 m deep, with velocities above 1.0 m s^{-1} at the surface, 0.5 m s^{-1} near 400 m, and southwestward flow (i.e. opposed to the surface movement) of up to 0.2 m s^{-1} below. Geostrophic calculations indicate that even in 4000 m of water the current does not reach much beyond 1500 m depth. Further downstream along the continental rise of Japan the Izu Ridge south of Honshu forms another obstacle to the flow. The current negotiates it along one of three paths. In the "large meander" path the current turns southeastward near 135°E and flows northward along the ridge before crossing it close to the coast (path 3 in Figure 8.16). In the "no large meander" path it alternates between a path that follows the coast closely (path 1) and a small meander across the ridge (path 2). As shown in Figure 8.16 the change from paths 1 or 2 to the large meander situation occurs every few years at irregular intervals. During those years when the Kuroshio does not follow the large meander path the current changes between paths 1 and 2 about every 18 months. What causes the Kuroshio to change its path remains to be fully explained. Observations show a distinct increase in velocity before the current changes from the large meander path to paths 1 or 2. This suggests some kind of hydraulic control exerted by the Izu Ridge.

In the discussion of geostrophy in Chapter 3 we noted that rule 2 is valid in western boundary currents as long as the hydrographic section is taken across the current axis. The Kuroshio is thus linked with a dramatic rise of the thermocline towards the coast, and a horizontal temperature map at (for example) 300 m depth cuts through the oceanic thermocline near the Kuroshio axis (Figure 8.17). The position of the 15°C isotherm on the 300 m or 200 m level is commonly used as an indicator of the Kuroshio's position. When the temperature map is compared with maps of curl(τ/f) (Figure 4.3) it is seen that

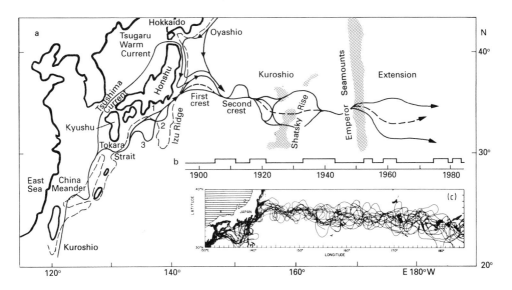

FIG. 8.16. Paths of the Kuroshio and Oyashio. (a) Mean positions of current axis, (b) time history of occurrence of the large meander path south of Japan (the current follows path 3 during the raised portions of the line), (c) individual Kuroshio paths observed during summer 1976 to spring 1980. The broken line is the 1000 m contour and indicates the shelf break. Adapted from Kawai (1972) and Mizuno and White (1983).

the region enclosed by the 15°C isotherm is a region of Ekman transport convergence (downwelling). Sinking motion in the surface layer removes nutrients and keeps biological productivity low despite high levels of sunlight. Ecologically, the central and western regions of the subtropical gyres are the oceanic equivalent of deserts. Devoid of detritus and other organic material, they display the deepest blue of all ocean waters and gave the Kuroshio (which carries their waters north) its name.

The separation point for the Kuroshio is reached near 35°N. It defines the transition from the Kuroshio proper to the Kuroshio Extension. Flow in the Extension is basically eastward, but the injection of a strong jet into the relatively quiescent open Pacific environment causes strong instability. Two regions of north-southward shift, the "First Crest" and the "Second Crest", are found between 140°E and 152°E with a node near 147°E. Both features, as well as the large meander path, are seen in the 300 m mean temperature (Figure 8.17) as features of the long-term mean circulation. East of the Second Crest the Shatsky Rise produces another region of alternative paths. On approaching the Emperor Seamounts the current breaks up into filaments which eventually form elements of the North Pacific Current.

Downstream from its separation point the Kuroshio continues into open water as a free inertial jet. Such jets create instabilities along their paths which develop into eddies or rings. In a map of eddy energy in the world ocean (Figure 4.8) the Kuroshio Extension therefore stands out as one of the regions with very high eddy energy. The process of ring formation is described in detail in Chapter 14 for the Gulf Stream, where more observations of the phenomenon are available. Kuroshio rings behave very much the same, so what is said in that chapter is relevant here as well. Observations over many years indicate that the Kuroshio

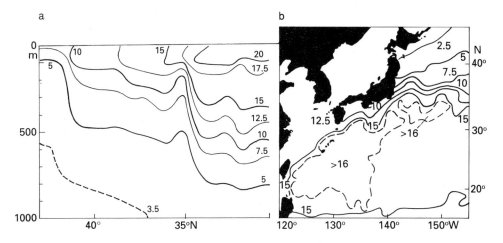

FIG. 8.17. The Kuroshio in the oceanic temperature field. (a) Vertical section of temperature (°C) across the Kuroshio Extension along 165°E, (b) temperature (°C) at 300 m depth. From Joyce (1987) and Stommel and Yoshida (1972).

forms about 5 rings every year when it flows along one of its stable paths and about 10 rings during years of transition. Kuroshio rings extend to great depth; analysis of long-term current meter measurements indicates coherence of kinetic energy virtually to the ocean floor. Eddy kinetic energy falls off across the Emperor Seamounts to one fifth of the amount observed in the west. Observations of deep flow over a one-year period (Schmitz, 1987) revealed unusually strong abyssal currents of 0.05 - 0.06 m s^{-1} below 4000 m depth just west of the Emperor Seamounts. These currents were directed westward near 165°E at either side of the Kuroshio Extension but eastward through a gap between two seamounts at 171°E, under the axis of the surface jet. The flow direction and strength was extremely stable and not reversed by eddies. In contrast, abyssal flow east of the Emperor Seamounts is so weak that it is regularly reversed by eddies, despite their lower energy levels there.

As in other western boundary currents, transport in the Kuroshio increases along its path, indicating entrainment of water from the subtropical gyre. In the Kuroshio Extension near 152°E and 165°E it has been estimated at 57 Sv, which is close to the 50 Sv estimated from closing the integrated Sverdrup flow. The current flows strongest during summer; seasonal variation of the sea level difference across Tokara Strait (0.6 m in the annual mean) indicates an increase of 13% from winter to summer. This apparently contradicts the idea that the Kuroshio is the continuation of the North Equatorial Current, which reaches maximum flow in winter. However, the seasonal variation of the wind field and the associated Ekman pumping is not uniform across the tropical Pacific Ocean; and while the variation of NEC transport is in phase with the variation of the wind in the central region, the seasonal wind variation in the western region is in phase with the Kuroshio transport variation. This suggests that a significant part of the Kuroshio transport is wind-generated in the western Pacific Ocean.

North of its separation point the Kuroshio is opposed by the *Oyashio*, the western boundary current of the subpolar gyre. Ekman flow diverges in the centre of this gyre, so the Oyashio carries cold water rich in upwelled nutrients and full of marine life - hence its name the "parent current". The two mighty streams meet south of Hokkaido, where the *Tsugaru Warm Current* also brings water from the Japan Sea into the Pacific Ocean (as described in detail in Chapter 10). This water proceeds partly southward along Honshu, while another part moves eastward against the advance of the Oyashio. As a result the Oyashio generally splits into two paths, called the First and Second Oyashio Intrusion (Figure 8.16), and the region to the east of Tsugaru Strait displays extremely complicated hydrography (Figure 8.18). Between one and two cyclonic (warm-core) eddies are formed each year in the region. Every six years or so one of them grows into a "giant eddy" which then dominates the area for nearly a year.

The southward boundary of the Oyashio with temperatures of 2 - 8°C defines the Polar Front; it is usually located at 39 - 40°N. Occasionally (for example during 1963, 1981, and 1984) the Oyashio pushes south as far as 36°N. This appears

FIG. 8.18. Satellite image of sea surface temperature of the Kuroshio/Oyashio frontal region taken on 21 May 1981. The dark area in the southeast indicates very warm water in the Kuroshio. Warm water from the Tsugaru Warm Current enters from the west, through the channel between Honshu in the south and Hokkaido in the north; as indicated by the grey tones it proceeds southward along the Honshu coast. Light tones represent cold water of the Oyashio; the First Intrusion is indicated by the coldest water. A large anticyclonic (warm) eddy is centred at 41°N, 147°E. The Second Oyashio Intrusion is seen east of the eddy. The complexity of the region is evident from the figure: Eddies, filaments, and meanders are seen in various stages of formation. In the centre of the observation area a jet-like intrusion from the Tsugaru Warm Current into Kuroshio water produces a "bipole", two small eddies of opposite rotation. (Bipoles are often found at straits or at outlets of strong-flowing rivers. Their eddies are smaller than eddies produced by western boundary currents and differ from them in that they contain the same water mass regardless of their sense of rotation - Kuroshio eddies contain warm water if they rotate anticyclonically but cold water if they have cyclonic rotation.) From Vastano and Bernstein (1984).

to occur when the region of zero wind stress curl moves southward, apparently extending the subpolar gyre southward by some 300 - 500 km. The southern edge of the Oyashio and the northern edge of the Kuroshio maintain their own frontal systems along the Kuroshio Extension. Thus, in the section shown in Figure 8.17a which is located about half-way between the Shatsky Rise and the Emperor Seamounts, the Kuroshio Front is seen at 35°N (identified by the 15°C isotherm) and the Oyashio or Polar Front at 41°N (the 5°C isotherm). Both fronts are associated with geostrophic flow, the front at 35°N with the 57 Sv of the Kuroshio and the front at 41°N with another 22 Sv as the continuation of the Oyashio. Between both flows and on either side movement is weakly westward.

The Oyashio is the continuation of two currents. The *Kamchatka Current* brings water from the Bering Sea southward. It is associated with quasi-permanent anticyclonic eddies on the inshore side which are caused by bottom topography and coastline configuration and result in countercurrents along the coast. The larger of the two Oyashio sources is the *Alaskan Stream*, the western boundary current along the Aleutian Islands. The distinction between the Alaskan Stream and the Alaska Current further to the east is gradual, and the two currents are sometimes regarded as one. They are, however, of different character, the Alaska Current being shallow and highly variable but the Alaskan Stream reaching to the ocean floor. This indicates that despite of its relatively modest speed of 0.2 - 0.3 m s^{-1} the latter is a product of western boundary dynamics, while the former belongs to the eastern boundary regime.

No transport estimates are available for the Alaskan Stream over its entire depth. Geostrophic estimates for the transport in the upper kilometre in the region 155 - 175°W vary between 5 and 12 Sv and show a width of the Stream of 150 - 200 km (Royer and Emery, 1987). South of the Stream the *North Pacific Current* continues from the Kuroshio and Oyashio Extension through the region between 30°N and some 200 km off the Alaskan coast, a broad band of eastward flow more than 2000 km wide. Because of the heavy distortion of distances in the subpolar parts of Figure 8.6, the figure cannot adequately portray this difference in width. Some authors distinguish between the North Pacific Current as the continuation of the Kuroshio Extension south of about 43°N and a Pacific Subarctic Current to the north representing the continuation of the Oyashio Front. Whether the clear separation of the Kuroshio and Oyashio Fronts in the west is maintained all across the Pacific basin, effectively suppressing exchange of water between the subtropical and subpolar gyres, is a matter of doubt, and it appears more appropriate to regard all eastward flow south of Alaska part of the same broad current.

A prominent western boundary current of the equatorial current system is the *Mindanao Eddy*. Captains of vessels that carry Australian wealth to Japan know it well and take advantage of it by following the Mindanao coast southward on their way from Japan to western Australia while travelling 100 - 200 km offshore on their way north. Its transport is estimated at between 25 and 35 Sv, with strong interannual variations. Observations in the western part of the eddy (the *Mindanao Current*) show that the equatorward flow does not extend beyond 250 m; flow in the depth range 250 - 500 m is poleward and carries some 16 - 18 Sv. The *Halmahera Eddy* and the *New Guinea Coastal Current* are seasonal boundary currents near the surface (the latter flows northwestward throughout the year below 200 m, see Fig. 8.11). The flow direction shown in Figure 8.6 reverses during a few months when the region of the Philippines, New Guinea, and northern Australia experiences the winds of the southern summer season: during December to March winds over the Philippines blow from the northeast, turning into northwesterlies south of the equator (Figures 1.2b and c).

The *East Australian Current* is the western boundary current of the southern hemisphere. Although it is the weakest of all boundary currents, carrying only about 15 Sv in the annual mean near 30°S, it is associated with strong instabilities. Its low transport volume is partly a consequence of flow through the Australasian Mediterranean Sea; models show that if the Indonesian passage were closed, flow from the Pacific into the Indian Ocean would be diverted through the Great Australian Bight, doubling the transport of the East Australian Current. The instabilities may result in part from the fact that the current first follows the Australian coast but has to leave it to continue along the eastern coast of New Zealand. The current therefore separates from the Australian coast somewhere near 34°S (the latitude of the northern end of New Zealand's North Island). The path of the current from Australia to New Zealand is known as the Tasman Front, which marks the boundary between the warm water of the Coral Sea and the colder water of the Tasman Sea. This front develops wave-like disturbances (meanders) and associated disturbances in the thermocline which eventually travel westward with Rossby wave speed. When the waves impinge upon the Australian coast they separate from the main current and turn into eddies.

Figure 8.19 shows the process of eddy formation. Because the meander closest to the coast always extends southward and thus can only trap water from the Coral Sea, the East Australian Current spawns many anticyclonic (warm core) but few cyclonic (cold core) eddies. Figure 8.20 shows the current extending to 37°S, flowing back past 34°S before turning eastward and forming the warm eddy "Maria" in the process. A band or "ring" of very warm water is clearly seen around the eddy, indicating the region of strongest currents. When the eddy becomes separated from the main current it will maintain its speed (1.5 - 2.0 m s[-1]) for many months while its hydrographic structure changes. Eddies that go through winter cooling and subsequent spring warming lose their surface signature and are no longer detectable in satellite observations of sea surface temperature (Figure 8.21). Eddy "Leo" in Figure 8.20 is such an eddy; its presence is revealed by the track of two drifting buoys but not visible in the temperature pattern. Remnants of eddies in the form of subsurface layers of uniform temperature and salinity are abundant south of the Tasman Front and are

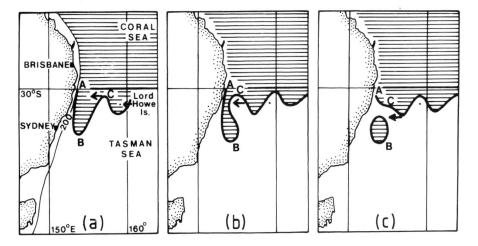

FIG. 8.19. A sketch of eddy formation in the East Australian Current through Rossby wave propagation along the Tasman Front. Point *C* moves westward towards *A*. In the process it pinches off the meander and releases current ring *B* which moves southward. Shading indicates warm Coral Sea water. From Nilsson and Cresswell (1981).

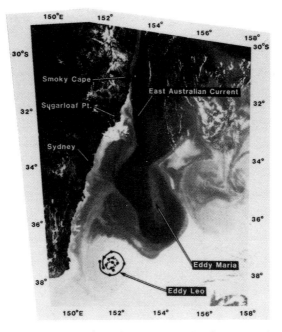

FIG. 8.20. The East Australian Current seen in the sea surface temperature distribution as observed by satellite on 20 December 1980. Dark is warm, light is cold. (Some small scale very light features are clouds.) The tracks in eddy "Leo" are the paths of two drifting buoys. Dots indicate noon positions for each day. From Cresswell and Legeckis (1986).

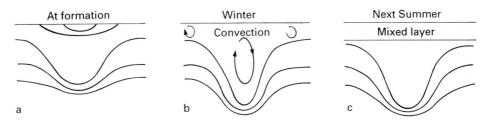

FIG. 8.21. Evolution of the hydrographic structure in an East Australian Current eddy: isotherms in a section across the eddy (a) at the time of formation, (b) during winter, (c) in the following summer. The original warm ring signature at the surface is destroyed by convection from cooling in winter. The eddy is then capped by the seasonal thermocline in the following spring.

characteristic of the upper 500 m of the Tasman Sea (Figure 8.22).

The East Australian Current spawns about three eddies per year, and some 4 - 8 eddies may co-exist at any particular time. Because the volume transport in the current is low, the eddies can contain more energy than the current itself. On occasions it is impossible to identify the path of the current; the western boundary current system then is a region of intense eddy activity without well defined mean transport. This is particularly true for the passage of the current from Australia to New Zealand, where an average location of the Tasman Front

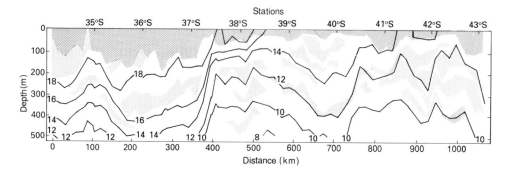

FIG. 8.22. A section of temperature (°C) through the Tasman Sea along a track about 150 km seaward of the continental shelf. The shaded regions indicate layers where the temperature changes by less than 0.1°C. In the permanent thermocline where the temperature usually changes by 0.1°C every 5 m, such layers indicate remnants of cores from East Australian Current eddies. From Nilsson and Cresswell (1981).

can only be defined in statistical terms (Figure 8.23). The current is stronger and reaches further inshore in summer (December - March) than in winter. This is evident from ship drift reports obtained from vessels sailing along the Australian coast between Bass Strait and the Coral Sea. These ships take advantage of the East Australian Current by proceeding along the shelf break on their voyage south; northbound vessels stay inshore of the current and remain over the shelf. Figure 8.24 shows that during winter northbound vessels occasionally experience an inshore countercurrent to assist their voyage. In summer they may encounter southward flow of more than 1 m s^{-1} even on the shelf.

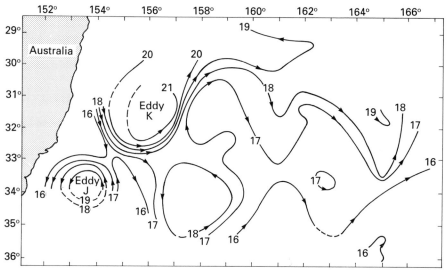

FIG. 8.23a. The Tasman Front. Dynamic topography (m^2 s^{-2}), or steric height multiplied by gravity, as observed in September - October 1979, showing the Tasman Front as a band of large steric height change along the 18 m^2 s^{-2} contour; to obtain approximate steric height in m, divide contour values by 10. Adapted from Stanton (1981).

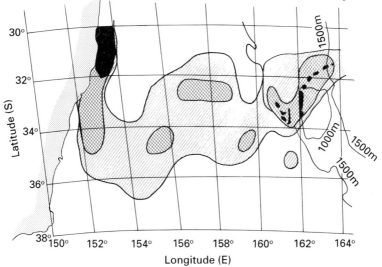

FIG. 8.23b. The Tasman Front. Mean position of the front as determined from satellite SST observations during March 1982 - April 1985; the front was found during more than 30% of the observation period in the lightly shaded area, more than 50% of the time in the dark region, and always in the black region. Adapted from Mulhearn (1987).

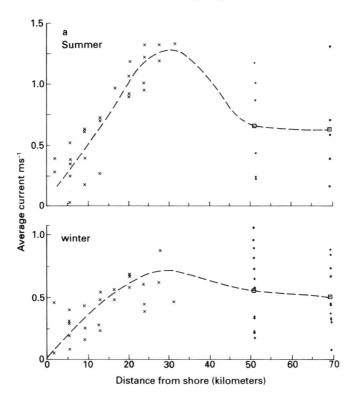

FIG. 8.24a. Seasonal variability of the East Australian Current. Mean current velocity for summer and winter near 30°S. From Godfrey (1973).

b

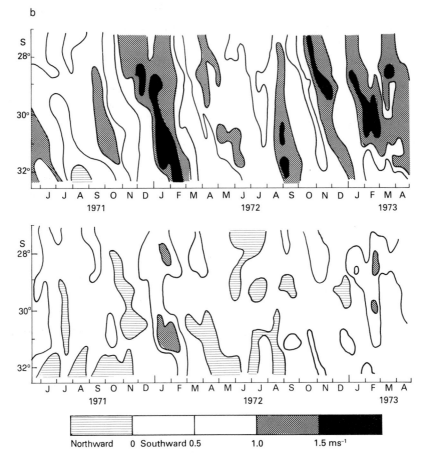

FIG. 8.24b. Seasonal variability of the East Australian Current. Current velocity deduced from the drift of southbound vessels (top; mean offshore distance 19 km) and from the drift of northbound vessels (bottom; mean offshore distance 6.5 km) as a function of time. From Hamon *et al.* (1975).

The continuation of the East Australian Current east of New Zealand is the *East Auckland Current*. It forms an anti-cyclonic eddy north of East Cape (near 37°S) similar in size and with the same homogeneous deep core as the eddies in the Tasman Sea; the eddy is, however, apparently topographically controlled, being found at the same location throughout the year. There is evidence that the East Auckland Current undergoes seasonal change; during summer most of its transport continues as the *East Cape Current* and follows the New Zealand shelf southward until it reaches the Chatham Rise, while during winter some of it separates from the shelf and continues zonally into the open ocean, forming a temperature front near 29°S. Another shallow front near 25°S, sometimes referred to as the Tropical Front, marks the northern limit of eastward flow in the subtropical gyre. At the surface, the westward flow of the South Equatorial Current rarely extends more than 300 km to the south of Fiji (i.e. 20°S); but the boundary between eastward and westward flow slopes down to the south, and at 800 m depth it is found more at 30°S (Roemmich and Cornuelle, 1990).

Eastern boundary currents and coastal upwelling

In the vertically integrated flow, the eastern part of the Pacific Ocean occurs as a region of broad and weak recirculation for the ocean-wide gyres. This is correct for the mean flow away from coastal boundaries but does not hold for currents over short periods or close to the shelf. The aspect of short term variability in the open ocean was already addressed through an example from the Atlantic Ocean (Figure 4.9); the same arguments apply to the Pacific or Indian Oceans. Closer to shore the dynamics are further modified as a result of the meridional direction of the winds. To understand the resulting circulation, known as coastal upwelling, it is necessary to review very briefly the balance of forces along eastern ocean boundaries.

The reason for the strong equatorward component of the Trades along eastern ocean coastlines is the strong difference in climatic conditions between the eastern and western coasts. In the west the Trades impinge on the land laden with moisture which they collected from evaporation over the sea. The coastal regions are therefore well supplied with rainfall; the eastern coastal regions of Madagascar, Brazil, Southeast Asia, New Guinea and the Cape York peninsula of Australia are all covered with luxurious rainforest. At the same latitudes in the east the Trades arrive depleted of moisture, having rained out over the land further east. Lacking the essential rain the coastal regions are deserts: the Simpson desert in Australia, the Atacama desert in South America, the Namib in South Africa, and the Californian desert regions in North America are all found at the same latitudes where rainforests flourish on the opposite side of the oceans. Over these desert lands the air is dry and hot in summer, creating low pressure cells (Figure 1.3). The resulting pressure difference between land and ocean gives rise to equatorward winds along the coast.

The effect of the wind on the oceanic circulation was already demonstrated in Figure 4.1 and is described in more detail in Figure 8.25: The Ekman transport E produced by equatorward winds is directed offshore; as a result the sea surface is lowered at the coast. The corresponding zonal pressure gradient which develops in a band of about 100 km width along the coast supports a geostrophic flow GF toward the equator, i.e. in the same direction as the wind. The water that is removed from the coast at the surface has to be supplied from below, hence upward water movement (of a few metres per week) occurs in a narrow region close to the coast. This upwelling water in turn has to be supplied from the offshore region. On a shallow shelf this can occur in a bottom boundary layer as indicated in Figure 8.25b, but outside the shelf flow toward the coast has to be geostrophic. In other words, in addition to the zonal pressure gradient produced by the Ekman transport, a meridional pressure gradient must exist as well, to support the supply of water for the upwelling process. This pressure

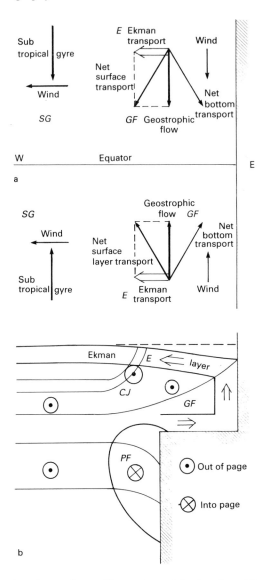

Fig. 8.25. Dynamics of coastal upwelling regions. (a) plan view, (b) vertical section (northern hemisphere). The thin lines in (b) represent isotherms or isopycnals. See text for details.

gradient is directed poleward; geostrophy implies that it is balanced by the Coriolis force linked with the onshore water movement. Close to the coast this movement runs into the shelf and comes to a halt; the associated Coriolis force goes to zero, and we encounter a situation already familiar from the dynamics of the Equatorial Undercurrent: In the absence of an opposing force the pressure field accelerates the water down the pressure gradient, until frictional forces are large enough to prevent further growth of the velocity. The result is poleward flow

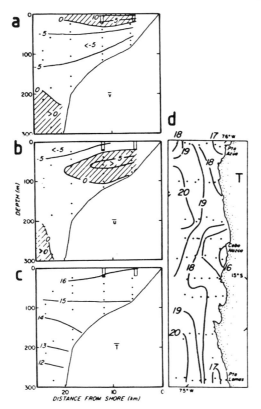

Fig. 8.26. The Peru/Chile upwelling system. (a) Mean alongshore velocity (cm s⁻¹, positive is equatorward), (b) mean cross-shelf velocity (cm s⁻¹, positive is shoreward); the offshore Ekman transport is indicated by negative values in the surface layer), (c) mean temperature (°C); all means are for the period 22/3 - 10/5/1977. Dots indicate current meters, the blocks at the surface special moorings. (d) Sea surface temperature during 20 - 23/3/1977. From Brink *et al.* (1983).

PF in a narrow band near and above the shelf break. This flow competes with the equatorward geostrophic flow driven by the zonal pressure gradient. It is therefore usually not observed at the surface but always seen as an undercurrent along the continental slope. Note that similar to western boundary currents or the equatorial undercurrent only the downstream pressure gradient is not in geostrophic balance. The cross-stream pressure gradient adjusts itself to geostrophic balance with the undercurrent, which can therefore be seen in a downward tilt of the isotherms and isohalines toward the shelf below the surface layer. At the surface, the upward tilt of the isotherms and eventual surfacing of the thermocline gives rise to a front which, through geostrophic adjustment, produces an intensification of the flow known as the coastal jet *CJ*.

The upwelling circulation just described is embedded in the much broader equatorward flow of the subtropical gyre circulation *SG*. At the surface both circulation systems combine to advect cold water towards the tropics, lowering the sea surface temperature along the eastern boundary. The result is a deflection of isotherms from zonal to meridional orientation along the eastern boundary (Figure 2.5a). The water warms as it moves equatorward and offshore; this is reflected in large net heat fluxes into the ocean (Figure 1.6).

The most impressive coastal upwelling system of the world ocean is found in the *Peru/Chile Current*. This current is strong enough to lower sea surface temperatures along South America by several degrees from the zonal average (Figure 2.5a). Embedded in the current is a vigorous upwelling circulation which lowers the temperatures within 100 km of the coast by another 2 - 4°C. The coastal upwelling band is too narrow to be resolved by the ocean-wide distribution of Figure 2.5a, so the low temperatures seen on the oceanic scale reflect advection of temperate water in the subtropical gyre. To see the effects of the coastal upwelling circulation we have to take a closer look at the inshore region. Figure 8.26 shows equatorward surface flow above 30 m, the poleward undercurrent between 30 m and 200 m, offshore Ekman transport above 30 m, and onshore movement of water mainly in the range 30 - 100 m with very little onshore or offshore movement below. Onshore transport in coastal upwelling regions does not extend below 400 m at most. The depth range for onshore movement in the Peru/Chile upwelling system is, however, particularly shallow, probably as a result of the extreme narrowness of the shelf. Upwelling is indicated by the shoaling of the 16°C isotherm, while the downward slope toward the coast of the isotherms below 100 m indicates geostrophic adjustment of the density field to the undercurrent along the slope.

Coastal upwelling systems are among the most important fishing regions of the world ocean because they offer optimum conditions for primary production. The basis for all marine life is photosynthesis in phytoplankton, which can only occur in a layer as deep as sunlight can reach (the so-called euphotic zone, which is less than 200 m deep even in very clear water and usually much shallower). The other requirement are nutrients to support phytoplankton growth. They are supplied by remineralization of dead organisms. In contrast to the nutrient cycle on land, where dead organisms are composted and the nutrients returned to the soil, the nutrient cycle in the ocean is not very efficient: Much of the nutrient reservoir is locked below the euphotic zone because dead organisms sink and escape from the euphotic zone before they can be remineralized. Coastal upwelling regions are among the few regions of the world ocean where nutrients are returned to the surface layer and made available for phytoplankton growth. This forms the basis for a marine food chain with high productivity. All coastal upwelling regions therefore support important fisheries.

The Peru/Chile upwelling system is the most productive coastal upwelling region of the world ocean. It extends from south of 40°S into the equatorial region where it blends into the equatorial upwelling belt. Despite its vast resources, human greed managed to destroy the basis of what before 1973 was the largest fishery in the world. Overfishing and natural variability of the upwelling environment brought about the end of an industry. This aspect of the Peru/

Chile Current System will be taken up again in Chapter 19.

Considerable uncertainty exists about the details of the flow field in the southern part of the Peru/Chile Current. The upwelling undercurrent is known to extend from at least north of 10°S to 43°S and possibly beyond, decreasing in strength from some 0.1 m s^{-1} in the north to barely more than 0.02 m s^{-1} in the south. Further offshore a conspicuous feature is a surface salinity minimum along 40°S (Figure 2.5b). It is known that rainfall along the coast is large in the region; but it can easily be shown that rainfall alone cannot explain the observed salinity minimum. Some researchers conclude that westward flow against the prevailing direction of the subtropical gyre circulation must occur in the region. This may explain why the Subtropical Front is displaced so far northward in the Pacific Ocean off South America and ill defined.

The corresponding coastal upwelling region in the northern hemisphere is found in the *California Current*. Its vast living resources are known from John Steinbeck's novel "Cannery Row" set in Monterey, the centre of the sardine fishery before it collapsed from overfishing in the 1930s. Winds along the coast are much more seasonal here than along the coasts of Peru and Chile (Figure 1.2). Equatorward winds prevail along the coast of Washington, Oregon, and California from April into September, while during the remainder of the year winds are variable and often southeasterly. As a result, poleward flow at the surface is observed during October - March over the shelf and even further offshore. This seasonal flow, which reaches its peak with 0.2 - 0.3 m s^{-1} in January - February, is often called the *Davidson Current*. Coastal upwelling with equatorward surface flow prevails during spring and summer, lowering the sea surface temperatures along the coast to 15°C and less at a time when only kilometres away the heat on land is hardly bearable. The associated cooling of the air leads to condensation, and a coastal strip usually less than a kilometre wide is nearly permanently shrouded in sea fog - the postcard photographs of the famous Golden Gate bridge spanning a blue San Francisco Bay in bright sunshine cannot be taken until October when the upwelling ends and sea surface temperatures reach their annual maximum. Even during the upwelling season poleward flow prevails along the coast of southern and central California in an inshore strip of up to 100 km width, apparently as part of a large cyclonic eddy between the California Current and the coast which has been observed to exist throughout the year except during March and April. Further north, inshore poleward flow with velocities in excess of 0.3 m s^{-1} can exist during periods of weak winds but is suppressed if the upwelling is strong.

High variability of winds and upwelling intensity are a characteristic feature of the Californian upwelling system. Figure 8.27 shows the circulation during a period of weak wind and a period where winds were particularly strong. The competing influences of wind-driven equatorward flow and poleward flow driven by the longshore pressure gradient are seen in the weakening of the undercurrent as the wind increases.

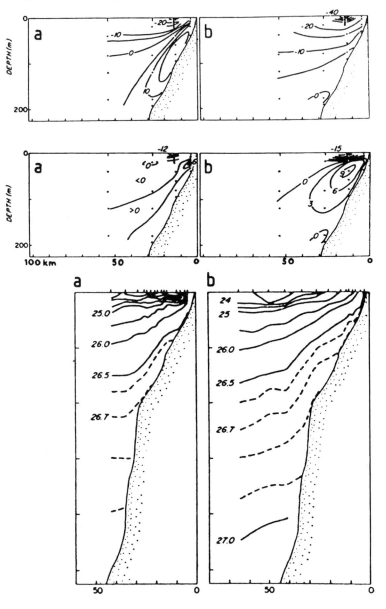

FIG. 8.27. Alongshore flow v (top; cm s^{-1}, positive is poleward), cross-shelf flow u (middle; cm s^{-1}, positive is shoreward), and density σ_t (bottom) in the Californian upwelling system. Left: During weak wind conditions, right: during strong wind conditions. Intensification of upwelling during strong winds is indicated by an increase in u and an increase in equatorward flow accompanied by a reduced undercurrent. The shallow pycnocline (near $\sigma_t = 24.5$) breaks the surface, forming a front some 20 km offshore during periods of strong upwelling; when the wind relaxes this front recedes towards the coast and may eventually disappear. From Huyer (1976).

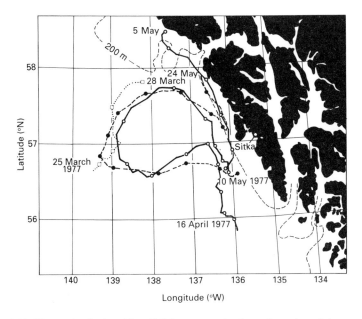

FIG. 8.28. The anticyclonic eddy off Sitka as seen in the trajectories of three satellite-tracked buoys during March - May 1977. Symbols indicate midnight positions. From Tabata (1982).

North of the Californian upwelling region is the *Alaska Current*, also called the Alaska Coastal Current, the eastern component of the subpolar gyre. Freshwater input from Alaska's rivers reduces the density in the upper layers near the coast, enhancing the pressure gradient across the current and constraining the current path to the coastal region. As a consequence, the current is concentrated on the shelf. It is strongest in winter when it shows speeds of up to 0.3 m s^{-1} and weakest in July - August when the wind tends to oppose its flow (Figure 1.2). During some years flow east of 145°W ceases altogether during these months and the subtropical gyre is dislocated some 700 km westward (Royer and Emery, 1987). Eddies may then dominate the region along the Canadian/Alaskan coast. A well defined anticyclonic eddy has been reported from buoy tracks and cruise data off Sitka (Figure 8.28) with average surface speeds exceeding 0.7 m s^{-1}. The eddy exists during spring and summer and possibly throughout the year. It reaches to at least 1000 m depth, although its speed is reduced by half at 200 m.

CHAPTER 9

Hydrology of the Pacific Ocean

The preceding chapter presented the currents of the Pacific Ocean essentially as wind-driven and treated temperature, salinity and other hydrographic properties as passive tracers, useful sometimes to demonstrate the effect of a current on hydrographic property fields but not responsible for the current's existence. This point of view is justified by the success of the Sverdrup model in explaining the major features of the oceanic circulation solely from the wind stress field, indicating that ocean currents are mainly wind-driven. On the other hand, our discussion of possible depths of no motion in Chapter 2 came to the conclusion that the Sverdrup circulation does not extend below 1500 m at most and that below that depth thermohaline forces must be responsible for water movement. We can use this fact to give our text some structure, by dividing the discussion of each ocean into two chapters. The first chapter concentrates on those elements of the circulation that are mainly wind-driven; thermohaline forces may modify details but are not the essential driving force. The second chapter addresses those elements of the circulation where thermohaline factors are responsible for the water movement or at least determine its direction.

This chapter discusses hydrographic property fields in the deep Pacific Ocean, using the distributions of temperature, salinity, and occasional other properties, to draw conclusions on currents below the direct influence of the wind. Since the driving agent for the thermohaline circulation is again the atmosphere, we begin the discussion with a brief review of surface conditions before turning to abyssal depths.

Precipitation, evaporation and river runoff

The region of highest precipitation in the Pacific Ocean as in all oceans is in the tropics where the southern and northern hemisphere Trade Winds meet, causing air to rise and release moisture. It occupies a narrow band in the vicinity of the equator and is characterized by weak and variable winds and heavy rainfall from cloud clusters associated with atmospheric convection and frequent electrical storms. This region is known in meteorology as the Intertropical Convergence Zone or ITCZ and as the Doldrums in the language of mariners.

In the Pacific Ocean the ITCZ was seen just north of the equator at 5°N in the last chapter, extending from South America to Indonesia and the Philippines. It is associated with annual mean rainfall in excess of 300 cm per year. Poleward of the ITCZ rainfall decreases towards the subtropics, reaching its absolute minimum of less than 10 cm per year in the coastal upwelling regions. As discussed in the last chapter, climatic conditions along the east and west coasts differ dramatically in the subtropics, and the subtropical rainfall minimum does not reach much below 200 cm per year in the west. In the northern hemisphere it cannot be identified west of 130°E where the band of high rainfall from the ITCZ bends northward to follow the Philippines Current and Kuroshio. The minimum is also very weak over the Coral and Tasman Seas. Further poleward, a rainfall maximum in both hemispheres is associated with the path of storm systems in the Westerlies (the Roaring Forties). In the northern hemisphere the band of high rainfall extends from Japan to Canada with values above 150 cm per year; in the southern hemisphere maximum values above 100 cm per year are found near 50°S.

A characteristic feature of the South Pacific Ocean is the existence of a second region of wind convergence in the tropics known as the South Pacific Convergence Zone or SPCZ. The last chapter showed it stretching southeastward from the ITCZ near the Philippines along New Guinea and the Solomon Islands towards Fiji and the Society Islands (Figure 8.5). Rainfall along the SPCZ is above 200 cm per year in the west and still above 150 cm per year between 140 - 160°W. The SPCZ is the result of strong monsoonal atmospheric motion across the equator over the Australasian Mediterranean Sea and in the extreme west of the equatorial Pacific Ocean, which causes the Trades to depart from their easterly direction on approaching Australia. During southern winter the southern hemisphere Trades are deflected toward the equator, establishing a convergence with winds coming in from the southeast (Figure 1.2b). During southern summer, when the Australasian monsoon extends weakly into the Coral Sea, air flow across the equator is from the north, and the southern Trades north of 20°S are deflected toward the pole (Figure 1.2c), again producing a convergence. The SPCZ is strongest during this season. It is seen that the SPCZ reflects the eastern limit for the influence of the monsoon system over the Indian Ocean on atmospheric conditions in the South Pacific Ocean.

Since over most of the Pacific Ocean evaporation varies much less than precipitation, the precipitation-evaporation balance (*P-E*; Figure 1.7) closely resembles the rainfall pattern. The SPCZ is evident as a ridge of high *P-E* values, as is of course the ITCZ. The contrast between the zonal uniformity in the northern hemisphere and the marked difference between the eastern and western South Pacific Ocean is remarkable. The increase of *P-E* values poleward of the subtropics is seen in the northern hemisphere. The corresponding increase in the southern hemisphere does not come out clearly in the figure due to lack of ship observations in this rarely travelled region.

Few rivers shed their waters into the Pacific Ocean, and the few that do have very small catchments. The largest rivers all enter the marginal seas along the western rim of the North Pacific basin, where they have a strong impact on the hydrography. This raises the freshwater input into of the marginal seas above the *P-E* values found in the open ocean at comparable latitudes; these aspects will be discussed in more detail in Chapter 10. In the Pacific Ocean proper, the only evidence for significant contributions to the freshwater balance from river runoff is seen along the Canadian coast. Although the rivers coming down from the mountain ranges are small they are numerous; their combined freshwater output of $23,000 \text{ m}^3 \text{ s}^{-1}$ is comparable to that of the Mississippi River and constitutes about 40% of all freshwater input into the northeast Pacific Ocean (Royer, 1982). In the southern hemisphere river catchments are restricted by the Andes in the east and the Great Dividing Range of Australia in the west, so river contributions are negligible.

A major contribution to Pacific rainfall comes from atmospheric moisture imported from the north Atlantic Trade Winds. In the southern hemisphere the moisture collected from evaporation over the Atlantic Ocean is released as rain over the Brazilian rainforest and returns to the sea through the Amazon river. The land barrier of Central America, while receiving plenty of rain and nurturing luxurious rainforest, is not wide enough to catch all the moisture collected by the northern hemisphere Trades. The Atlantic Ocean therefore suffers a net freshwater loss across the land barrier, while the Pacific Ocean experiences a freshwater gain. This increases the salinity of the North Atlantic and decreases the salinity of the North Pacific Ocean. The resulting salinity difference between the two oceans has important consequences for the world climate. This aspect will be discussed in detail in Chapters 18 and 20.

Sea surface temperature and salinity

The distribution of sea surface salinity (SSS, Figure 2.5b) mirrors the *P-E* distribution closely. The ITCZ is seen as a band of low salinity along 5 - 10°N. The SPCZ is evident as an extension of the tropical low salinities from New Guinea to Fiji. Minimum salinities in the ITCZ/SPCZ system occur in the Gulf of Panama, where salinity drops below 33.0 in the annual mean. This water originates from the North Equatorial Countercurrent, which crosses the Pacific Ocean estward under the heavy rains of the ITCZ and arrives in the east with substantially reduced salinities. Weak currents in the Gulf of Panama expose the water further to the heavy rains produced by the import of Atlantic moisture. Eventually the diluted water exits into the North Equatorial Current and - via the Kuroshio - into the northern subtropical gyre. As a result the surface layer of the North Pacific Ocean is significantly less saline than that of the South Pacific Ocean.

The centres of the subtropical gyres coincide with the region of negative *P-E* or strong evaporation and are therefore characterized by high salinities. The salinity maxima are shifted somewhat to the west from the *P-E* minimum. This can be explained by the observation that winds in the centre of the gyres are weak and currents sluggish, so evaporation occurs over a longer time span than in the faster flowing eastern boundary currents.

Poleward of the subtropical gyres SSS decreases again, particularly in the northern hemisphere where the effect of ice melting from glaciers is felt along the Canadian and Alaskan coast. Salinities are also very low in the marginal seas along the western coastline. The influence of the Arctic Mediterranean Sea is seen in very low SSS values in Bering Strait.

The outstanding feature in the sea surface temperature (SST) is the expanse of very warm water in the western equatorial region. The contouring chosen for Figure 2.5a shows it with SST values higher than 28.0°C. Highest temperatures are encountered between eastern New Guinea and the equator where the annual mean is above 29.5°C. Annual mean temperatures are above 29.0°C between 10°N and New Guinea and the Solomon Islands and extend eastward along the SPCZ to 170°W. The thermal equator follows the ITCZ from 5°N in the west to 15°N in the east, with annual mean SST values generally above 27.0°C. The very high temperatures associated with the SPCZ and ITCZ are remarkable because they occur in a region of maximum cloud cover (Figure 8.5) which limits direct atmospheric heating. However, this effect is outweighed by the fact that the winds are weak over this region during most of the year (Figure 1.2), so there is little evaporative cooling.

The general poleward decrease of surface temperature produces a zonal orientation of most isotherms. Marked departures from zonal SST distribution are observed in the coastal upwelling regions. The Peru/Chile upwelling region stands out in this regard because it prevails strongly throughout the year. In the northern hemisphere the effect of coastal upwelling on the climatological mean SST is much weaker because the Californian upwelling is restricted to spring and summer. It is worth remembering when looking at Figure 2.5a that it is based on heavily smoothed data and cannot display features on scales much smaller than 700 km. The meridional orientation of the isotherms along the eastern coast is therefore more a result of equatorward advection of cold water in the Peru/Chile and California Currents than a direct consequence of coastal upwelling. But the cold water that upwells on and near the shelf is eventually transported offshore in the subtropical gyre and thus contributes indirectly to the lowering of offshore temperatures seen in the large scale SST distribution.

In the western Pacific Ocean the isotherms depart from strictly zonal orientation as a result of advection in the western boundary currents. The effect is clearly seen in the northern hemisphere where the convergence between the Kuroshio and the Oyashio produces a crowding of isotherms at the Polar Front. Advection by the East Australian Current can also be seen in the SST distribution; but the effect is much weaker, since the current has a much smaller transport and most of the current's energy is contained in its eddies.

Abyssal water masses

Although the Ross Sea, an important formation region of bottom water for the world ocean, is located in the Pacific sector of the Southern Ocean, the abyssal waters of the Pacific Ocean are renewed very slowly. This is primarily a result of topography: Arctic Bottom Water access is blocked by the very shallow Bering Strait, while most of the Antarctic Bottom Water produced in the Ross Sea is prevented from flowing north by the combined action of the Circumpolar Current and the Pacific-Antarctic Ridge and escapes through Drake Passage. The deep basins of the Pacific Ocean are therefore filled from the west. This is evident from the distribution of near bottom potential temperature in Figure 9.1 which shows *Antarctic Bottom Water* (AABW) entering south of New Zealand with potential temperatures below 0°C and moving eastward. Formation of AABW in the Ross Sea is indicated by a region of water colder than -0.5°C. The figure indicates three entry routes of AABW into the Pacific basins. All originate from the northern flank of the Circumpolar Current and therefore carry AABW in the form of Circumpolar Water; some authors therefore prefer that term for the bottom water of the Pacific Ocean. The western route, northward flow east of Australia, is blocked by topography near 20°S and of little importance outside the Tasman and Coral Seas. The eastern route, inflow along the East Pacific Rise east of 110°W, is blocked by the Chile Rise near 40°S and thus, too, does not contribute to bottom water renewal of most basins. This leaves the central route as the major point of supply. AABW enters the Southwest Pacific Basin east of the New Zealand Plateau and Chatham Rise and spreads gradually northward until it enters the Northwest Pacific Basin through the Samoa Passage (about 10°S, 169°W, east of the Tokelau Islands) and finds its way into the basins west of the Mariana Ridges through narrow deep passages. Most of this flow takes the form of narrow western boundary currents below 3500 m depth. In contrast, the vast expanse of the Northeast Pacific Basin is most likely renewed by uniform sluggish eastward flow.

The existence of western boundary currents above the ocean floor is demonstrated more convincingly in a vertical east-west section across the South Pacific Ocean (Figure 9.2). The observed rise of the isotherms against the western coast at depths below 4000 m is consistent with a northward "thermal wind" in which speed increases with depth (Rule 2a of Chapter 3). The feature is seen both in the Tasman Sea (150°E) and along New Zealand (165°W). The boundary current associated with the East Pacific Rise cannot be seen in the section, 43°S being too close to the Chile Rise to allow strong meridional flow there.

Detailed investigation of the circulation in Drake Passage reveals that circumnavigation of Antarctica is not the only route of AABW into the Pacific Ocean. The most direct route, from the Weddell Sea *westward* through Drake Passage, is taken by some 2 Sv of Weddell Deep Water. The water enters the Scotia Sea through a depression in the South Scotia Ridge near 40°W and follows the bottom topography along Antarctica in a narrow westward flow. This is clearly seen in the distribution of hydrographic properties (Figure 9.3). The differences between values west and east of the Shackleton

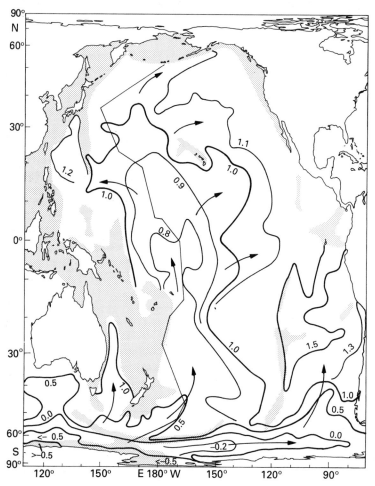

FIG. 9.1. Near bottom potential temperature (°C) in the Pacific Ocean. Arrows indicate the flow of Antarctic Bottom Water. Regions shallower than 3000 m are shaded. The thin line in the western basins gives the location of the section of Fig. 9.4. After Mantyla and Reid (1983).

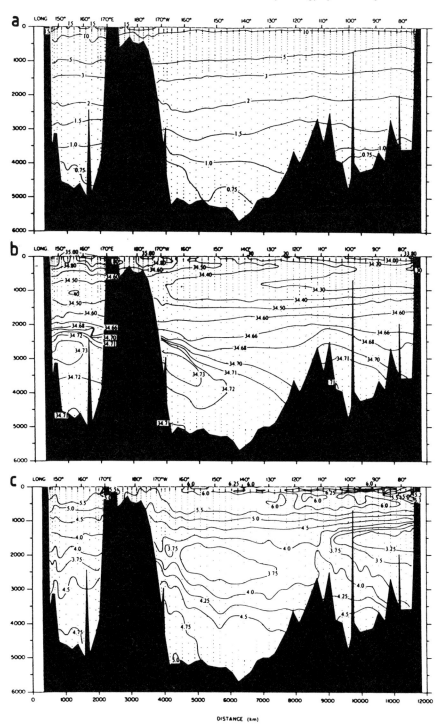

FIG. 9.2. A zonal section across the South Pacific Ocean along 43°S. (a) Potential
temperature (°C), (b) salinity, (c) oxygen (ml/l). From Reid (1986).

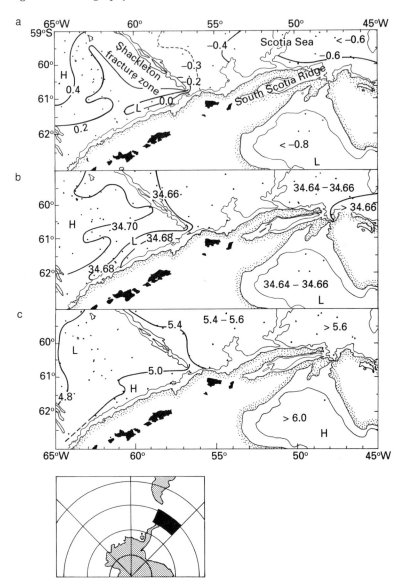

FIG. 9.3. Evidence for westward flow through southern Drake Passage. (a) Near bottom potential temperature (°C), (b) near bottom salinity, (c) near bottom oxygen content (ml/l). The 3000 m contour is shown as a thin line. Stippling indicates depths less than 2000. After Nowlin and Zenk (1988).

Fracture Zone indicate a warming of 0.5°C, a salinity increase of 0.04, and an oxygen loss of 0.4 ml/l for a circumnavigation; but cold, fresh, and oxygen-rich water is seen penetrating westward through a gap south of 61°S. The flow has been confirmed by direct current measurements which gave persistent speeds of up to 0.2 m s^{-1}.

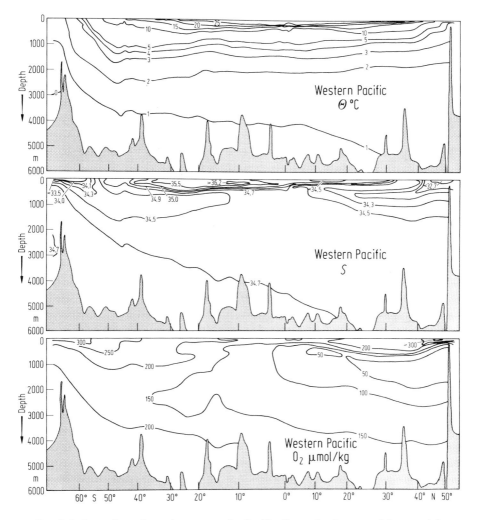

FIG. 9.4. A north-south section across the Pacific Ocean in the west. (a) potential temperature (°C), (b) salinity, (c) oxygen (μmol/kg; an approximate conversion to ml/l, correct near Θ = 5°C, S = 34.45, is 100 μmol/kg = 2.24 ml/l). From Craig *et al.* (1981). See Fig. 9.1 for location of section.

The three basins along the South American coast are separated from the remainder of the Pacific Ocean by topography and are therefore not reached by the AABW circulation. Figure 9.1 shows near bottom potential temperatures in the Peru and Chile Basins some 0.3 - 0.5°C warmer than elsewhere. This is the temperature above the Chile Rise, indicating that below the sill depth the Chile Basin is filled with water crossing the ridge from the south. The water then follows the Peru/Chile Trench along South America into the Peru Basin and continues into the Panama Basin, a small basin of near 3800 m depth blocked on all sides at the 2300 m level, with a sill depth in the trench of approximately 2900 m just south of the equator. Observations 100 m above the sill gave a mean inflow speed over a 17 day period of 0.33 m s^{-1}, which gives a rough transport estimate of 0.2 Sv, with a potential temperature of 1.6°C. By the time this amount of water leaves the Panama Basin through upwelling and outflow above the 2300 m level it has acquired a potential temperature of 1.9°C, while its salinity remains unchanged. This indicates that the heat required to raise the temperature by 0.3°C is not so much derived from mixing with the waters above (which would affect the temperature and salinity) but from geothermal heating in the Galapagos sea floor spreading centre. This places the Panama Basin amongst the few ocean regions where geothermal heating provides an important contribution to the heat budget. Geothermal heating has also been verified in the Northeast Pacific Basin (Joyce *et al.*, 1986), but the increase in bottom temperature does not exceed 0.05°C in that case.

Figure 9.4 gives a meridional hydrographic section through the western Pacific Ocean. The outstanding feature is the uniformity of water properties below 2000 m depth. The circulation in this region is very sluggish. AABW is slowly advected from the south, mixing with the water above, its aging being indicated by the northward decrease of oxygen content. The oldest water is found in the northern hemisphere, just below the tongue of well oxygenated low salinity Intermediate Water (to be discussed below), where oxygen values fall below 50 µmol kg^{-1} (about 1.1 ml/l). Measurements of ^{14}C indicate that more than 1000 years elapsed since this water was in contact with the atmosphere. In contrast to the North Atlantic Ocean which contributes to the formation of the oceans' abyssal water masses, no deep or bottom water is formed in the North Pacific Ocean. This is not only the result of Bering Strait blocking deep communication with the Arctic Mediterranean Sea; in the North Atlantic Ocean, Deep Water would still be produced if the connection between the Arctic Mediterranean Sea and the Atlantic Ocean were closed, by deep winter convection in the Greenland and Labrador Seas. In the North Pacific Ocean such deep winter convection is inhibited because its surface salinity is much lower than that of the North Atlantic Ocean (Figure 2.5b). This reduces the surface density sufficiently to prevent it from exceeding the density below the mixed layer even in winter. Surface salinity in the northern Pacific Ocean is thus of immense importance for the global oceanic circulation and consequently for climate. This aspect will be taken up again in the last chapters.

The distribution of oxygen (Figure 9.4c) indicates the penetration of AABW into the northern hemisphere below 3000 m and active circulation associated with the spreading of Intermediate Water above 1000 m depth. In the northern hemisphere, water in the depth range 1000 - 3000 m does not participate much in the circulation; its properties are determined nearly entirely through slow mixing processes. This water is usually called *Pacific Deep Water* (PDW), in analogy to the North Atlantic and Indian Deep Waters which occupy the same depth range. Identification of the oldest water in the Pacific Ocean as a distinct water mass is justified but requires some explanation in view of our water mass definition of Chapter 5. Figure 9.5 compares Deep Water properties of the three oceans. The different character of Atlantic, Indian, and Pacific Deep Water comes out clearly. North Atlantic Deep Water is formed at the surface in a region of very high surface salinity and is therefore seen in the T-S diagram as a salinity maximum between Bottom and Intermediate Water. The characteristics of Indian Deep Water will be discussed in Chapter 12; in essence, it is North Atlantic Deep Water (NADW) advected into the Indian Ocean and therefore given a different name. In the T-S diagram it is again present as a salinity maximum, with T-S values very close to those of North Atlantic Deep Water east of 40°E, the maximum being eroded further east and north through mixing with the waters above and below (Figure 9.5 shows maximum salinity just above 34.7 near 70°E). In contrast, the T-S values of Pacific Deep Water do not depart very much from the mixing line between Bottom and Intermediate Water except in the vicinity of the Southern Ocean, where a faint salinity maximum indicates that traces of North Atlantic Deep Water, having crossed the Indian Ocean under the name of Indian Deep Water, are entering the Pacific Ocean

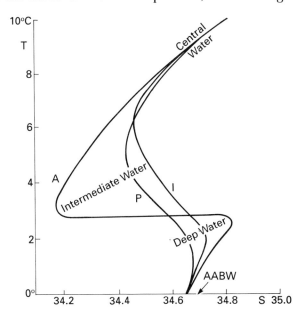

FIG. 9.5. A comparison of T-S diagrams from the three oceans in the southern hemisphere. A: Atlantic Ocean (at 41°S), I: Indian Ocean (at 32°S, 70°E), P: Pacific Ocean (near 43°S, 120°W). AABW: Antarctic Bottom Water.

from the Great Australian Bight. It is thus seen that the constituents of Pacific Deep Water are Antarctic Bottom Water, North Atlantic Deep Water, and Antarctic Intermediate Water. This mixing process constitutes the common formation history of all Pacific Deep Water.

Evidence for a contribution of NADW to Pacific Deep Water can be seen in Figure 9.2b which shows a weak salinity maximum at and below 3000 m depth, indicating that some NADW enters in western boundary currents along the continental rises of Australia and New Zealand. Lateral mixing must be important along the path, as the horizontal extent of the maximum is far too large to indicate the true width of the boundary currents. Figure 9.6 traces NADW from its formation region through the Atlantic and Indian Oceans into the Pacific Ocean. The salinity maximum in the Pacific Ocean can be followed nearly to the equator where the extinction of NADW through conversion into PDW is complete. The path of NADW shown in the figure differs significantly from the recirculation path discussed in Chapter 7. It suggests NADW propagation into the Indian and Pacific Oceans without significant upwelling and water mass conversion (from NADW into AAIW) in the Southern Ocean. Since there is no exit for PDW from the north Pacific basins, it has to upwell into the overlying Intermediate Water; so conversion of NADW into Intermediate Water does occur eventually, but the details of the heat and salt budgets involved are quite different for the two pathways. Figure 9.6 leaves no doubt that some NADW reaches the Indian and Pacific Oceans with the abyssal circulation. On the other hand, the discussion of Chapter 7 showed that deep upwelling and associated conversion of NADW

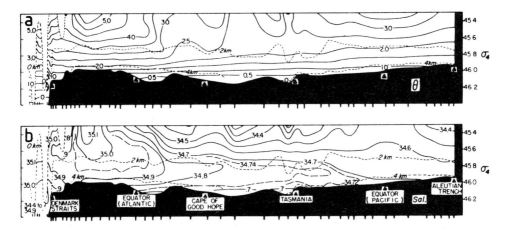

Fig. 9.6. A vertical section from the North Atlantic Ocean near Denmark Strait southward around southern Africa, then eastward through the southern Indian Ocean past Tasmania and New Zealand, then northward into the North Pacific Ocean. (a) potential temperature (°C), (b) salinity. The ordinate uses σ_4, the density a parcel of water would have if moved adiabatically to a pressure of 40,000 kPa (approximately 4000 m depth). Actual depths are indicated by broken lines. Tick marks along the bottom indicate station positions. From Reid and Lynn (1971).

into AAIW in the Southern Ocean is a dynamic necessity of the circumpolar circulation. The necessary conclusion is that the NADW recirculation is shared between both paths.

The salinity minimum near 800 - 1000 m depth in both hemispheres indicates the presence of Intermediate Water. At this depth the movement of water no longer follows the deep circulation (northward movement in western boundary currents with slow return circulation through the eastern basins) but is part of the upper layer circulation which shows symmetry around the equator (broad equatorward flow in the subtropical gyres and poleward flow in the western boundary currents). Until very recently it was thought that *Antarctic Intermediate Water* (AAIW) originates mainly from subduction along the Polar Front. Recent results (England *et al.*, in press) indicate that most of the AAIW in the world ocean may form west of South America through convective overturning of surface waters during winter. It is then injected into the subtropical gyre, filling the southern subtropics and tropics from the east. *Subarctic Intermediate Water* (SIW) originates from the Polar Front formed between the Kuroshio and Oyashio in the western North Pacific Ocean, where it is formed mainly by mixing of surface and deeper waters (i.e. without direct contact with the atmosphere) and subducted into the subtropical gyre, filling the northern hemisphere south of 40°N again from the east. A map of salinity at the depth of the minimum (Figure 9.7) is somewhat indicative of Intermediate Water movement. Strong contrast between minimum salinities in the east and west is seen in the South Pacific Ocean. Numerical models of the oceanic circulation attribute this to a strong injection of new AAIW from the winter convection region west of southern South America which is transported westward with the subtropical gyre, opposing the eastward flow of old AAIW advected from the west. The latter component is traced back to the winter convection region in the Atlantic Ocean (east of southern South America) by the models. Verification of these ideas in the field is required before we can be more definite on the reasons for the east-west salinity gradient at the level of the salinity minimum. Minimum salinities are much more uniform near the equator where AAIW and SIW meet and dissipate through upwelling, leaving only a faint trace of a minimum (Figure 9.4).

Movement of AAIW in the western South Pacific Ocean is strongly influenced by the topography. The presence of New Zealand results in AAIW entering the Tasman Sea along two paths. AAIW with minimum salinity less than 34.4 enters from the south, taking the shortest route from the Polar Front. It does not spread very far northward, however, being opposed by a second source of AAIW supply in the Eastern South Pacific Ocean. This water spreads westward through the Coral Sea, enters the Tasman Sea from the north and leaves it with the East Australian Current extension around the northern tip of New Zealand's North Island (It can be traced, through its

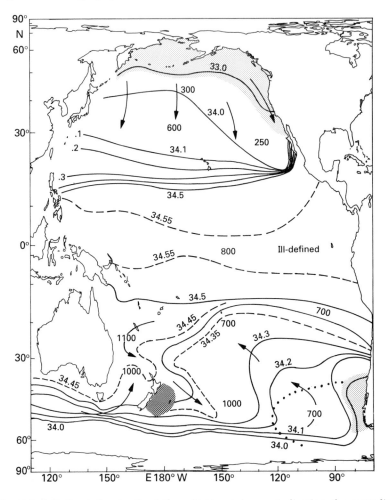

FIG. 9.7. Salinity at the depth of the salinity minimum, indicating the spreading of Intermediate Water. The depth of the minimum is also indicated; light shading indicates regions where the minimum is at the surface. The dotted line in the southeast marks a region where surface salinities are lower than those shown for the minimum at 700 m. East of New Zealand the water depth is too shallow for Intermediate Water to occur. Note that salinities between 33 and 34 are not contoured.

slightly higher minimum salinity, along the northern flank of the Chatham Rise to about 160°W). T-S diagrams from the Tasman Sea often do not show the Intermediate Water as a broad salinity minimum in a smooth T-S curve but reveal a high level of finestructure and interleaving (Figure 9.8). The phenomenon has been observed in all oceans north of the Antarctic Polar Front (Piola and Georgi, 1982); but in the Tasman Sea it is found well north of the formation region of AAIW and further north than in all observations elsewhere. A possible explanation is that AAIW formation takes place in a number of regions along the Polar Front (e.g. in the South Atlantic and

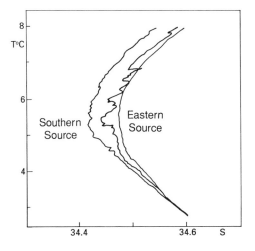

FIG. 9.8. T-S diagrams from the northern Tasman Sea near 28°S, showing stations with and without finestructure and interleaving at the level of Antarctic Intermediate Water. Interleaving is high over the Lord Howe Rise, which suggests increased turbulent mixing between the two sources over shallow ridges.

eastern South Pacific Oceans) and that its T-S properties vary slightly between different formation regions. The small inversions in the T-S curve would then indicate interleaving of AAIW from different source regions. Similar interleaving occurs in the formation region of SIW, suggesting Intermediate Water formation in distinct regions of the Polar Front of the northern hemisphere as well.

Since Intermediate Water originates from the Polar Fronts and joins the subtropical gyre circulation, it is not found in the subpolar gyre of the North Pacific Ocean. The water that occupies the depth range of Intermediate Water north of the Arctic Polar Front is called the Subarctic Upper Water. It is not restricted to subsurface layers but occupies the upper few hundred metres of the water column. Its discussion is therefore taken up again in the next section.

Water masses of the thermocline and surface layer

Chapter 5 explained how the water masses of the thermocline are subducted in the Subtropical Convergence (STC) and fill the upper kilometre of the ocean by spreading on isopycnal surfaces. The name Central Water was introduced in that chapter, and it was pointed out that these water masses are characterized by temperature-salinity (T-S) relationships that span a large range of T-S values in a well defined manner (see Figsures 5.3 and 5.4). The principle is easily applied to the Indian and Atlantic Oceans, where it serves as a valuable guide to a physical interpretation. The thermocline of the Pacific Ocean, on the other hand, displays a variety of T-S relationships, and some care is required to identify associated water masses and processes. The existence of different T-S relationships for different parts of the Pacific Ocean is most likely due to its large size, in comparison to the other two ocean basins, which leads to variations of atmospheric conditions and therefore surface T-S relationships in the Subtropical Convergence.

Six thermocline water masses can be distinguished in the Pacific Ocean (Figure 9.9). *Western South Pacific Central Water* (WSPCW) is the most saline; its T-S properties are virtually identical to those of Indian and South Atlantic Central Water, indicating identical atmospheric conditions in their formation regions. WSPCW is formed and subducted in the STC between Tasmania and New Zealand. Its occurrence is restricted to the region west of 150°W and south of 15°S (Figure 9.10). The transition to the fresher *Eastern South Pacific Water* (ESPCW) is gradual; but east of the transition, between 145°W and 110°W, the T-S properties of the thermocline are quite uniform and very close to the T-S curve labelled ESPCW in Figure 9.9, indicating that ESPCW is indeed a Central Water variety with its own formation history. Comparison with surface T-S diagrams across the Subtropical Convergence indicates that it is formed between 180° and 150°W (Sprintall and Tomczak, 1993). East of 110°W the Subtropical Convergence weakens considerably, and advection of Subantarctic Upper Water in the Peru/Chile Current reduces the salinity in the thermocline. Salinities as low as 34.1 are found in the upper thermocline east of 90°W. This region, identified in Figure 9.9 as a

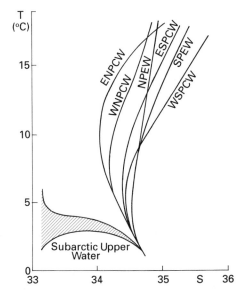

Fig. 9.9. Temperature-salinity relationships in the Pacific Ocean thermocline. ENPCW: Eastern North Pacific Central Water, WNPCW: Western North Pacific Central Water, SPEW: South Pacific Equatorial Water, NPEW: North Pacific Equatorial Water ESPCW: Eastern South Pacific Central Water, WSPCW: Western South Pacific Central Water.

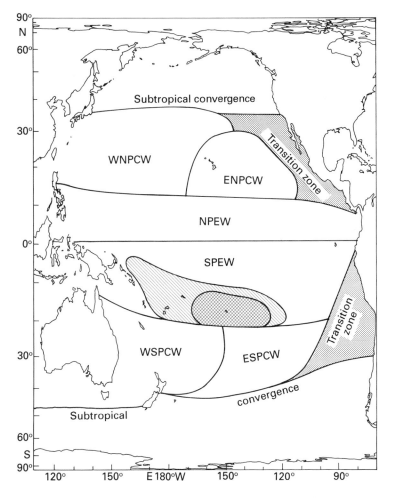

FIG. 9.10. Regional distribution of thermocline water masses in the Pacific Ocean. Shading in the region of South Pacific Equatorial Water indicates the areas where mean climatological salinity during August is above 36 at the surface (dark shading) and at 200 m depth (light shading). See Fig. 9.9 for more details and abbreviations.

"transition zone" for lack of a better name, requires more study before its water masses can be properly described.

The dominant water mass of the thermocline in the northern subtropical gyre is the *Western North Pacific Central Water* (WNPCW). It is formed and subducted in the northern STC. Sea surface salinities there are lower than those found in the southern STC, and WNPCW is thus significantly fresher than the Central Waters in the south. Again, a second variety of Central Water can be distinguished in the east. *Eastern North Pacific Central Water* (ENPCW) is fresher than WNPCW in the temperature range below 17°C but more saline in the upper thermocline. This and the fact that the boundary between the two water masses near 170°W is quite distinct, indicates that ENPCW has its own formation history. Its low salinities at the lower temperatures are probably the result of mixing with Subarctic Upper Water; but the high salinities above 17°C are well above those of all water masses in the vicinity and can therefore only be generated at the surface. This places the formation region in the region of the surface salinity maximum just south of 30°N, where salinities in excess of 35 are found throughout the year (Figure 2.5b).

The largest volume of the Pacific thermocline is occupied by Pacific Equatorial Water, a water mass without equivalent in the other two oceans. At temperatures above 8°C it displays two varieties; both have T-S properties intermediate between those of Central Water found in the two hemispheres (Figure 9.9). Below 8°C their T-S properties merge into a single curve which eventually reaches T-S combinations outside the range of Central Water. The high salinities at these temperatures reflect the absence of Intermediate Water in the tropics and indicate that Equatorial Water has contact and mixes with Deep Water.

The formation region of *North Pacific Equatorial Water* (NPEW) is at the boundary between the subtropical gyres and involves mixing in the Equatorial Countercurrent and in the Equatorial Undercurrent. Since both currents originate in the west, the water masses involved in the mixing are of the western variety. The T-S diagram indicates that NPEW is a mixture of WNPCW and SPEW, with the larger Central Water contribution in the upper thermocline. Mixing in the core region of the eastward components of the equatorial current system thus has to be regarded the common formation mechanism for all elements of this water mass; in other words, NPEW is one of the few water masses not formed through air-sea interaction. *South Pacific Equatorial Water* (SPEW), on the other hand, is partly formed by convective sinking of surface water in the tropics: South of the equator and east of 170°W (from the Polynesian islands to South America) evaporation exceeds precipitation (Figure 1.7), and sea surface salinity exceeds 36 throughout the year, making Polynesia the region with the highest salinity of the Pacific Ocean. The extent of the area for August is shown in Figure 9.10, which also includes the area where the salinity is above 36 at 200 m depth. It is seen that the high salinities reach much further west in the thermocline than at the sea surface. The origin of these high thermocline salinities must be at the sea surface, since the highest salinity reached in the Subtropical Convergence and therefore found in Central Water does not reach 35.8. This leads to the conclusion that convective sinking through evaporation occurs in the Polynesian region, at sea surface temperatures of 26°C and above. Maximum temperatures and salinities are reduced during sinking and spreading by mixing; but even in the Coral Sea where maximum salinities have fallen off to 35.8 and less, SPEW can still be clearly identified at temperatures above 20°C by its higher salinity when compared with WSPCW

errata sheet
for
Regional Oceanography: an Introduction

Page		reads	should read
13	Fig. 1.6	. . . (W m^{-2}). From . . .	. . . (W m^{-2}). Minimum values in the Kuroshio exceed -150 W m^{-2}, in the Gulf Stream -200 W m^{-2}. From . . .
37	line 11	The ratio $-(H/$ t$)/(H/$ x$)$	The ratio $-(\partial H/\partial t)/(\partial H/\partial x)$
43	Fig. 4.3	curl(t/f)	curl(τ/f)
57	line 5	Figure 16.27).	Figure 10.9).
132	Fig. 8.17b	150°W	150°E
165	Fig. 9.10	Subtropical convergence	Subtropical Front (*twice*)
	line 19	Subtropical convergence	Subtropical Front
171	line 7	Bradley *et al.*	Godfrey *et al.*
174	line 42	175°W	165°W
216	line 2	offshore,	offshore as the "Great Whirl",
217	Fig. 11.17b	. . . during June. From . . .	. . . during June. This eddy is often called the "Great Whirl". From . . .
218	line 41	monsoon winds.	monsoon winds combined with seasonal cooling.
225	Fig. 12.2a	■	7.5
		0.1	1.0
231	Fig. 12.8b	μmol g	μmol/kg
249	Fig. 13.7	μmol/l	μmol/kg
273	Fig. 14.16	the poleward jet	the equatorward jet
289	line 37	south to north	north to south
294	line 13	increase	decrease
311	line 35	Figure 16.16	Figure 16.15
312	Fig. 16.13a	>34.7	<34.7
322	line 11	19°E	9°E
355	line 17	in Figure 18.4	in Figure 18.5
376	line 18	Bradley *et al.*	Godfrey *et al.*
392	Fig. 20.10	crosses indicate	hatching indicates

References

Nelson (1989)	. . . Barber (editors) . . .	. . . S. J. Neshyba, C. N. K. Mooers, R. L. Smith and R. T. Barber (editors) . . .
Peters H (1976)	*Forschungs-Berichte*	*Forschungs-Ergebnisse*

Index entry

31	eqn. 3.2	should read	$$M = \int_{Z_1}^{Z_2} \rho v\, dz$$

37	eqn. 3.3	should read	$$M' = \int_{A}^{B} \rho v_n\, dl$$

| 37 | eqn. 3.9 | should read | $g H \Delta\rho \, \Delta H \left(\dfrac{1}{f(y_1)} - \dfrac{1}{f(y_2)} \right) = g H \Delta\rho \, \Delta H \dfrac{\beta \, \Delta y}{f^2(y_1)}$ |

| 37 | eqn. 3.10 | should read | $\rho_o \dfrac{\partial H}{\partial t} = \dfrac{g H \Delta\rho \, \Delta H \beta \, \Delta y}{f^2(y) \Delta x \, \Delta y}$, or |

$$\frac{\partial H}{\partial t} = \frac{\beta \, g \, H}{f^2(y)} \frac{\Delta\rho}{\rho_0} \frac{\partial H}{\partial x}$$

59	Fig. 5.3	the 27.04 σ_t isopycnal in the right diagram should go through point A
86	Fig. 6.16	swap *CWB* at the top with *Antarctic Divergence*
92	Fig. 7.3b	change *1015* on the second contour from right at the bottom to *1010*
175	Fig. 10.1	replace the figure by the enclosed version
177	Fig. 10.3	replace the figure by the enclosed version
181	Fig. 10.7	replace the figure by the enclosed version
185	Fig. 10.11	replace the figure by the enclosed version
195	Fig. 11.1	add the following to the caption:
		"South of Australia the Indian Ocean extends to Tasmania (146°55′E). See Figure 8.3 for the topography east of 125°E."
244	Fig. 13.5	replace the figure by the enclosed version

colour section   The longitude scale above and below Fig. 4.8 is wrong. It should be replaced as follows:

for	180°W	150°W	120°W	90°W	60°W	30°W	0°
read	0°	30°E	60°E	90°E	120°E	150°E	180°
for	30°E	60°E	90°E	120°E	150°E	180°	
read	150°W	120°W	90°W	60°W	30°W	0°	

Fig. 13.5.

Fig. 10.1.

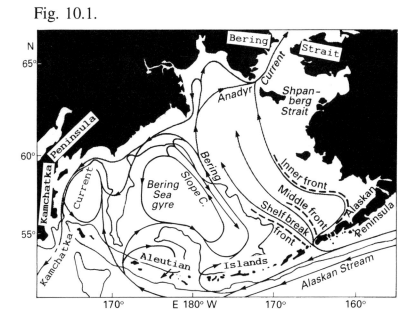

Fig. 10.3.

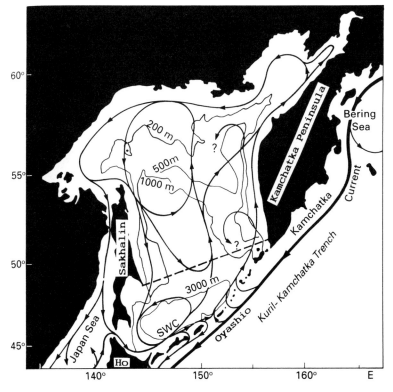

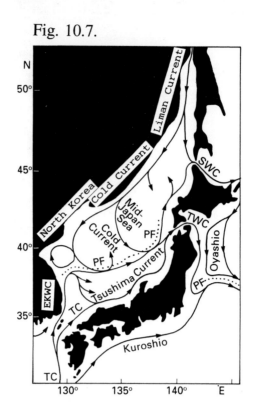

Fig. 10.7.

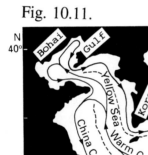
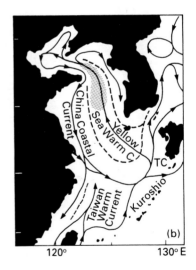

Fig. 10.11.

(Figure 9.11). Below 20°C SPEW appears to be a mixture of WSPCW and ESPCW.

Although the differences in T-S values between SPEW and WSPCW are small, the boundary between the two water masses along 20°S in the central Pacific Ocean and about 15°S in the Coral Sea is quite distinct. This is borne out in Figure 9.11 which shows the small salinity differences between the two groups of stations as being well outside the standard deviation, i.e. the natural variability, of salinity in each water mass. Steric height and salinity distributions indicate that south of 15°S WSPCW is advected from the south-east, while north of 15°S SPEW enters the region from the east. It is worth noting also that at the upper level (where the temperature is well above 20°C) the presence of SPEW is indicated by a salinity maximum (Figure 9.12c), while at the lower level (where temperatures are close to 17°C) SPEW salinity is lower than WSPCW salinity (Figure 9.12d).

The lack of communication between the circulation of the two hemispheres, already demonstrated by last chapter's T-S diagrams of Figure 8.10, produces a distinct separation between SPEW and NPEW along the equator, where the change from one water mass to the other occurs within less than 250 km distance. The change from NPEW to Central Water north of 10°N is more gradual and characterized by a front in which Central Water is found above Equatorial Water, gradually expanding downwards as one proceeds toward north. Observations along 170°W (Figure 9.13) show the transition from NPEW to WNPCW to occur at 200 - 300 m depth near 12°N and at 400 - 600 m near 18°N; north of 20°N the entire thermocline is taken over by WNPCW.

The water of the surface layer south of the southern hemisphere Subtropical

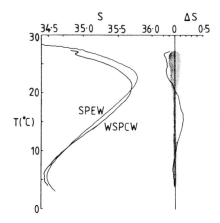

Fig. 9.11. T-S diagrams for WSPCW and SPEW derived from 42 CTD stations (37 for SPEW, 5 for WSPCW) in the eastern Coral Sea. The curve at the right is the salinity difference ΔS between the two water masses. The shaded area gives the standard deviation for salinity in both water masses. From Tomczak and Hao (1989).

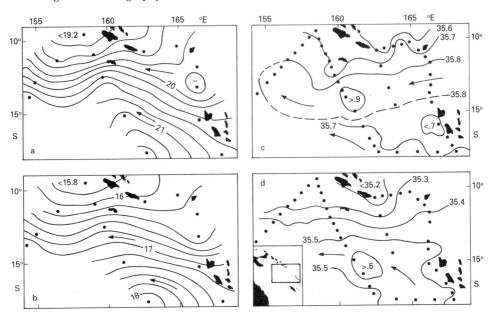

FIG. 9.12. Manifestation of the boundary between SPEW and WSPCW in the eastern Coral Sea. Dynamic height, or steric height multiplied by gravity, relative to an assumed level of no motion of 1200 m, at 148 m (a) and 248 m depth (b), and salinity at 148 m (c) and 248 m depth (d). Arrows give the inferred flow direction. Dots indicate station positions used in drawing the maps; the inset shows the location of the region. From Tomczak and Hao (1989).

Convergence is the Subantarctic Upper Water already discussed in detail in Chapter 6. The equivalent water mass in the northern hemisphere is the *Subarctic Upper Water*. Both water masses are characterized by very low salinities. Some of their water is carried toward the tropics in the eastern boundary currents of the subtropical gyres and mixes with Central Water. The range of T-S combinations produced in the process can be imagined by looking at the range of values spanned by the T-S curves of the Subarctic Upper Water and of ENPCW. Water properties in the transition regions of Figure 9.10 are found in that range. They are the only regions in the world ocean where salinity increases with depth over large parts of the thermocline (usually salinity decreases downwards until the salinity minimum of the Intermediate Water is reached). Examples of T-S curves are given in Figure 9.14.

Main aspects of the hydrographic structure above the permanent thermocline were already discussed in an earlier section of this chapter. A major aspect of surface layer dynamics in the tropical Pacific Ocean is the existence of a barrier layer, particularly in the western region where the layer is a permanent feature. Inspection of Figure 5.7 shows that its existence is closely linked with the Intertropical and South Pacific Convergence Zones (ITCZ and SPCZ), i.e. the location of maximum precipitation. This indicates that in the western Pacific Ocean the mechanism that creates and maintains

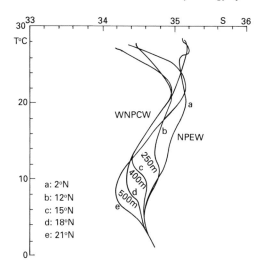

FIG. 9.13. T-S diagrams along 170°W showing the change from NPEW to WNPCW. Data from Osborne *et al.* (1991).

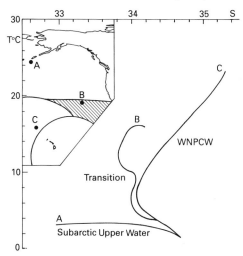

FIG. 9.14. T-S diagrams of Central and Subpolar Upper Water and from the "transition zone" of Fig. 9.10. Data from Osborne *et al.* (1991).

the barrier layer is freshening of the surface layer by local rainfall. Figure 9.15 gives an example of the resulting temperature and salinity structure.

The existence of the barrier layer has only been noted recently and has caused oceanographers and meteorologists to revise some of their ideas about ocean-atmosphere coupling mechanisms. What happens between the ocean and the atmosphere in the western equatorial Pacific Ocean is of utmost importance for the dynamics of the ENSO phenomenon, a major fluctuation of climatic conditions over one half of the globe. The dynamics of ENSO will be discussed in Chapter 19; in the context of barrier layer dynamics it

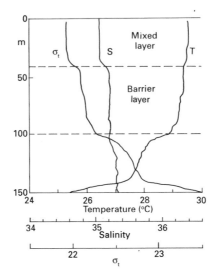

Fig. 9.15. An observation of the barrier layer in the western equatorial Pacific Ocean; T: temperature (°C), S: salinity. The halocline is above the tropical thermocline, the depth of the mixed layer indicated by the shallower of the two. From Lukas and Lindstrom (1991).

is sufficient to note that winds tend to blow towards regions of highest sea surface temperature (SST; Figure 2.5a shows that the ITCZ and SPCZ are regions of maximum SST). These high temperatures were traditionally believed to be the product of a net heat flux into the ocean. The heat gained from the atmosphere would raise the temperature of the surface mixed layer until a balance is achieved between heat gain at the surface and heat loss at the bottom of the mixed layer from mixing with colder water. This concept reflects the dynamics of the mixed layer in many regions of the subtropics and temperate zones rather well. It fails in the presence of a barrier layer because the temperature gradient at the bottom of the mixed layer is zero (Figure 9.15), which excludes heat loss through mixing. Horizontal temperature gradients in the western tropical Pacific Ocean are also very small, so countering the heat gain at the surface by bringing in cold water with the currents is also not possible. The conclusion has to be that the net heat flux into the ocean at the surface must be close to zero. This should be compared with the distribution of Figure 1.6 which shows a net heat flux of 40 W m^{-2} for the region in question. This is a low value - earlier heat flux maps give 80 W m^{-2} and more - but still too large to close the oceanic heat budget in the presence of a barrier layer.

   The problem lies in the calculation of the various components of the heat budget from observations of so-called bulk parameters, i.e. parameters that do not themselves represent heat fluxes but are easy to measure or estimate, such as air temperature, cloudiness, or wind speed. The formulae used to derive the contributions to the heat budget from long wave back-radiation, evaporation, and direct heat transfer to the atmosphere from bulk parameters are based on semi-empirical arguments and calibrated at moderate to high wind speeds. Winds over the western tropical Pacific Ocean are, however, very light during most parts of the year (Figure 1.2), and the formulae used successfully for other ocean regions may not apply to such calm conditions. Recent measurements of the heat flux terms under low wind speed conditions in the field (Bradley *et al.*, 1991) indicate indeed that under the cloudy sky of the ITCZ and SPCZ in the western equatorial Pacific Ocean the evaporative heat loss is large enough to counter the heat gained by direct solar radiation, and that the net heat transfer between ocean and atmosphere is very small. It is thus very likely that in regions of very low wind speeds the net heat flux values given in Figure 1.6 are much too high. A more detailed discussion of the exchange of properties between the ocean and the atmosphere will be presented in Chapter 18.

CHAPTER 10

Adjacent seas of the Pacific Ocean

Although the adjacent seas of the Pacific Ocean do not impact much on the hydrography of the oceanic basins, they cover a substantial part of its area and deserve separate discussion. All are located along the western rim of the Pacific Ocean. From the point of view of the global oceanic circulation, the most important adjacent sea is the region on either side of the equator between the islands of the Indonesian archipelago. This region is the only mediterranean sea of the Pacific Ocean and is called the Australasian Mediterranean Sea. Its influence on the hydrography of the world ocean is far greater in the Indian than in the Pacific Ocean, and a detailed discussion of this important sea is therefore postponed to Chapter 13. The remaining adjacent seas can be grouped into deep basins with and without large shelf areas, and shallow seas that form part of the continental shelf. The Japan, Coral and Tasman Seas are deep basins without large shelf areas. The circulation and hydrography of the Coral and Tasman Seas are closely related to the situation in the western South Pacific Ocean and were already covered in the last two chapters; so only the Japan Sea will be discussed here. The Bering Sea, the Sea of Okhotsk, and the South China Sea are also deep basins but include large shelf areas as well. The East China Sea and Yellow Sea are shallow, forming part of the continental shelf of Asia. Other continental shelf seas belonging to the Pacific Ocean are the Gulf of Thailand and the Java Sea in South-East Asia and the Timor and Arafura Seas with the Gulf of Carpentaria on the Australian shelf.

The Bering Sea and the Sea of Okhotsk

The two seas at the northern rim of the Pacific Ocean are characterized by subpolar conditions. Both are surrounded by land masses on three sides and separated from the main ocean basins by island arcs with deep passages, allowing entry of Pacific Deep Water. Another feature these two marginal seas have in common is their nearly equal division into deep basins and regions belonging to the continental shelf or rise. The *Bering Sea* is set between the Siberian and Alaskan coasts and approximates the shape of a sector with a radius of 1500 km, the circular perimeter being described by the Alaska Peninsula and the Aleutian Islands. It is the third largest marginal sea (after the Arctic and Eurafrican Mediterranean Seas), with a total area of $2.3 \cdot 10^6$ km^2 and a total volume of

$3.7 \cdot 10^6$ km^3. Northwest of a line from the Aleutian islands near 166°W to the Siberian coast near 179°E the Bering Sea is shallower than 200 m and forms part of the vast Siberian-Alaskan shelf which continues through Bering Strait into the Chukchi Sea. Southeast of that line depths fall off rapidly, reaching 3800 - 3900 m over most of the region. The Shirshov Ridge runs along 171°E with depths between 500 m and 1000 m. The slightly shallower Bowers Ridge forms a submarine arc from the Aleutian islands along 180° and then 55°N. Together, these ridges divide the western Bering Sea into three basins (Figures 10.1 and 8.3).

Knowledge of the circulation in the Bering Sea is still incomplete, and circulation schemes proposed by different authors show considerable variation. With one exception near 180°, sill depths between the Aleutian islands east of 171°E are generally less than 1000 m, and although tidal currents between the islands are strong - 1.5 m s^{-1} are common, and 4 m s^{-1} have been reported - net transport through most passages appears to be small. The major water exchange between the Pacific Ocean proper and the deep basins of the Bering Sea is believed to occur between 168°E and 172°E where the sill depth is 1589 m. A significant part of the Alaskan Stream enters the Bering Sea through this passage, turning east almost immediately and driving a cyclonic gyre in the deep part of the Bering Sea (Figure 10.1). Velocities in the inflow are near and above 0.2 m s^{-1}; in the gyres they are closer to 0.1 m s^{-1}. As explained during the discussion of the Antarctic Circumpolar Current in Chapter 6, the water temperature in subpolar ocean regions (i.e. regions poleward of the Subtropical Front) varies little with depth and currents reach very deep. The current therefore experiences the Shirshov and Bowers Ridges as obstacles to its progress, and a system of two eddies over the two basins is set up. Current shear between the gyre interior and the current axis appears to be strong; large eddies have been observed separating from the Bering Slope Current (the gyre section over the steep continental rise) into the gyre interior. The Bering Slope Current is associated with a countercurrent attached to the slope. Maximum velocities exceed 0.25 m s^{-1} and are usually found at 150 - 170 m depth. The current appears to be an eastern boundary current in a subpolar gyre circulation, i.e. the dynamics of eastern boundary currents explained in Chapter 8 apply here as well, if poleward and equatorward directions are reversed.

An amount of water nearly equivalent to that carried by the inflow from the Alaskan Stream leaves the Bering Sea with the Kamchatka Current (also known as the East Kamchatka Current), with some leakage (0.6 - 1.5 Sv, see Chapters 7 and 18) through Bering Strait. Typical velocities in the Kamchatka Current are 0.2 - 0.3 m s^{-1}.

Currents in the shallow eastern Bering Sea draw on the surface waters of the Alaskan Stream only and therefore receive their inflow through a shallow but broad passage at 175°W. Observed speeds in the passage are about 0.1 m s^{-1}, while over most of the shelf long-term mean velocities do not exceed 0.03 m s^{-1}. For

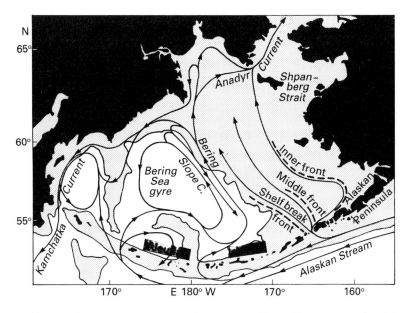

Fᴵɢ. 10.1. Surface currents in the Bering Sea. Shading indicates water depth less than 3000 m; in the region of the Bering Slope Current the 200 m isobath runs close to the 3000 m isobath. The Shirshov Ridge is seen near 171°E, the Bowers Ridge north of the Aleutian Islands near 180°.

reasons related to coastal and shelf dynamics - a topic outside the scope of this text - they are coupled with a system of fronts, along which most of the transport occurs. They are also strongly influenced by the local winds and therefore strongest in August and September when the Bering Sea is ice-free. (Ice begins to form in river mouths during October. In early November sea ice is found south of Bering Strait, and by January ice covers the entire shelf. Ice coverage during this time is usually 80 - 90%. Off Kamchatka the inflow of very cold air from Siberia results in ice coverage well beyond the shelf. Disintegration of the ice sheet starts in April and continues into July, when the Bering Sea is again free of ice.)

Currents in the northernmost section of the Bering Sea are relatively strong despite shallow water depths, being driven by sea level differences across Bering Strait. Flow through the 45 m deep Bering Strait varies between 0.1 m s^{-1} in summer and 0.5 m s^{-1} in winter. Most of its water is supplied by the Anadyr Current which flows at about 0.3 m s^{-1} and varies little with season. To compensate for the seasonal difference, flow through Shpanberg Strait is northward in winter but reverses to weakly southward in summer (Muench *et al.*, 1988).

The water mass structure is controlled by advection of water from the Pacific Ocean proper and modification of water properties on the shelf. Station data show a pronounced temperature minimum at or below 100 m depth, a rapid rise of salinity within the upper 300 m from low surface values, and generally low oxygen concentration (Figure 10.2). They indicate the presence of three water masses. The water above the temperature minimum is surface water from the area south of the Aleutian Islands imported by the Alaskan Stream. The water below the minimum is Pacific Deep Water also transported by the Alaskan Stream. As Pacific Intermediate Water is formed well south of the Alaskan Stream and does not enter the Bering Sea (compare Figures 9.4 and 9.7), Pacific Deep Water fills the entire water column below about 250 m depth where it mixes with the water of the temperature minimum. This water originates on the shelf during winter as a result of convection under the ice. Its salinity of about 33 corresponds to the highest salinities found on the shelf during the year. (The range of surface temperatures and salinities on the shelf covers -1.6 - 10°C and 22 - 33, respectively.) It sinks to 100 - 200 m depth and joins the general circulation of the deeper western part. It can be traced well into the western gyre (Figure 10.2) and into the recirculation from the Kamchatka Current to the Pacific inflow in the south.

The *Sea of Okhotsk* is set between the Siberian coast in the west and north, the Kamchatka Peninsula in the east, and the Kurile Islands in the south and southeast. The distinction between a deep and shallow region is not quite as straightforward as in the case of the Bering Sea, the main division being along the 1000 m isobath which runs diagonally through the sea from south of the Kamchatka Peninsula toward northwest (Figures 10.3 and 8.3). To the northeast of this line the depth gradually shallows to 500 m in the vicinity of 54°N and to 200 m near 57°N, although departures from this rule occur west of Kamchatka. Typical depths in the basin to the southwest of the dividing line are around

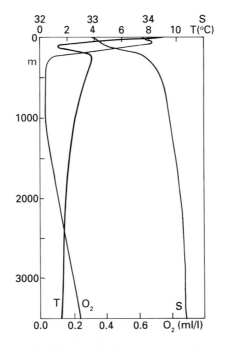

Fig. 10.2. Temperature T (°C), salinity S, and oxygen O2 (ml/l) as functions of depth in the Bering Sea, at a station near the centre of the western gyre (57°N, 167°E).

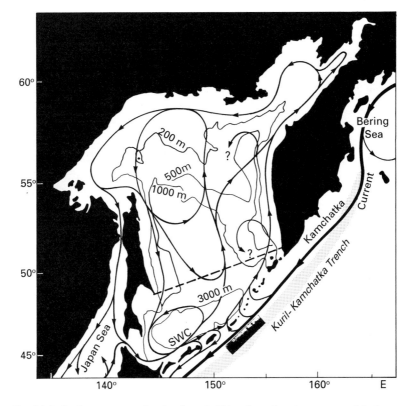

Fig. 10.3. Surface currents in the Sea of Okhotsk and major topographic features. Shading indicates regions deeper than 6000 m. The heavy broken line indicates the location of the section of Fig. 10.5. Ho: Hokkaido, SWC: Soya Warm Current.

1500 m or less. South of 49°N the ocean floor falls off further to 3000 m and more in the Kurile Basin. Numerous deep passages between the Kurile Islands connect this basin with the Pacific Ocean proper, the most important ones being Boussole Strait near 46.5°N which accounts for 43% of the total cross-sectional area and has a sill depth of 2318 m, and Kruzenshtern Strait near 48.5°N which accounts for 24% with a sill depth of 1920 m. Two additional passages connect the Sea of Okhotsk with the Japan Sea in the south. Tatarskyi Strait between Siberia and Sakhalin Island has a sill depth of less than 50 m and provides a very restricted exit for cold water from the northern shelves. Soya Strait (also known as La Pérouse Strait) between Sakhalin Island and the island of Hokkaido is less than 200 m deep and dominated by strong inflow of warm water from the Japan Sea.

Atmospheric conditions over the northern Okhotsk Sea are similar to those over the Bering Sea, and most of the region is covered with drift ice during 6 - 7 months every year. The effect of the monsoon system that dominates the climate of the marginal seas further south is felt in the southern part. The combination of winter monsoon conditions in the south and polar conditions in the north produces strong northerly or northwesterly winds blowing out of

(a)

(b)

Fig. 10.4. Eddies spawned by the Soya Warm Current. (a) A composite of two radar images obtained at two coastal stations on Hokkaido; (b) a photograph of the eddy marked by the arrow in a) taken from an aircraft at 3500 m altitude. In both figures the eddies are made visible by ice belts composed of uniform ice floes with about 10 m diameter. The diameter of the eddy in (b) is about 20 km. From Wakasutchi and Ohshima (1990).

the atmospheric high pressure cell over Siberia from October to April, often reaching storm conditions and causing waves to reach up to 10 m in height. In contrast, the southeasterly winds of the summer monsoon from May to September are rather weak, and calm conditions are encountered during 30% of the time. Both wind systems support cyclonic circulation of the surface waters along the coast with moderate velocities (0.1 - 0.2 m s^{-1}). Currents in the inner parts of the Okhotsk Sea are weaker and irregular; the limited observational data available indicate some closed circulation features particularly in the northwest and over the Kurile Basin (Figure 10.3).

An important element of the surface circulation is the Soya Warm Current, an extension of the Tsushima Current from the Japan Sea which passes through the southern part of the Sea of Okhotsk. It has the character of a boundary current with velocities reaching 1.0 m s^{-1} and traverses the Okhotsk Sea rapidly, staying close to the coast along its way. Strong current shear between the fast-flowing inshore waters and the offshore region persistently produces eddies, typically of 10 - 50 km diameter, which are easily seen when the sea is partly covered with ice (Figure 10.4).

The hydrographic structure shows strong similarities with the Bering Sea, indicating similar layering of water masses (Figure 10.5). The temperature minimum at or above 100 m is again the result of winter convection on the shelf, particularly those parts which extend deep into the Siberian land mass; as a result, water temperatures at the minimum are much lower here than in the Bering Sea. The waters above and below the minimum are again advected from the Pacific Ocean.

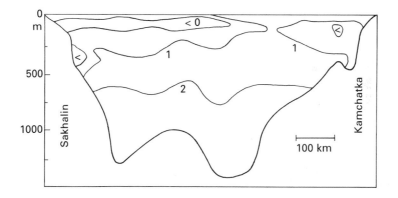

FIG. 10.5. A temperature section (°C) through the southern Okhotsk Sea. Note the lower minimum temperatures in the west, a result of the cyclonic circulation which brings the cold shelf water to the western part first. See Fig. 10.3 for the location of the section.

The Japan Sea

The Japan Sea or Sea of Japan consists of an isolated deep sea basin and its connections to the East China Sea in the south, the Sea of Okhotsk in the north, and the Pacific Ocean proper in the east. Exchange with the surrounding seas is through mostly narrow passages with sill depths not exceeding 100 m. North of about 40°N bottom depths generally exceed 3500 m; this region is known as the Japan Basin. South of 40°N the Yamato Ridge separates the Yamato Basin to the east, which is somewhat deeper than 2500 m, from the Japan Basin to the west which in this part shows a complicated topography with depths of 1000 - 2500 m.

Given the topographic features of great depth, shallow sills, and restricted communication with the open ocean, the conclusion that the Japan Sea is a mediterranean sea does not seem far-fetched. However, the geographic location at the crossroads between two mighty western boundary currents prevents the establishment of mediterranean characteristics, and the term mediterranean sea is not applied to the Japan Sea. The influence of the western boundary currents is seen clearly in the distribution of sea surface temperature which shows a distinct frontal region between central Korea and Tsugaru Strait with salinities well above 34 to the south but around and below 34 to the north (Figure 10.6). The situation looks remarkably similar to the situation found east of Japan where large horizontal temperature and salinity gradients are produced by the Polar Front through the confluence of the Kuroshio and Oyashio. The Japan Sea is indeed a meeting place for warm currents from the south and cold currents from the north; its separation into a warm part on the Japanese side and a cold part on the Siberian and Korean side indicates that the Polar Front does not terminate at the east coast of Japan but continues in modified form into the Asian mainland.

Figure 10.7 shows how the various currents combine to shape the hydrography of the Japan Sea. Warm water is brought in by the *Tsushima Current*, a branch of the Kuroshio, through Korea Strait. The branching of the north Pacific western boundary current caused by the islands of Japan pushes the position of the Polar Front in the Japan Sea much further north than in the Pacific Ocean east of Japan. Warm water from the subtropics can thus enter the Pacific Ocean proper with the *Tsugaru Warm Current* and meet the cold subpolar Oyashio as far north as 42°N. (The identification of currents as warm or cold is an east Asian tradition; elsewhere these currents would simply be called the Tsugaru and Soya Currents.) It can even proceed to 45°N, pass through the Sea of Okhotsk with the *Soya Warm*

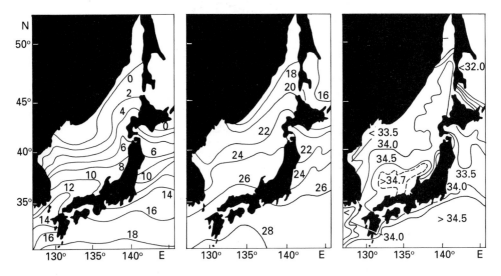

Fɪɢ. 10.6. Hydrographic conditions at the surface of the Japan Sea. (a) Temperature (°C) in February, (b) temperature (°C) in August, (c) annual mean salinity.

Current, and encounter the Oyashio some 800 km north of the latitude where the Polar Front is found in the Pacific Ocean. The complexity of the region east of Japan, seen in the satellite observations of Figure 8.18, is thus partly the result of the existence of islands in the path of the north Pacific western boundary current which allow water from the Kuroshio to bypass the Oyashio in the west and enter the region from the northwest.

Cold water enters the Japan Sea with the *Liman Current* from the Sea of Okhotsk; some continues southward along the western coast to northern Korea as the *North Korea Cold Current* before it joins the northward flow in the Polar Front. The central part of the Japan Sea is dominated by slow southward cold water movement into the Polar Front; this flow is known as the *Mid-Japan Sea* or *Maritime Province Cold Current.*

The Tsushima Current separates into two branches around the Tsushima Islands which divide Korea Strait near 35°N into a western and an eastern channel. It flows strongest in summer (August) when it carries about 1.3 Sv (about 2% of the total Kuroshio transport) with speeds of up to 0.4 m s^{-1} and weakest in winter (January) when its transport amounts to just 0.2 Sv and speeds are below 0.1 m s^{-1}. Most of the summer transport passes through the western channel and follows the Korean coast until it separates near 37 - 38°N and follows the Polar

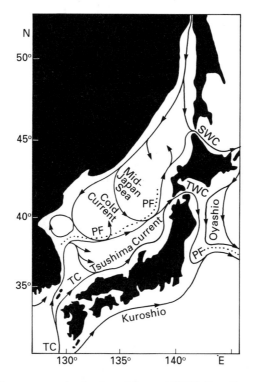

FIG. 10.7. Surface currents in the Sea of Japan. EKWC: East Korea Warm Current, PF: Polar Front, SWC: Soya Warm Current, TC: Tsushima Current, TWC: Tsugaru Warm Current.

Front. Flow through the eastern channel, which is weak throughout the year, follows the Japanese coast closely. Northeastward transport in the central Japan Sea is fairly steady at 2.5 Sv throughout the year; this incorporates the transport of cold water brought in by the North Korea and Mid-Japan Sea Cold Currents.

The separation of the Tsushima Current from the Korean coast is accompanied by instabilities typical for western boundary currents. This includes the formation of large eddies and major shifts in the paths of the two branches. Figure 10.8 (in the colour section of the book) shows that in early 1982 the situation depicted schematically in Figure 10.7 was observed. The western branch extends northward as the East Korea Warm Current and establishes the Polar Front north of Ulleuing Island (*U* in the Figure). Eddy shedding is evident between Ulleuing Island and 35°N. In contrast, during early 1981 the East Korea Warm Current did not proceed beyond 36°N and rejoined the main Tsushima Current along the Japanese coast, producing strong deformations of the Polar Front. This situation was observed to persist for six months.

The seasonal variability of the Tsushima Current is associated with strong seasonal changes of the hydrography. The sea surface salinity in Korea Strait is comparable to open ocean salinities during winter, with values close to 35 (Figure 10.9). These values fall to below 32.5 during summer when the Tsushima Current takes in large amounts of Yellow Sea water which is diluted by river runoff during the Summer Monsoon. The dilution effect does not reach much below 50 m depth and in most areas does not extend down to 30 m. Going north, the annual range of salinity is reduced by mixing; off Hokkaido surface salinity varies between 33.7 and 34.1. Seasonal changes in the Japan Sea also play an important

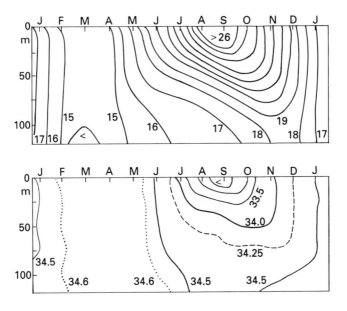

FIG. 10.9. Seasonal variation of the hydrography in Korea Strait. (a) Temperature (°C), (b) salinity (full lines use a contour interval of 0.5). From Inue *et al.* (1985)

role in the heat transfer between ocean and atmosphere. As Figure 10.6 shows, the sea surface temperature rises by 14 - 18°C from winter to summer, a warming that is almost entirely a result of increased inflow of subtropical Kuroshio water. The heat advected from the tropics is transferred to the atmosphere during winter by cold strong winds from Siberia.

Below the surface water is what is known as the *Japan Sea Middle Water* (again an Asian tradition; elsewhere this water would be called Intermediate Water). It occupies the depth range 25 - 200 m and is characterized by a rapid drop of temperature from 17°C to 2°C. Compared with the major oceans it takes the place of both Central and Intermediate Water; but its depth distribution is much more restricted. The warmer layers of Middle Water are advected into the Japan Sea from the Kuroshio, while the colder layers are formed through a combination of sinking at the Polar Front and on the northern shelf; an oxygen maximum of 8 ml/l near 200 m depth indicates recent contact of this water with the atmosphere.

Japan Sea Deep Water, usually known as *Japan Sea Proper Water*, occupies all depths below 200 m (84% of the volume of the Japan Sea). Its hydrographic properties are remarkably uniform (temperature 0 - 1°C, salinity 34.1), a result of the isolation from all other ocean basins by the shallow sills. The water mass is formed by winter convection north of 43°N and in the region 41° - 42°N, 132° - 134°E. Details of the formation process are not well known but it seems likely that salt advection from the Tsushima Current is an important factor, since deep convection will be inhibited by low densities. Instabilities of the Polar Front such as those seen in Figure 10.8 play an important role in transferring salt from the Tsushima Current into the northern regions and may thus influence the rate of formation of Japan Sea Proper Water. Compared to the same depth range

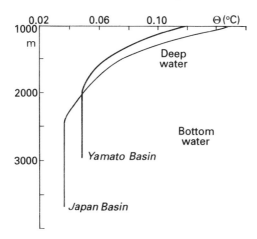

FIG. 10.10. Potential temperature (°C) in the northern Japan Basin (41.5°N, 138°E) and in the Yamato Basin (38.5°N, 135.5°E) showing different thermal gradients in Japan Sea Deep Water and Japan Sea Bottom Water. Note the extremely expanded temperature scale. From Gamo *et al.* (1986)

in the open North Pacific Ocean, the water in the deep basins of the Japan Sea is extremely well ventilated. Tritium, a product of bomb testing that entered the ocean in vast quantities some 30 years ago, had not yet reached the north Pacific waters below 1000 m depth in 1985 but was present below 2000 m depth in the Japan Sea. Oxygen levels below the thermocline are also much higher in the Japan Sea than in the open North Pacific Ocean, the Sea of Okhotsk, and the Bering Sea which are typically 1 - 2 ml/l (Figure 9.4). Oxygen values in the Japan Sea Proper Water are near 6 ml/l above 2000 m, falling off only slightly to 5.5 ml/l below. The difference in oxygen content below and above the 2000 m level is most likely a reflection of the existence of two formation regions. More recently high quality CTD data have shown a change in the gradient of potential temperature at the same depth (Figure 10.10). Some authors use these observations to differentiate between two variants of Japan Sea Proper Water, which they call Japan Sea Deep Water (200 - 2000 m) and Japan Sea Bottom Water (2000 m - bottom). The consistent temperature difference of 0.01°C between Bottom Water in the Japan and Yamato Basins has been used to infer deep winter convection in the Yamato Basin. The residence time of Bottom Water has been estimated at 300 years. At these depths there are some obvious similarities between the Sea of Japan and true mediterranean seas.

The East China Sea and the Yellow Sea

South of Tsushima Strait and adjoining the Japan Sea is a vast expanse of continental shelf which reaches from the Chinese mainland to Taiwan and stretches as far south as Vietnam. The East China and Yellow Seas encompass the region to the north of Taiwan (the southern shelf belonging to the South China Sea). Both seas form a hydrographic and dynamic unit but are distinguished by tradition. The East China Sea is usually defined as reaching from the northern end of Taiwan Strait to the southern end of Kyushu, where according to some it adjoins the Yellow Sea along a line just north of 33°N; others draw the line from Kyushu to Shanghai (the mouth of the Yangtze River). To the east the East China Sea is bordered by the Ryukyu and Nansei Islands, while the Yellow Sea continues northward between China and Korea. Its innermost part, which is fully enclosed by Chinese provinces and separated from the Yellow Sea proper by the Shandong and Liaodong peninsulae, is known as the Bohai Gulf. The Yellow Sea derives its name from the huge quantities of sediment discharged into the Bohai Gulf by the Yellow River.

With the exception of the Okinawa Trough west of the Ryukyu Islands which reaches 2700 m depth, the East China and Yellow Seas are part of the continental shelf. A complete analysis of their hydrography and dynamics is therefore only possible in the framework of coastal and shelf oceanography which is beyond the scope of this book. The following brief discussion concentrates on aspects relevant and interpretable in the context of dynamics on oceanic scales.

Two factors determine the characteristics of the East China and Yellow Seas, their proximity to the Kuroshio, and the monsoon winds which bring northerly winds during winter and southerly or southeasterly winds during summer to the entire region (Figure 1.2). Advection of warm saline Kuroshio water in the *Yellow Sea Warm Current* (Figure 10.11) raises the sea surface temperature of the central Yellow Sea several degrees above those of the coastal waters (Figure 10.12). Current speeds are generally below 0.2 m s^{-1} and decrease rapidly with depth; water temperatures below the 50 m isobath remain below 10°C during most of the summer. (This water is known as the Yellow Sea Bottom Cold Water; Figure 10.11.) The *China Coastal Current* brings water of low salinity from the northern Yellow Sea southward. A narrow coastal current along the west coast of Korea brings low salinity water from the Bohai Gulf. The *Taiwan Warm Current* carries water of oceanic properties northward, some of it as an offshoot from the Kuroshio and some through Taiwan Strait. The second path has been well documented for the period of the summer monsoon; but there is some evidence that supply from Taiwan Strait continues through winter. More observations are required to clarify the situation. Further north the path of the Taiwan Warm Current overlaps partly with that of the China Coastal Current, particularly in winter when

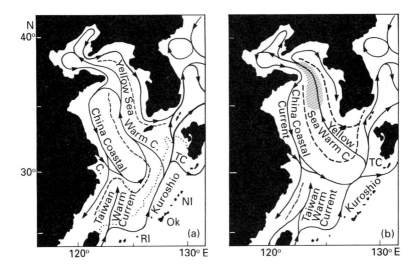

FIG. 10.11. Circulation of the East China and Yellow Seas. (a) During the winter monsoon, (b) during the summer monsoon. TC: Tsushima Current, Ky: Kyushu, NI: Nansei Islands, Ok: Okinawa, RI: Ryukyu Islands, YR: Yangtze River. The shaded area in (b) indicates the region of the Yellow Sea Bottom Cold Water.

it flows against the wind and submerges, leaving the upper 5 m of the water column to the southward flowing China Coastal Current, and during all seasons near the mouth of the Yangtze River where it is flooded by diluted water of low density (Figure 10.13).

The alternating southward and northward flows are separated by frontal regions (the broken lines in Figure 10.11). The current system exists throughout the year, the Yellow Sea Warm Current heading into the northerly monsoon winds

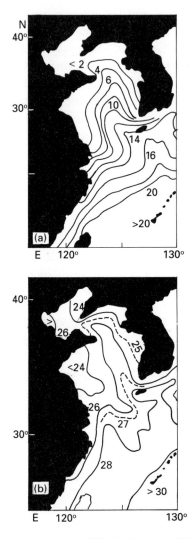

FIG. 10.12. Sea surface temperature (°C) in the East China and Yellow Seas. (a)
During the winter monsoon, (b) during the summer monsoon.

during winter, and the coastal currents opposing the southerly winds of the
summer monsoon. Unlike the Yellow Sea Warm Current, which is much weaker
when it is opposed by the monsoon winds, the China Coastal Current is
strengthened by river runoff from monsoonal rainfall in summer. The current
therefore continues unabated against the weak but opposing winds and extends
southeastward. Taking in most of the waters of the Yangtze River, it contributes
greatly to the increased summer transport of the Tsushima Current.

From the point of view of global climate the East China and Yellow Seas can
be described as a radiator. Water is withdrawn from the oceanic circulation
through the Yellow Sea Warm Current, circulated through a region with a very

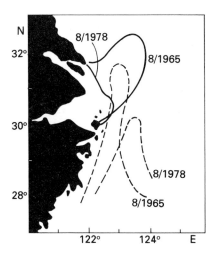

FIG. 10.13. Evidence for subsurface flow of Taiwan Warm Current water underneath low-salinity water from the Yangtze River. Data are from August of 1965 and 1978. Full lines give the 26 isohalines near the surface, broken lines the 20°C isotherms near the bottom. The two-layer structure of the flow is particularly clear during 1965. From Weng and Wang (1988).

large surface to volume ratio where it is exposed to increased air-sea interaction, and returned to the oceanic circulation in the coastal currents. The two seas also serve as a huge mixing bowl, blending large quantities of freshwater into the oceanic environment. Recent estimates derived from radiocarbon measurements (Nozaki *et al.*, 1989) put the shelf water contribution to the Tsushima Current at 20% and the residence time of the shelf water at 2.3 years.

The South China Sea

Continuing south in the sequence of marginal seas in the western Pacific Ocean, the South China Sea begins with Taiwan Strait and ends some 700 km south of Singapore. It includes within its boundaries large shelf regions and deep basins. The major basin between the Philippines and Vietnam is around 4300 m deep; in its eastern part it contains numerous seamounts studded with coral reefs. To the east of this basin is a moderately wide shelf which narrows southwards to about 50 km along the coast of Vietnam between 12° - 15°S. Further south the shelf widens to one of the largest shelf areas of the world ocean, covering the region between eastern and western Malaysia and Indonesia west of 109°E and south of 5°N. By convention this shelf region, known as the Sunda Shelf, is included in the South China Sea, with the exception of the Gulf of Thailand which will be addressed in the next section.

The only connection between the South China Sea and the Pacific Ocean proper is the Bashi Channel between Taiwan and Luzon, which has a sill depth of about 2600 m. Mindoro Channel and Balabac Channel connect the region

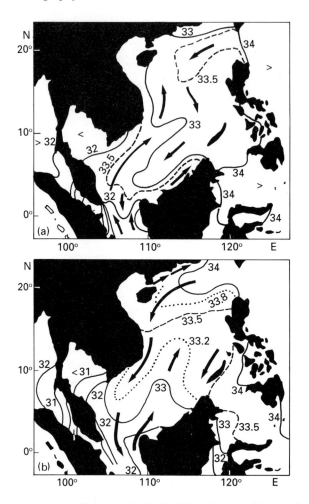

Fig. 10.14. Sea surface salinity in the South China Sea. (a) During the southwest monsoon (August), (b) during the northeast monsoon (February). Arrows indicate the inferred direction of flow. After Wyrtki (1961).

with the Australasian Mediterranean Sea to the east and have sill depths of 450 m and 100 m. The connection to the Java Sea in the south is through Karimata Strait and Gasper Strait, which are simply openings of the shallow shelf between islands without sills. Taiwan Strait in the north, the connection to the East China Sea, has a sill depth of about 70 m. Malacca Strait, the only connection to the Indian Ocean, is extremely restricted in cross-section; it has a sill depth of 30 m and a width of only 32 km. It is dominated by large tidal currents which produce periodically shifting sand dunes of 4 - 7 m height and 250 - 450 m wave length at the bottom of the Strait.

The entire region of the South China Sea is under the influence of the monsoon system, and in the absence of major oceanic inflow the currents undergo a seasonal reversal of direction. This is particularly true for currents on the shelf which are easily forced by pressure gradients established through coastal sea level

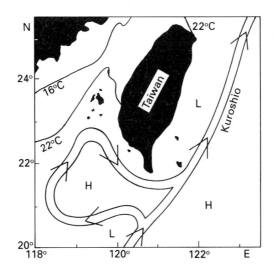

FIG. 10.15. Sea surface temperature and inferred flow direction in Taiwan Strait on 8 January 1986. Observed differences in steric height between high (*H*) and low (*L*) sea level are of the order of 0.15 m. Adapted from Wang and Chern (1988).

set-up. Direct current measurements are rare but some inferences can be made from the distribution of salinity. During May to September the southwest monsoon pushes the shelf water northward; this is believed to result in some compensatory southward movement over the deep basins (Figure 10.14). High rainfall during this season lowers salinities on the eastern shelf. During November to March the northeast monsoon reverses the direction of flow and the salinity adjusts accordingly. Along the coast of Vietnam this may develop into a strong boundary current. Further north, observations show that at least in the area north of 18°N poleward flow persists throughout winter in the inshore zone (Guan, 1986); detailed analysis of shallow water dynamics would be required to discuss this feature further.

By convention, Taiwan Strait is considered part of the South China Sea, so some words on the flow through this strait are included here (it could have been included just as well in the discussion of the East China Sea, on which it exerts considerable influence). For a long time it was believed that water movement along the west coast of Taiwan is towards north throughout the year, fed by an offshoot from the Kuroshio. More recent observations have shown that this flow is interrupted by the northeast monsoon, which holds the warm tropical Kuroshio water back behind a front (Figure 10.15). The Kuroshio water then passes to the south of Taiwan and rejoins the main Kuroshio path. The front is broken during periods of weak winds, when large parcels of Kuroshio water manage to escape through Taiwan Strait into the East China Sea. There is thus still a net supply of water from the South China Sea during winter, but it occurs sporadically rather than continuously and is related to variations in the strength of the northeast monsoon.

The Australasian shelf seas

The last group of marginal seas to consider is found in the regions to the west, and to the southeast, of the Australasian Mediterranean Sea. Both regions belong to continental shelves and thus cannot be discussed in detail without an understanding of coastal and shelf dynamics. We therefore conclude this chapter with a very brief summary of their features without going into much detail of what brings those features about.

The seas to the west and northwest of the Australasian Mediterranean Sea form part of the largest shelf region of the world ocean, which consists of the Gulf of Thailand, the Sunda Shelf, Malacca Strait, and the Java Sea. With depths in the range 40 - 80 m this shelf is shallower than most shelves bordering the oceans.

The *Gulf of Thailand* has a bowl-shaped topography with average depth of 45.5 m and maximum depth of 83 m in the centre. It is separated from the South China Sea by a sill with 58 m sill depth and can be considered a large estuary or mini-mediterranean sea with negative E - P balance (see Chapter 7 for a discussion of mediterranean sea dynamics; precipitation P here includes river runoff). Its hydrography thus shows a two-layer system with low-salinity water leaving the Gulf near the surface and colder, more saline water entering near the bottom. Average surface salinities are in the range 31 - 32 throughout the year. The inflowing water has a temperature below 27°C and a salinity above 34. This water fills the Gulf below about 50 m depth. Currents are variable, responding to the seasonal cycle of the monsoon winds which are generally weak and variable over the Gulf. The weak mean flow is clockwise during summer, anti-clockwise during winter.

The Sunda Shelf forms part of the South China Sea; its circulation and hydrography were addressed in the last section. At 3°S it connects through Karimata Strait with the *Java Sea*, a shallow region with average depths around 40 - 50 m. The Java Sea was formed by the drowning of two large river systems which now form shallow channels in the otherwise flat sea floor. Its circulation and hydrography are determined by the monsoon winds, which in this region show the same annual cycle as the winds over the Australasian Mediterranean Sea (Chapter 13). Currents flow westward from June to August and eastward during the remaining eight months. A tongue of high salinity from the South China Sea (Figure 10.14b) then penetrates deep into the Java Sea, pushing the 32 isohaline as far east as 112°E.

To the south of the Sunda Island arch, the southern boundary of the Australasian Mediterranean Sea, is the extensive shelf of the Australian continent which embraces the Timor and Arafura Seas and the Gulf of Carpentaria. The *Timor Sea* between the island of Timor and northern Australia is characterized by a narrow trench on its northern side and a broad shelf in the south. The shelf is generally less than 50 m deep but contains a large central depression, the Bonaparte Basin with a maximum depth of 140 m. Maximum depths in the Timor Trough are near 3200 m. To the southwest the trough is closed to the

Indian Ocean by a sill with about 1800 m sill depth; towards the east it is connected with the Aru Basin (which belongs to the Arafura Sea) via a sill with about 1400 m sill depth. Deep water renewal therefore occurs from the Indian Ocean.

The *Arafura Sea* south of the island of New Guinea is mostly a vast expanse of shelf generally 50 - 80 m deep, rising in its northwest to the Aru Islands. These islands are located close to the shelf break, which forms the base of many coral reefs before it falls off into the Aru Basin, a small isolated deep basin with maximum depths around 3650 m (Figure 13.5). Even though the sill depth to the Seram Basin in the north is slightly deeper than the sill depth in the south, a section of potential temperature (Figure 10.16) demonstrates that deep water renewal is from the Timor Trough (see also Figure 13.10).

Currents in the Timor and Arafura Sea are influenced by the winds and the throughflow from the Pacific Ocean through the Australasian Mediterranean Sea (see Chapter 13). There is therefore a steady westward flow along the southern side of the Sunda Islands. Further south and on the shelf currents are variable. This is the region of the shifting boundary between the Monsoon winds and the Trades; and the variability of the winds is reflected in the oceanic circulation.

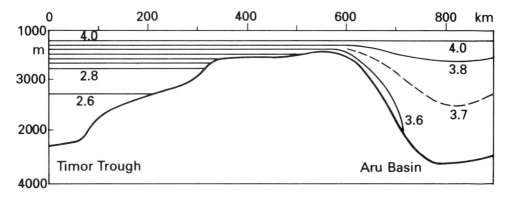

FIG. 10.16. Potential temperature (°C) below 1000 m depth along the axis of the Timor Trough and the Aru Basin. Contouring interval is 0.2°C; potential temperatures >4°C are not contoured.

CHAPTER 11

The Indian Ocean

We now turn to the Indian Ocean, which is in several respects very different from the Pacific Ocean. The most striking difference is the seasonal reversal of the monsoon winds and its effects on the ocean currents in the northern hemisphere. The absence of a temperate and polar region north of the equator is another peculiarity with far-reaching consequences for the circulation and hydrology.

None of the leading oceanographic research nations shares its coastlines with the Indian Ocean. Few research vessels entered it, fewer still spent much time in it. The Indian Ocean is the only ocean where due to lack of data the truly magnificent textbook of Sverdrup *et al.* (1942) missed a major water mass - the Australasian Mediterranean Water - completely. The situation did not change until only thirty years ago, when over 40 research vessels from 25 nations participated in the International Indian Ocean Expedition (IIOE) of 1962 - 1965. Its data were compiled and interpreted in an atlas (Wyrtki, 1971, reprinted 1988) which remains the major reference for Indian Ocean research. Nevertheless, important ideas did not exist or were not clearly expressed when the atlas was prepared, and the hydrography of the Indian Ocean still requires much study before a clear picture will emerge. Long-term current meter moorings were not deployed until two decades ago and notably during the INDEX campaign of 1976 - 1979; until then, the study of Indian Ocean dynamics was restricted to the analysis of ship drift data and did not reach below the surface layer.

Bottom topography

The Indian Ocean is the smallest of all oceans (including the Southern Ocean). It has a north-south extent of 9600 km from Antarctica to the inner Bay of Bengal and spans 7800 km in east-west direction between southern Africa and western Australia. Without its Southern Ocean part it covers an area of $48 \cdot 10^6 \, \text{km}^2$. If the Southern Ocean part is included, the area increases to $74.1 \cdot 10^6 \, \text{km}^2$. The only large shelf area is the Northwest Australian Shelf, a region of strong tidal dissipation. It is part of the large expanse of continental shelf between Australia and south-east Asia that continues as the Timor and Arafura Seas and the Gulf of Carpentaria; however, these latter regions are considered part of the Pacific Ocean. The Northwest Australian Shelf itself is insufficient in size to have much impact on the mean depth of the Indian Ocean, which with 3800 m is between that of the Pacific and Atlantic Oceans. Most basins show depths well below 5000 m; in the east, the *Wharton Basin* exceeds 6000 m depth. The *Arabian Sea* reaches depths below 3000 m over most of its area, while the depth in the *Bay of Bengal* decreases gradually from 4000 m south of Sri Lanka to 2000 m and less at 18°N.

Three mediterranean seas influence the hydrographic properties of Indian Ocean water masses. The Persian Gulf is the smallest of the three; with a mean depth of 25 m, a maximum depth of only 90 m, and a sill which rises barely above the mean depth its impact is not felt much beyond the Gulf of Oman. The Red Sea is a very deep basin with maximum depths around 2740 m and a mean depth near 490 m; its sill depth is about 110 m. The Australasian Mediterranean Sea, a series of very deep basins with depths exceeding 7400 m, communicates with the Indian Ocean through various passages between the Indonesian islands where the depth is in the range 1100 - 1500 m.

Two ridge systems run through the Indian Ocean in a roughly meridional direction, dividing it into three parts of about equal size (Figure 11.1). The *Central Indian Ridge* between the western and central part is a northward extension of the interoceanic ridge system (Figure 8.1) and connects with the ridge systems of the Atlantic and Pacific Oceans at a bifurcation point in the south. Similar to the East Pacific Rise and Mid-Atlantic Ridge it is a broad structure which rises consistently above 4000 m; again and again it reaches above 3000 m (in the region of the Mascarene Plateau it rises high enough to influence upper ocean currents), but numerous fractures and depressions make the blockage for water flow incomplete at that level. The Eastern Indian or *Ninety East Ridge* stretches in nearly perfect north-south orientation from the Andaman Islands to 33°S. It is much narrower and less fractured than the other ridges. South of 10°S it reaches the 3000 m level consistently and the 2000 m level frequently. South of 30°S it rises above 1500 m and merges with the *Southeast Indian Ridge* (a continuation of the Central Indian Ridge) at the 4000 m level. The *Broken Plateau* branches off to the east near 30°S, also reaching above 1500 m. North of 10°S the Ninety East Ridge shows occasional gaps in the 3000 m contour. The passages are apparently deeper than the merging depth of the ridge systems in the south and therefore provide a bottom water entry point for the Central Indian Basin in the northeast.

In contrast to the central and eastern parts of the Indian Ocean which are dominated by single basins of large meridional extent, the western part is subdivided by secondary ridges and the island of Madagascar into a series of smaller deep basins. The Arabian Basin is closed in the south by the merger of the *Carlsberg, Central Indian,* and *Chagos-Laccadive Ridges.* The merger region shows extremely complicated topography with multiple fracture zones; but the closure is complete somewhere near the 3500 m level, somewhat higher than the deepest passage in the Owen Fracture Zone. As a consequence the bottom water entry point for the Arabian Basin is also found in the north rather than in the south.

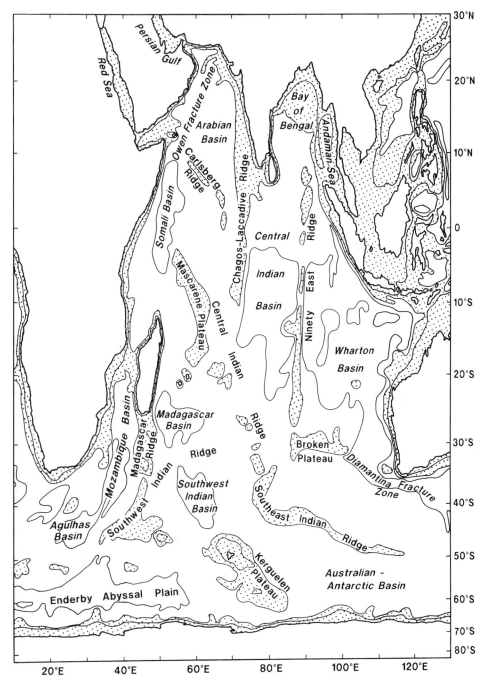

Fig. 11.1. Topography of the Indian Ocean. The 1000, 3000, and 5000 m isobaths are shown, and regions less than 3000 m deep are stippled.

The wind regime

Monsoonal climate dominates the northern Indian Ocean, and its effects are felt far into the subtropics of the southern hemisphere. Annual mean distributions of atmospheric and oceanic parameters are therefore of only limited use. Instead, we define two mean states and discuss both separately. The information is again contained in Figures 1.2 - 1.4. A summary of the monsoon cycle is given in Figure 11.2. An introductory but comprehensive account of monsoon meteorology is given by Fein and Stephens (1987).

The *Northeast* or *Winter Monsoon* determines the climate of the northern Indian Ocean during the northern hemisphere winter (December - March). It is

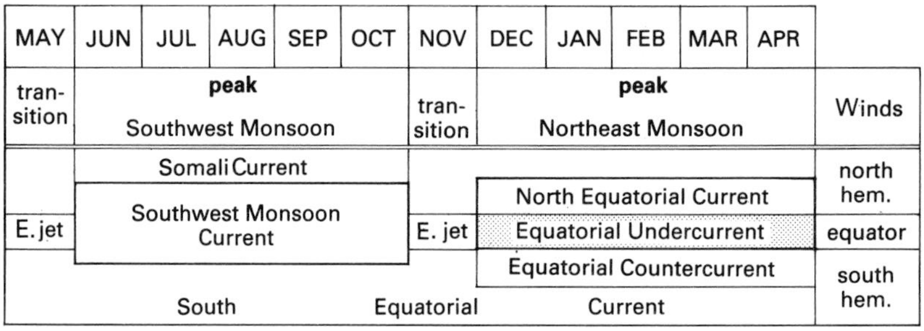

MAY	JUN	JUL	AUG	SEP	OCT	NOV	DEC	JAN	FEB	MAR	APR	
tran-sition	peak					tran-sition	peak					Winds
	Southwest Monsoon						Northeast Monsoon					
	Somali Current											north hem.
E. jet	Southwest Monsoon Current					E. jet	North Equatorial Current					
							Equatorial Undercurrent					equator
							Equatorial Countercurrent					south hem.
	South			Equatorial			Current					

FIG. 11.2. A summary of the monsoon system in the Indian Ocean. The top part indicates the wind cycle, the lower part shows the major currents that develop in response to the wind.

characterized by high pressure over the Asian land mass and northeasterly winds over the tropics and northern subtropics. The situation resembles the annual mean wind circulation over the Pacific Ocean, except that the Intertropical Convergence Zone (ITCZ) and the Doldrums are located south of the equator (near 5°S) rather than north. The wind over the northern Indian Ocean represents the Trades, but because of its seasonality it is known as the Northeast Monsoon. (The word monsoon is derived from the Arabic, meaning seasonally reversing winds.) Since most of the air pressure gradient is retained behind the Tibetan Plateau, air pressure gradients over the ocean are small. This protects the ocean from the full force of the winds blowing off the Mongolian high pressure region and results in a wind of moderate strength, comparable to the Pacific Northeast Trades which are also relatively weak at this time of year. The wind also carries dry air, and the Winter Monsoon season is the dry season for most of southern Asia.

The situation in the southern hemisphere is determined by the pressure gradient between the tropical low and the subtropical high pressure belts. The axis of low pressure in the tropics is near 10°S, while the subtropical high pressure belt is dominated by an air pressure maximum 1000 km north of the Kerguelen Islands. The resulting Southeast Trades are rather uniform and somewhat stronger than in the Pacific Ocean. As in the other oceans the Trades are southerly along the eastern coast, but over Australia's Northwest Shelf winds become south-westerly, skirting the heat low over that continent. This brings summer rain to northern Australia. Rain also occurs throughout the Doldrums, with maximum rainfall in the Indonesian region.

The *Southwest* or *Summer Monsoon* determines the climate of the northern Indian Ocean during the northern hemisphere summer (June - September). A deep heat low develops over northern Arabia and Pakistan. The Australian heat low of the southern summer is replaced by a centre of high pressure, while the atmospheric high north of the Kerguelen Islands is shifted westward towards southern Africa. Whereas during the winter monsoon season the north-south pressure gradient from Arabia to Madagascar barely exceeds 6 hPa, there is now a gradient of 22 hPa acting in the opposite direction. As a result the winds in the northern Indian Ocean reverse completely and are no longer like the Trades anywhere. A wind jet, believed to be an atmospheric version of a western boundary current, develops along the high east African topography. Winds blow steadily at Beaufort 6 or more over the entire western Indian Ocean north of the equator. Further east along the equator the winds weaken, bringing moderate rainfall; but there is still a southerly component throughout the entire Indian Ocean everywhere north of 30°S. On the Northwest Shelf the wind blows directly offshore with moderate strength.

The Southwest Monsoon, as this wind is called in the northern hemisphere, is the continuation of the southern hemisphere Trades, which between 10 and 20°S are stronger, during this time of year, than anywhere else in the world. The Southwest Monsoon skirts the low over Pakistan to deposit rain on the Himalayas, thus bringing with it the monsoon rains and floods that are so crucial to Asian agriculture.

The position of the ITCZ over the ocean changes little between the seasons, but during the Southwest Monsoon frequent disturbances break away from it to settle south of the Himalayas, bringing with them most of the season's rain. Winds at the equator change direction but remain weak throughout the year. Because of their meridional orientation the associated Ekman transport does not develop a divergence at the equator. There is therefore no equatorial upwelling in the Indian Ocean. Strong equatorial downwelling occurs during the transition months (May and November) when the winds turn eastward at the equator, producing Ekman transport convergence.

Conditions for coastal upwelling are also markedly different in the Indian Ocean. Along the eastern coastline, where the most important upwelling regions of the Pacific and Atlantic Oceans are found, winds favourable for upwelling along the Australian coast are weak during the Northeast Monsoon season and absent during the Southwest Monsoon season. A small upwelling region can be expected in the latter season along the coast of Java. But the strongest upwelling of the Indian

Ocean occurs along its western coastline when the Southwest Monsoon produces strong Ekman transport away from the coasts of Somalia and Arabia.

We saw in Chapter 4 that the Sverdup relation, which compares circulation patterns derived from annual mean integrated steric height and annual mean wind stress, gives very good agreement in the Indian Ocean despite the seasonal reversals of the monsoon winds. It is therefore instructive to look at the annual mean atmospheric circulation over the Indian Ocean as well. The southern hemisphere Trades stand out as particularly strong, in the annual mean, when compared with the other oceans. As we have seen, the larger contribution to this comes from the Summer Monsoon season. The stronger winds of this season dominate the annual mean in the northern Indian Ocean as well. This produces a net southwesterly stress over the northwestern Indian Ocean and in particular along the East African coast. Mean winds in the northeast and along the equator are weak and westerly.

The integrated flow

We begin the comparison between the depth-integrated flow (Figs 4.4 - 4.7) and the observed surface circulation (Figure 11.3) with the southern Indian Ocean which does not experience monsoonal winds. The Sverdrup stream function (Figure 4.7) shows a particularly strong subtropical gyre, a result of large wind stress curl between the annual mean Trades and Westerlies (Figure 1.4). South of 10°S and away from the Australian coast the pattern agrees well with the observations. Both show the South Equatorial Current being fed by the throughflow of Pacific water through the Indonesian seas and strengthening westward. The bifurcation east of Madagascar and again near the African coast is also seen in the Sverdrup flow, as is the joining of the Mozambique and East Madagascar Currents into the Agulhas Current. Furthermore, the Sverdrup relation implies net southward flow across 10°S in response to strong negative wind stress curl; this generates the eastward flowing Equatorial Countercurrent north of 10°S as its source of supply. To the best of our knowledge, the annual mean transports of the open-ocean currents are well predicted by the Sverdrup relation, and the transports of the East Madagascar, Mozambique, and Agulhas Currents follow quite accurately from mass continuity.

Significant discrepancies between the Sverdrup circulation and observations are seen in the Agulhas Current extension region, for reasons discussed in Chapter 4, and in the Leeuwin Current. The net depth-averaged flow along the Australian coast is in fact very close to that given by the Sverdrup circulation; but the shallow Leeuwin Current is accompanied by an undercurrent of nearly equal transport (to be discussed below), and the surface circulation is therefore not representative for the depth-averaged flow.

In contrast, the annual mean flow in the northern Indian Ocean is almost featureless, both in the Sverdrup circulation maps and in observations. The stream function map does not show a single streamline for the northern hemisphere,

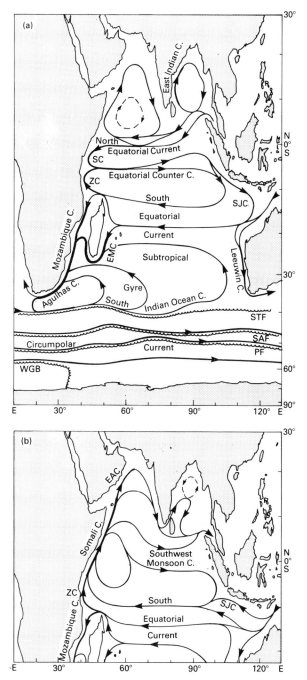

Fig. 11.3. Surface currents in the Indian Ocean. Top, late Northeast Monsoon season (March - April); bottom, late Southwest Monsoon season (September - October; the circulation south of 20°S remains unchanged). Abbreviations are used for the East Arabian (EAC), South Java (SJC), Zanzibar (ZC), East Madagascar (EMC), and Somali (SC) Currents. Other abbreviations denote fronts: STF: Subtropical Front, SAF: Subantarctic Front, PF: Antarctic Polar Front, WGB: Weddell Gyre Boundary.

so any mean circulation that does develop must be weaker than 10 Sv (the contour interval used for Figure 4.7). This is, of course, a somewhat theoretical result since it represents the mean of two strong monsoonal current systems. The annual mean is still of use when it comes to the calculation of heat fluxes between the hemispheres, and we shall come back to this in Chapter 18. But in a discussion of monsoonal surface currents the Sverdrup relation has its limits, and we now turn to a description based on current and hydrographic observations.

During the Northeast Monsoon season the current system resembles closely those of the Pacific and Atlantic Oceans. The subtropical gyre dominates the southern hemisphere. For a long time it was believed that eastward flow in the south is achieved by the Circumpolar Current. Following the recent discovery of the South Atlantic Current as a distinct element of the south Atlantic subtropical gyre (Chapter 14), Stramma (1992) showed the existence of a similar South Indian Ocean Current north of the Circumpolar Current. The current follows the northern flank of the Subtropical Front, carrying some 60 Sv in the upper 1000 m southeast of Africa and gradually releasing its waters into the gyre. Southwest of Australia its transport is reduced to about 10 Sv. Zonal flow immediately south of the Subtropical Front is usually weak, indicating that the South Indian Ocean and the Circumpolar Currents are two distinct features of the circulation.

The subtropical gyre of the northern hemisphere is not well-defined; most of the water carried in the North Equatorial Current returns east with the Equatorial Countercurrent near 5°S, leaving little net transport for the weak currents in the northern Indian Ocean. The Equatorial Countercurrent continues into the South Java Current, which during this season feeds its waters into the Indonesian seas and southward into the South Equatorial Current. Currents flow in the opposite direction north of 10°S during the Southwest Monsoon season, when the South Equatorial Current intensifies and feeds part of its flow into the Somali Current, the western boundary current of the anticyclonic circulation that develops in the northern hemisphere. The North Equatorial Current disappears, and the Equatorial Countercurrent becomes absorbed into the Southwest Monsoon Current. Its broad eastward flow dominates the northern Indian Ocean.

A feature of the circulation which is independent of the monsoon cycle is the strong separation of the northern and southern hemisphere flow fields along the ITCZ east of 60°E. Little flow crosses the Doldrums east of the western boundary current. The semi-annual reversal of the intense flow across the ITCZ along the east African coast north of 10°N has been well documented during the last two decades. However, it is obvious that at least some of the water that enters the northern hemisphere from June to October in the west must leave it during that season. Likewise, at least some of the water that is withdrawn during December to April has to enter it somewhere in the east during these months. Our knowledge of the circulation near Java is, however, incomplete, and the details of the mass balance on seasonal time scales are a topic for the future.

The Indonesian throughflow from the Pacific Ocean enters the Indian Ocean during both seasons as a narrow band of low salinity water. It is embedded in generally westward flow and therefore apparently does not develop the strong lateral shear necessary to induce much instability. It continues westward, providing the core of maximum westward flow in the equatorial current system, and can be followed over the entire width of the Indian Ocean.

The subtropical gyre of the southern hemisphere is seen with two western boundary currents, one along eastern Madagascar and one along the coast of Mozambique. Whether both feed into the Agulhas Current - as indicated by the depth-integrated flow - or whether the East Madagascar Current feeds back into the gyre independently has been the subject of debate for some time and will be discussed in more detail later on. Here we only note the difference between the flow fields based on 1500 m and 2500 m levels of no motion (Figs 4.5 and 4.6) south of Madagascar; they indicate that the circulation near the end of the East Madagascar Current reaches deeper than 1500 m, not unlike the situation found in separation regions of other western boundary currents. The fact that deep flow is not restricted to the Agulhas Current but extends eastward away from Madagascar makes the estimation of transports in the gyre difficult and affects in particular our estimates for the South Equatorial Current.

The equatorial current system

Cutler and Swallow (1984) used the records of ship drifts collected by the British Meteorological Office from daily log book entries of merchant ships, to compile an atlas of surface currents. Currents in the Indian Ocean are stronger than in the Pacific or Atlantic Oceans during most of the year, which makes ship drift estimates reasonably reliable. Given the paucity of other information, this atlas is the best source of information on surface currents near the equator. Information on the subsurface structure became available with the long-term current meter moorings and time series of vertical current profiles taken during INDEX; unfortunately it is restricted to the region west of 62°E.

The evolution of surface currents through the seasons is shown in Figure 11.4. The *North Equatorial Current* is prominent in January and March when the Northeast Monsoon is fully established. It runs as a narrow current of about 0.3 m s^{-1} from Malacca Strait to southern Sri Lanka, where it bends southward and accelerates to reach 0.5 - 0.8 m s^{-1} between 2°S and 5°N in the region between 60°E and 75°E. The *South Equatorial Current* occupies the region south of 8°S with velocities rarely exceeding 0.3 m s^{-1}. Between these westward flows runs the *Equatorial Countercurrent* with 0.5 - 0.8 m s^{-1} in the west but getting weaker in the east; in January it does not reach beyond 70°E, being opposed in the east by weak westward flow.

The transition from Northeast to Southwest Monsoon (Figure 11.4c) is characterized by the intense *Indian Equatorial Jet* first described by Wyrtki (1973a). The long-term mean distributions derived from the ship drift data show it from

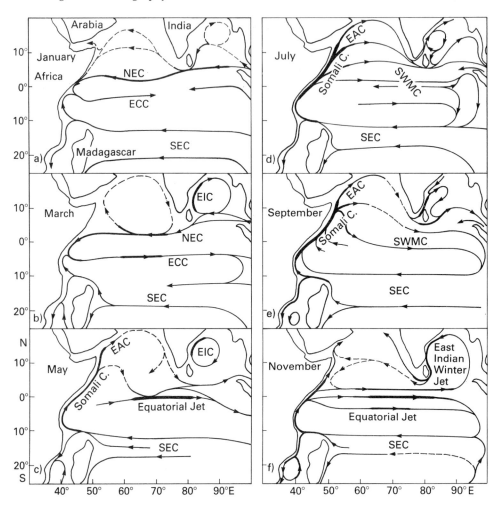

Fig. 11.4. Surface currents in the northern Indian Ocean as derived from ship drift data. SEC: South Equatorial Current, NEC: North Equatorial Current, ECC: Equatorial Countercurrent, SWMC: Southwest Monsoon Current, EAC: East Arabian Current, EIC: East Indian Current. Adapted from Cutler and Swallow (1984).

early April until late June with velocities of 0.7 m s^{-1} or more. It is possible that in any particular year the jet appears within the three-month window April - June as a feature of shorter (one month) duration with higher peak velocities. The averaging employed with the ship drift data would spread it over the three months as a weaker feature. The jet is easily observed with drifting buoys since the current converges at the equator, keeping drifting objects trapped near its core. Away from the equator the current speed falls off to less than 0.2 m s^{-1} at 3°S or°N.

When the Southwest Monsoon is fully established during July and September, the entire region north of 5°S is dominated by the eastward flow of the *Southwest Monsoon Current*, the only exception being a narrow strip along the equator in July to which we shall return in a moment. Velocities in the Southwest Monsoon

Current are generally close to 0.2 - 0.3 m s^{-1}, but an acceleration to 0.5 - 1.0 m s^{-1} occurs south and southeast of Sri Lanka. The South Equatorial Current expands slightly towards north, reaching 6°S in September. The transition before the onset of the Northeast Monsoon (Figure 11.4f) is again characterized by the Equatorial Jet. Concentrating all eastward flow in a 600 km wide band along the equator it reaches its peak in November with velocities of 1.0 - 1.3 m s^{-1} and disappears in early January, when the annual cycle is repeated.

A remarkable feature of the equatorial current system is the dominance, within the annual monsoon cycle, of a semi-annual flow reversal in a narrow band along the equator. Figure 11.5 compares $2^1/_2$ years of wind data near the equator with average currents in two layers above the thermocline. Winds at the equator are generally light, the meridional wind component being dominated by the annual monsoon cycle. The zonal wind component, on the other hand, shows westerly winds during the transition periods and exerts a strong semi-annual signal on the ocean. The associated change of current direction produces semi-annual variations in thermocline depth and sea level (Figure 11.6). The 20°C isotherm rises off Africa and falls off Sumatra during periods of eastward flow, indicating a net transport of 20 Sv in the equatorial band. The same transport in the opposite direction occurs during periods of westward flow when the 20°C isotherm rises off Sumatra and falls off Africa. The associated zonal slope of the thermocline finds its mirror image (as expressed in our Rule 1a of chapter 3) in the slope of the sea surface, as indicated by the sea level at Sumatra.

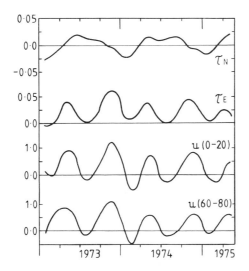

FIG. 11.5. Monthly mean winds and currents during 1973 - 1975 near the equator at 73°E: Meridional wind stress component τ_N (Pa), zonal wind stress component τ_E (Pa), and zonal velocities (m s^{-1}) averaged over 20 m for 0 - 20 m and 60 - 80 m depth. All directions indicate where currents and winds are going, north and east are positive. After McPhaden (1982).

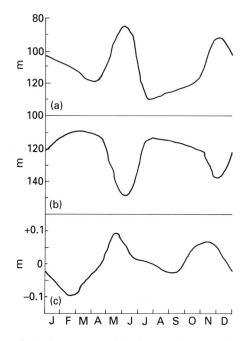

FIG. 11.6. Climatological mean monthly thermocline depth and sea level at the equator. (a) Depth of 20°C isotherm off Africa, (b) depth of 20°C isotherm off Sumatra, (c) sea level at the west coast of Sumatra. After Wyrtki (1973a)

Figure 11.5 suggests a direct relationship between the zonal current and the zonal wind. However, the current is not purely wind-driven. An array of current meter moorings was deployed along the equator between 48°E and 62°E during INDEX (1979 - 1980), with instruments at 250 m, 500 m, and 750 m depth. The semi-annual current reversal occurred at all depths, much deeper than could be explained by direct wind forcing (Luyten and Roemmich, 1982). It appears that the wind reversal in the west is only the triggering mechanism for a wave phenomenon peculiar to the equatorial region known as a Kelvin wave. The same process is responsible for interannual climate variations in the equatorial Pacific Ocean which will be discussed in detail in Chapter 19. We therefore leave the explanation of Kelvin waves to that chapter and note here only the essence of the process, which is as follows. The arrival of westerly winds in the west lifts the thermocline up near the African coast. The resulting bulge in the thermocline travels eastward, accompanied by strong eastward flow, until it reaches Sumatra. Thus, eastward flow does not occur simultaneously but propagates along the equator. This explains why we find both eastward and westward flow at the equator on occasions, as in Figs. 11.4a and d.

The limited information available on subsurface flow near the equator indicates a complicated and unique current regime. Observations with profiling current meters show that the currents reach to great depth with only slightly reduced velocities (Figure 11.7). Several layers of alternate flow direction were found, with

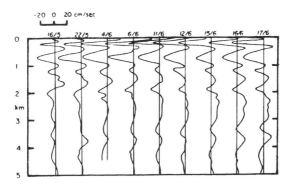

Fɪɢ. 11.7. Eastward components of velocity on the equator at 53°E. After Luyten
and Swallow (1976).

velocities reaching 0.12 m s^{-1} at 4000 m depth. The uppermost layer of westward
flow near 200 m depth is commonly referred to as the *Equatorial Undercurrent*,
but it is obvious that westward transport occurs at much greater depths as well.
During the Southwest Monsoon of 1976 mean flow at moorings in the area
52 - 58°E was westward with a northward component into the Somali basin at
500 m and 1500 m depth.

The oscillations of the undercurrent known from the Pacific and Atlantic Oceans
occur in the Indian Ocean as well. At 200 m depth they were seen from August
1979 until January 1980 (in a record which covered the period March 1979 -
June 1980). They were of 1300 km wavelength and 0.2 m s^{-1} meridional amplitude
and shifted the undercurrent axis back and forth by about 150 km either side
of the equator. Drifting buoy data for the period 1979 - 1982 indicate that at
the peak of the Southwest Monsoon in August and September the oscillations
reach into the surface layer with meridional flow amplitudes up to 0.8 m s^{-1}.

Circulation in the Arabian Sea and Bay of Bengal

Seasonal reversal of the currents dominates the two major subdivisions of the
northern Indian Ocean as well, but the opposing flows occupy periods of very
different length, and the transitions are less well defined than in the equatorial
zone. Weak westward flow, an extension of the North Equatorial Current with
velocities rarely exceeding 0.2 m s^{-1}, dominates in the Arabian Sea at the peak
of the Northeast Monsoon season. Northwestward flow along the western Indian
shelf begins as early as November (Figure 11.4f) and persists into January, with
a width of some 400 km and a depth of about 200 m in the south (10°N), getting
narrower and deeper as it flows along the continental slope. This current flows
against the prevailing Northeast Monsoon and thus cannot be wind-driven. The
East Indian Winter Jet supplies fresh, low density water from the Bay of Bengal
at this time, while on the north Indian coast cold continental winds result in
cooling and convective overturn (Shetye et al., 1991). The resulting gradient
of steric height overwhelms the wind forcing. These dynamics are discussed in
more detail with the Leeuwin Current below.

Westward flow prevails south of 15°N and west of 65°E until late April, while in the remaining area currents are less and less well defined and change gradually into the weak anticyclonic pattern of Figure 11.4b. The Somali Current responds quickly to the onset of the Southwest Monsoon in April; northward flow develops, strengthening the pattern in the west. By mid-May (Figure 11.4c) the *East Arabian Current* is fully established with velocities of 0.5 - 0.8 m s^{-1}. At the same time the anti-cyclonic pattern breaks up from the east where the flow joins the Equatorial Jet around southern India and Sri Lanka. Moderate eastward flow, an extension of the Somali and Southwest Monsoon Current, dominates the region during the next 4 - 5 months. During its peak in June and July it reaches 0.3 m s^{-1} and more but weakens rapidly in October when the second occurrence of the Equatorial Jet concentrates most eastward transport in the equatorial zone and outflow from the Bay of Bengal begins to oppose eastward flow around Sri Lanka. By mid-November currents are again diffuse; south of 15°N they are weak but already westward. General westward flow is again established by early December.

A notable feature of the Arabian Sea circulation is the occurrence of strong coastal upwelling in the East Arabian Current. As in other coastal upwelling regions it owes its existence to an offshore transport direction in the Ekman layer (the Southwest Monsoon blowing parallel to the coast with the coast on its left). Positive curl(τ/f) over a 400 km wide strip along the coast adds to the upwelling through Ekman suction. From May to September coastal temperatures are lowered by 5°C and more (Figure 11.8). The fact that the upwelling is embedded in a western boundary current reduces its effectiveness for primary production - the swift current removes much of the additional biomass from the system before it can be utilized. Compared to upwelling regions in the Pacific and Atlantic Ocean, zooplankton levels in the Arabian Sea upwelling are much less exceptional. Nevertheless, an abundance of sea birds off the Arabian coast during the upwelling period indicates that it is sufficient to support a significant marine resource.

An increase of zooplankton biomass associated with a drop in sea surface temperatures reminiscent of upwelling has also been documented for the western Indian shelf during the Southwest Monsoon season (Figure 11.9). Since the Southwest Monsoon cannot drive the surface waters offshore on that coast, the reasons for the increased productivity must be found elsewhere. Most likely the phenomenon is related to nearshore advection of river water associated with the monsoon rains and not to the oceanic circulation.

The circulation in the Bay of Bengal is characterized by anticyclonic flow during most months and strong cyclonic flow during November. In January currents are weak and variable. In the west, the *East Indian Current* strengthens as the Northeast Monsoon becomes stronger, exceeding 0.5 m s^{-1} in March (Figure 11.4b) and remaining strong (0.7 - 1.0 m s^{-1}) until May/June. Throughout this time the current runs into the wind, apparently as an extension of the North Equatorial Current. The fact that it persists during May when flow south of the Bay turns eastward is remarkable and still requires an explanation. During the Southwest

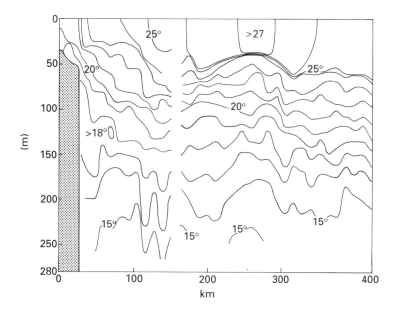

Fig. 11.8. A temperature section across the Arabian upwelling region during July 1983, from the Kuria Muria Islands (18°N) towards southeast. The break in the isotherm slope about 150 km from the coast separates the coastal upwelling to the left from the Ekman suction region to the right. After Currie *et al.* (1973).

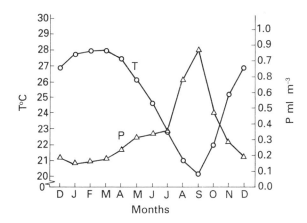

Fig. 11.9. Monthly mean temperature at 50 m depth (T) and zooplankton biomass (P) on the western Indian shelf between 8°N and 15°N. From Murty (1987).

Monsoon season currents in the entire bay are weak and variable again. The highest velocities (around 0.5 m s^{-1}) are found in the East Indian Current; flow along the eastern coast rarely exceeds 0.2 m s^{-1} but is often directed into the wind.

An indication of a current reversal in the west is seen in September (Figure 11.4e). Currents are consistently southwestward and strong (0.5 m s^{-1} and more) north of 15°N, and close to the shelf southwestward flow prevails. Complete reversal of the East Indian Current into the *East Indian Winter Jet* is not achieved until late October, when water from the Equatorial Jet enters the Bay in the east and a cyclonic circulation is established. The East Indian Winter Jet is a powerful western boundary current with velocities consistently above 1.0 m s^{-1}. It follows the topography south of Sri Lanka and feeds its water into the Arabian Sea. Very little exchange occurs with the Equatorial Jet south of Sri Lanka; currents in the separation zone between the two jets (near 3°N) are weak and variable. The East Indian Winter Jet fades away from the north in late December, its southern part merging with the developing North Equatorial Current.

Western boundary currents

The story of the western boundary currents begins east of Madagascar, where both the integrated flow (Figure 4.7) and the ship drift currents show a separation of the South Equatorial Current into a northern and southern branch. The distribution of their transports - 30 Sv in the northern branch, 20 Sv in the southern branch - varies little over the year (Swallow *et al.*, 1988). The contribution of the northern branch to the circulation in the southern hemisphere is the Mozambique Current; it is maintained throughout the year. The contribution to the circulation in the northern hemisphere ceases during the Northeast Monsoon season. The southern branch feeds the *East Madagascar Current.* The current field of this small but well-defined western boundary current reaches to the 2000 m level. Below 3100 m some 4 - 5 Sv are carried northward, with little movement inbetween. Having passed the southern tip of Madagascar the current apparently alternates between three flow patterns. A first path (*a* in Figure 11.10a) continues north along the west coast to return south with the Mozambique Current (*a'*). In the second path (*b*) the current flows directly west. Both paths feed into the Agulhas Current (*c*), going through a cyclonic loop (*d*) on their way. The third path, complete retroflection south of Madagascar (*c'*), cuts the East Madagascar Current off from the Agulhas Current; it is rarely followed by drifting buoys but occasionally seen in satellite data (Figure 11.10b).

The contribution of the *Mozambique Current* to the Agulhas Current is comparatively small. Although southward transport along the African shelf through Mozambique Strait was estimated at 30 Sv near 15°S, the Mozambique Current contributed only 6 Sv to this flow, the remaining 24 Sv coming from the northward looping East Madagascar Current (Zahn, 1984); in other words, the net southward transport near 15°S was only 6 Sv. Entrainment of water from the loop increases the net southward flow to 15 Sv near 20°S. These estimates are based on a depth of no motion of 1000 m and do not include flow on the shelf, so they almost certainly underestimate the true transport of the Mozambique Current; but they leave little doubt that the East Madagascar Current is the more important source for the Agulhas Current.

The region south of Mozambique Strait is characterized by the frequent occurrence of cyclonic eddies (*d* in Figure 11.10), spawned by the passage of

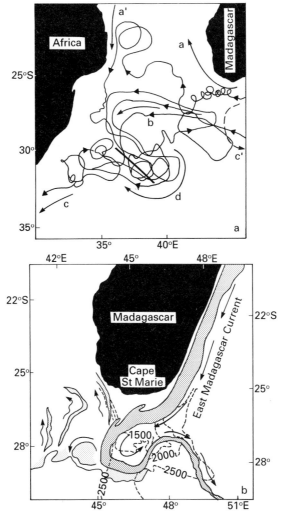

Fig. 11.10. Paths of the East Madagascar Current extension. (a) from tracks of satellite-tracked drifters, (b) from a satellite image of sea surface temperature on 13 June 1984. The warm core of the East Madagascar Current is indicated by dark shading. Light shading indicates intermediate water temperatures. The heavy line in (a) is the location of the section shown in Fig. 11.11. See the text for the meaning of letters in (a). From Lutjeharms *et al.* (1981), Gründlingh (1987) and Lutjeharms (1988).

the joint flow from the Mozambique and East Madagascar Currents over the Mozambique Ridge. This ridge does not reach much higher than 1500 m; but the current reaches deep enough to be influenced by it. Figure 11.11 shows that the eddies are also deep-reaching energetic features, with life spans of many months and transports of 15 - 30 Sv.

South of 30°S the flow continues as the Agulhas Current, one of the strongest currents of the world ocean. In contrast to other western boundary currents it shows little seasonal variation. Mean speeds are 1.6 m s⁻¹ throughout the year,

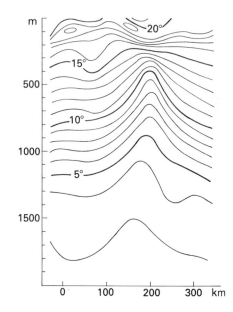

FIG. 11.11. A temperature section through a cyclonic eddy near the Mozambique Ridge. The form of the 3°C isotherm indicates that the eddy reaches deeper than 2000 m. See Fig. 11.10 for the location of the section. From Gründlingh (1985a).

and peak speeds exceed 2.5 m s^{-1} in most months. Transport estimates from observations give 70 Sv near 31°S and an increase of 6 Sv for every 100 km (as in the Gulf Stream). On approaching the shallow Agulhas Bank near 35°S it carries 95 - 135 Sv. The current occasionally floods the bank, lowering the inshore temperatures by several degrees (Figure 11.12). This upwelling is a result of the thermocline slope across the current (Rule 2 of Chapter 3) and not related to the wind. An increase in current speed increases the thermocline slope; since isotherm depths on the oceanic side of the current cannot change, the result is an uplift of cold water onto the shelf.

Transport estimates for the Agulhas Current agree well with the depth-integrated estimate of Chapter 4 until the current rounds the Cape of Good Hope. Further south the current encounters the circumpolar eastward drift, and most of its transport turns back into the Indian Ocean (Figure 11.13), quite in contrast to the prediction of Sverdrup theory which indicates a continuation of the current toward South America (Figs. 4.4 - 4.7). The reason for the striking departure from Sverdrup dynamics is that the current enters the Atlantic Ocean as a free jet and develops instabilities accompanied by eddy shedding, in a manner similar to the generation of East Australian Current eddies: The retroflection loop moves westward until its western part pinches of and the loop retreats to its most eastern position. (The two positions are indicated in Figure 11.13a by the crowding of front locations near 15°E and 19°E.) Most eddies are ejected into the Benguela Current and drift away toward northwest; in Figure 11.13c and d two eddies are

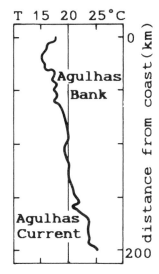

Fɪɢ. 11.12. Upwelling over the Agulhas Bank induced by the Agulhas Current. (a) Satellite image of sea surface temperature (dark is warm; small light patches are clouds. Numbers next to triangles are temperature in °C), (b) Temperature section along the track labelled *A* in (a). From Walker (1986).

seen by the depression of the thermocline in their centres (as explained with Figure 3.3). The eddies are among the most energetic in the world ocean and are believed to have life spans of many years. The associated transport of Indian Central Water into the Atlantic Ocean is an important element in the recirculation of North Atlantic Deep Water (see Chapter 7). Observations show that net transfer of water between the two oceans is westward in the thermocline (above 1500 m) but eastward underneath. Estimates for the time-averaged transfer of thermocline water from Agulhas Current eddies range from 5 to 15 Sv.

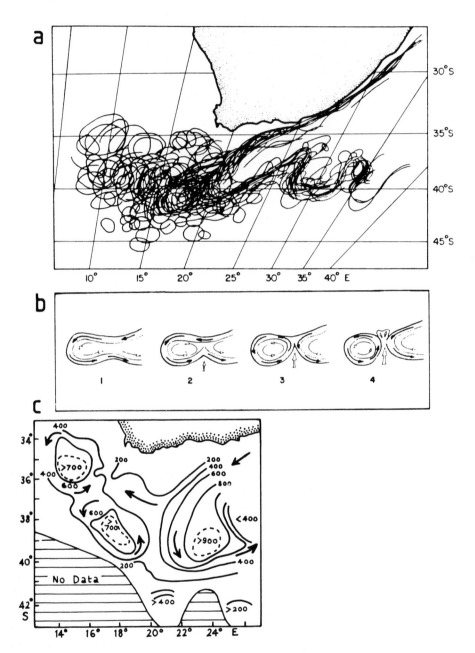

Fig. 11.13a–c. The retroflection region of the Agulhas Current. (a) Positions of the temperature front along the Agulhas Current for a 12 month period 1984/85; (b) sketch of the eddy shedding process at the retroflection; (c) depth of the 10°C isotherm (representative for the thermocline) in November/December 1983. From Lutjeharms and van Ballegooyen (1988) and Gordon (1985).

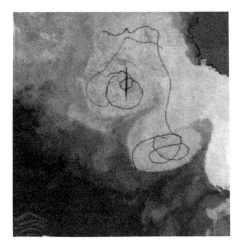

Fig. 11.13d. The retroflection region of the Agulhas Current. Sea surface temperature during 7 - 9 December 1983 and tracks of drifting buoys in November/December 1983. The area shown is approximately 32° - 42°S, 10° - 20°E. To minimize cloud disturbance the picture is a composite of three images. Lighter tones indicate warmer water. From Gordon (1985).

Near 25°E the Agulhas Plateau which reaches above 2500 m depth causes a northward excursion and further instability in the path of the retroflection current. The feature is seen in the frontal positions of Figure 11.13a; it was observed as early as 1899 when the steamer *Waikato* broke down on the Agulhas Bank and drifted in the current for 100 days; it rounded the Agulhas Plateau, looped through two eddies, and drifted eastward, reaching 65°E after a journey of 5000 km.

Winds near southern Africa are westerly to southwesterly throughout the year, bringing cold unsaturated maritime air over the Agulhas Current system. Unlike the Gulf Stream and Kuroshio the Agulhas Current therefore loses heat during all seasons. On average annual net heat loss amounts to $75 \, W \, m^{-2}$, less than half of the values found in the northern hemisphere. A unique aspect of the region is that the Mozambique, East Madagascar, and Agulhas Currents all run against the prevailing wind direction during the Southwest Monsoon season. This produces a steepening of the wind waves. Further south the current runs into the mature swell of the Southern Ocean. The resulting steepening of the swell produces some of the most dangerous waves of the world ocean; with wave heights of 20 m and more and a steepness that can cause ordinary wind waves of the open ocean to turn into breakers they are capable of inflicting severe damage on the largest vessels.

North of 10°S the East African Coastal or *Zanzibar Current* flows northward, fed by the northern branch of the South Equatorial Current. Because of its permanent character it is well resolved in the vertically integrated flow (Figure 4.7). During the Northeast Monsoon season it runs against light winds and is opposed by the southward flowing Somali Current. The point where northward and southward flows meet at the surface moves from 1°N at the beginning of the season to 4°S during the February peak, when the Zanzibar Current is at its weakest. Soon after the peak it starts moving north again, reaching the equator by early April. Throughout this period the current feeds into the Equatorial Countercurrent. Below the surface flow the Zanzibar Current flows northward across the equator at all times, taking the form of an undercurrent under the southward flowing Somali Current during the Northwest Monsoon season (Figure 11.14). During the Southwest Monsoon season the strength of the Zanzibar Current increases considerably; INDEX observations from April/May 1979 show it with speeds of 2.0 m s^{-1} and a transport of 15 Sv. The current now feeds the northward flowing Somali Current; some consider it part of the Somali Current during these months.

Southward flow in the *Somali Current* during the Northeast Monsoon is limited to the region south of 10°N. It first occurs in early December south of 5°N and expands rapidly to 10°N in January (Figure 11.4a) with velocities of 0.7 - 1.0 m s^{-1}. In March the southward flow contracts again to 4°N, until the surface flow reverses in April. During the Southwest Monsoon the Somali Current develops into an intense jet with extreme velocities; INDEX observations gave surface speeds of 2.0 m s^{-1} for mid-May and 3.5 m s^{-1} and more for June.

South of 5°N the Somali Current is extremely shallow; below 150 m depth southward flow is maintained throughout the year (Figure 11.15). Further north the jet deepens, eventually embracing the permanent thermocline. The current structure on the equator is extremely complex and shows layering similar to the equatorial flow further east (Figure 11.7) but oriented northward-southward.

The period of northward flow can be divided into two phases of different dynamics. During the transition in May, flow in the Equatorial Jet is eastward, and westward flow develops only slowly during the following two months. From the point of view of mass continuity there is not much need for a strong western boundary current until the monsoon reaches its peak in August, and the Somali Current is first established as a response to the wind reversal along the African coast. Winds are southerly but light in late April and May and strengthen abruptly in late May or June. This drives northward flow across the equator; but the flow turns offshore near 3°N, and a coastal upwelling regime develops between 3°N and 10°N. Figure 11.16a displays this division into two separate northward flows in the sea surface temperature for May/June 1979. The dynamic structure of the upwelling region is identical to that of other coastal upwelling regimes (see for example the Peru/Chile upwelling in Chapter 8); the southward flow beneath 150 m mentioned above is in fact the continuation of its undercurrent. The increase in wind speed four weeks later strengthens the oceanic circulation, without destroying the two-gyre structure (Figure 11.17); current speeds can reach 3.5 m s^{-1} at the surface, and coastal temperatures in the two upwelling centres are lowered dramatically (Figure 11.16b). Reported transport estimates are 27 Sv above 100 m and 80 Sv between the surface and 700 m for the southern gyre,

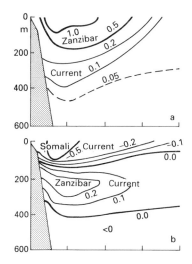

FIG. 11.14. Mean meridional current velocity (m s^{-1}) in the western boundary currents at the equator, derived from 2 years of observations during 1984 - 1986. (a) June - September, (b) December - February. After Schott et al. (1990).

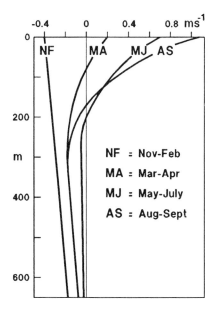

FIG. 11.15. Mean alongshore velocity (positive = northeastward) in the Somali Current at 5°N, from 30 months of current meter records. After Quadfasel and Schott (1983).

and 22 Sv above 100 m in the northern gyre. As the monsoon reaches its peak in August, the temperature front at 4°N is pushed northward along the coast until it merges with the Ras Hafun front; by September the Somali Current is established as a continuous western boundary current, from the Zanzibar Current

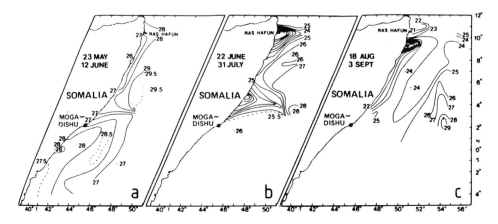

FIG. 11.16. Development of the Somali Current during the onset of the Southwest Monsoon as seen in the sea surface temperature during 1979. (a) Advection of warm water south of 3°N and coastal upwelling north of 3°N, (b) two gyres with upwelling regions near 4°N (minimum temperature <19°C) and Ras Hafun (minimum temperature <18°C), (c) continuous northward flow and upwelling near Ras Hafun (minimum temperature <17°C). From Evans and Brown (1981).

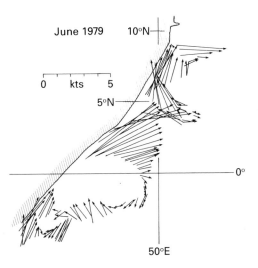

FIG. 11.17a. The two-gyre phase of the Somali Current. Surface currents in June 1979 along the track of *Discovery*. From Swallow and Fieux (1982).

in the south to the East Arabian Current in the north. The Ras Hafun gyre survives offshore, keeping mixed layer depths greater than 200 m near 9-10°N, 53-54°E during the November transition and early Southwest Monsoon season when mixed layer depths are near 50 m elsewhere.

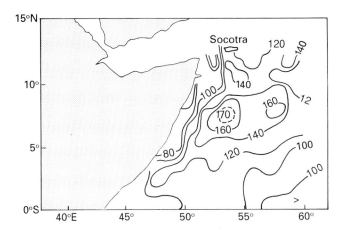

FIG. 11.17b. The two-gyre phase of the Somali Current. The northern gyre as seen in the mean depth of 20°C isotherm (m) during June. From Bruce *et al.* (1980).

Eastern boundary currents

The eastern boundary of the tropical Indian Ocean is different from the Atlantic and Pacific eastern boundaries in several respects. First, the mean temperature and salinity stratification is less developed than off Peru and West Africa, so when upwelling occurs it generally does not generate marked surface cooling. Secondly, the monsoon wind reversal is so complete along this coast that the annual mean longshore wind stresses here are close to zero, from the northern Bay of Bengal south to Java (Figure 1.4) and the upwelling in one season is counteracted by downwelling in the following season. Thirdly, twice a year (around May and November) the Equatorial Jet feeds warm water towards Sumatra, generating a pulse of current that flows poleward in both hemispheres. The seasonal cycle thus has strong semiannual as well as annual components, and is quite complicated. The strongest westward currents along the southern coast of Java, the seasonally reversing *South Java Current*, occur in August, when the monsoon winds are easterly and the Equatorial Jet is inactive. Surface cooling occurs off South Java at this time. This is also the time when the sea level difference from Java to Australia is largest, implying maximum strength in the Indonesian throughflow (see Chapter 13) and suggesting that at least some of the water for the South Java Current is then supplied from the Pacific Ocean.

The dynamics of the eastern boundary current along the western Australian coast, known as the *Leeuwin Current*, are very unusual and require some explanation. In the Pacific and Atlantic Oceans equatorward winds along the eastern boundary produce coastal upwelling, an equatorward surface flow, and a poleward undercurrent. In the Indian Ocean, annual mean winds along western

Australia do blow towards the equator, but at the surface a vigorous poleward flow runs against the wind, and the undercurrent is equatorward. The reason is that eastern boundary currents are driven by the combined effects of longshore winds and longshore pressure gradients in the upper ocean. Figure 2.8b demonstrates that the upper ocean pressure distribution in the Indian Ocean differs substantially from those of the other oceans. While variations in dynamic height in the subtropics do not exceed $0.1 \, m^2 \, s^{-2}$ along most eastern coastlines, this difference is $0.5 \, m^2 \, s^{-2}$ (equivalent to about 0.5 m steric height) along western Australia. In the open ocean it drives an eastward geostrophic flow; but closer to the coast eastward flow becomes impossible, and the water accelerates down the pressure gradient. In the Indian Ocean the resulting poleward flow is strong enough to override the wind-driven equatorward current, and the onshore geostrophic flow is strong enough to override the offshore Ekman flow.

The large drop in steric height along western Australia is apparently related to the connection between the Pacific and Indian Oceans through the Australasian Mediterranean Sea. The free connection from the Pacific to the Indian Ocean permits steric height to have similar values at either end of the channel — i.e.the steric height off Northwest Australia is essentially the same as in the western equatorial Pacific Ocean (Figure 2.8b). Steric height is about 0.5 m larger than that off Peru, because the easterly winds blowing along the equatorial Pacific Ocean maintain the steric height gradient along the equator that can be seen in Figure 2.8b. The high steric height at 15°S off northwest Australia cannot be maintained at 34°S — steric height is primarily a depth-integral of temperature through eqn (2.3), and to maintain the same surface steric height at 34°S would require maintenance of an average temperature of about 25°C over the top 200m, in a region where strong winds and air temperatures of 12°C occur in winter. Strong cooling and convective overturn occurs (Figure 11.18), to bring water temperatures down nearer to air temperatures throughout a mixed layer typically 150m deep, near 34°S in winter. This heat loss can be seen in Figure 1.6, which shows that, unlike the other eastern boundaries, a net heat loss to the atmosphere occurs near the western Australian coast poleward of 20°S. It extends down to 150 m and results in relatively low steric height at the southern end of western Australia. The resulting southward flow in turn feeds the warm water polewards to maintain the surface heat loss. The process is self-perpetuating, as southward flow is accompanied by surface cooling, and surface cooling produces southward flow.

As mentioned before, the dynamics of the currents along the western Indian shelf during the Northeast Monsoon are similar to those of the Leeuwin Current. In the northern hemisphere case the large longshore steric height gradient is the result of differences in water temperature and salinity produced by the monsoon winds.

The annual mean transport of the Leeuwin Current is estimated at 5 Sv, with average current velocities of $0.1 - 0.2 \, m \, s^{-1}$. However, the intensity and southward

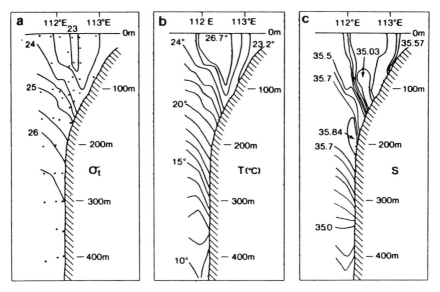

FIG. 11.18. Hydrographic properties in the Leeuwin Current near 25°S. (a) Density, (b) temperature (°C), (c) salinity. Note the deep mixed layer produced by convective cooling between 112°E and 113°E. From Thompson (1984).

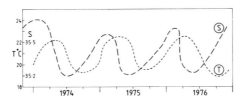

FIG. 11.19. Seasonal variation of surface temperature (T) and salinity (S) on the central shelf off Perth (32°S). Note the rapid drop of salinity between March and June; the Leeuwin Current is at its peak around April/May, bringing warm, fresh water from the tropics. From Cresswell and Golding (1980).

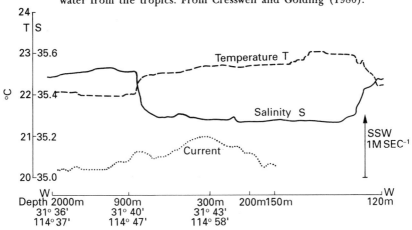

FIG. 11.20. A section across the Leeuwin Current off Perth (31°40'S), showing temperature and salinity fronts near 900 m and 120 m water depth on either side of the current. From Cresswell and Golding (1980).

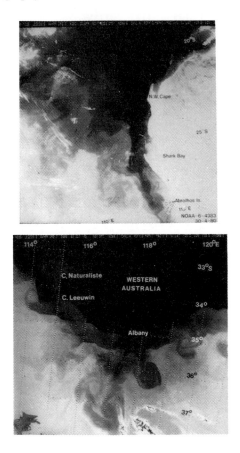

Fig. 11.21. Satellite images of sea surface temperature in the Leeuwin Current and associated eddies. Warm is dark. (a) nigth time image of 30 April 1979; the Leeuwin Current is seen as a narrow band of warm water along the north-eastern coastline. (b) day time image of October 1979 showing large eddies near Cape Leeuwin.

extent of the current vary strongly with the seasons, in response to variations in the southerly wind and perhaps also the longshore pressure gradient. The southerly wind is weakest in May. At this time the Leeuwin Current passes round Cape Leeuwin at speeds of up to 1.5 m s^{-1} and enters the Great Australian Bight. The associated advection of warm low salinity water produces a distinct seasonal cycle in the water properties along the western Australian coast (Figure 11.19) and hydrographic fronts on the offshore and inshore edges of the current (Figure 11.20). Eddies produced by the strong current shear across the fronts are clearly visible in satellite images of sea surface temperature (Figure 11.21).

CHAPTER 12

Hydrology of the Indian Ocean

The distribution of hydrological properties in the Indian Ocean is much less affected by the seasonal monsoon cycle than the near-surface current field. Direct monsoonal influence is restricted to the surface mixed layer and the western boundary currents. The two most important factors which make Indian Ocean hydrology different from the hydrology of the other oceans are the closure of the Indian Ocean in the northern subtropics and the blocking effect of the equatorial current system for the spreading of water masses in the thermocline. In the Pacific and Atlantic Oceans, mixing of water in the tropics and subtropics occurs mainly on isopycnal surfaces (the exception being the Equatorial Undercurrent). This is not so in the northern Indian Ocean, because conservation of mass under isopycnal conditions is impossible, given the geographical and dynamic constraints.

Precipitation, evaporation, and river runoff

The most striking characteristic of the rainfall distribution over the Indian Ocean is the anomalous difference between the eastern and western regions in the north. Annual mean precipitation varies between 10 cm per year in the west (on the Arabian coast) and 300 cm per year or more in the east (near Sumatra and over the Andaman Sea). This is the reverse of the situation usually encountered in the subtropics, where the Trades bring dry continental air out over the sea in the east and rain to the western coast. The normal situation of little rain in the east and high rainfall in the west prevails in the southern Indian Ocean; western Australia receives less than 50 cm per year but Madagascar some 200 cm per year. As a result, contours of equal precipitation show more or less zonal orientation in the south but meridional orientation in the north.

Given the small variation of evaporation over the region, the precipitation-evaporation balance (P-E; Figure 1.7) reflects the rainfall distribution closely. The change from zonal to meridional direction of the P-E gradient occurs near 10°S in the west and closer to the equator in the east. The continuation of the Pacific Intertropical Convergence Zone (ITCZ) and associated rainfall region into the Indian Ocean along 5°S brings rain to the seas west of Sumatra throughout the year; it dominates the annual mean distribution between 10°S and the equator. North of the equator the annual mean is a poor representation of the actual situation. P-E values vary from 600 cm per year along the western Indian coast and in the eastern Bay of Bengal during the Summer Monsoon to -150 cm per year during the Winter Monsoon. As with the winds and the currents, the Summer Monsoon produces the stronger signal and dominates the annual mean *P-E* distribution, which shows the eastern Bay of Bengal as a region of freshwater gain comparable in importance to the ITCZ.

Although the land drainage area of the Indian Ocean is rather small - the west coast of Africa (the Sambesi being the most important freshwater source), Madagascar, the coastal strip of western Australia, Sumatra, Java, and the Indian and Indochinese subcontinent - the influence of the Asian rivers is amplified by the monsoonal climate. The summer floodwaters which the Ganges and Brahmaputra empty into the Bay of Bengal, the Indus into the Arabian Sea, and the Irrawady and Salween Rivers into the Andaman Sea, influence the salinity of the surface waters over thousands of kilometres offshore.

Sea surface temperature and salinity

As in the other oceans, sea surface salinity (SSS; Figure 2.5b) follows the *P-E* distribution (Figure 1.7) outside the polar and subpolar regions closely. The *P-E* minimum near 30°S is reflected by a SSS maximum. The decrease of SSS values further south continues into the Southern Ocean, reflecting freshwater supply from melting Antarctic ice. However, the lowest surface salinities are found in the northern subtropics, where they reach values of 33 and below on annual mean; during the Summer Monsoon season, surface salinity in the inner Andaman Sea is below 25.

Surface salinity in the eastern tropical region is rather uniform near and below 34.5, close the the values found in the western tropical Pacific Ocean. SSS values increase towards the African coast and north into the Arabian Sea, where the annual mean reaches its maximum with values above 36. Higher salinities are reached in the Red Sea and Persian Golf, two mediterranean seas with extreme freshwater loss from evaporation (see Chapter 13).

When it comes to the distribution of sea surface temperature (SST) the entire northern Indian Ocean appears as a continuation of the western Pacific "warm pool" (the equatorial region east of Mindanao which is generally regarded the warmest region of the open ocean). The contouring interval chosen for Figure 2.5a displays it with temperatures above 28°C; over most of the region, annual mean temperatures are in fact above 28.5°C. Only the Somali Current region shows annual mean temperatures below 28°C, a result of upwelling during the Southwest Monsoon which brings SST down to below 20°C during summer (Figure 11.16). A remarkable feature of the seasonal SST cycle in the northern Indian Ocean is that the SST maximum does not occur during summer but during the spring transition from Northeast to Southwest Monsoon. May SST values are above 28°C everywhere north of the equator and north of 10°S in the east. As the Southwest Monsoon develops, advection of upwelled water reduces summer SST values to 25 - 27°C.

Lack of upwelling along the western Australian coast means that surface isotherms in the southern hemisphere show nearly perfect zonal orientation. Small deviations along both coastlines reflect the poleward boundary (Agulhas and Leeuwin) currents. There is also no upwelling along the equator, so the equatorial SST minimum which is so prominent in the Pacific and also visible in the Atlantic Ocean is not found in the Indian Ocean.

Abyssal water masses

Antarctic Bottom Water (AABW) fills the Indian Ocean below approximately 3800 m depth. By the time it leaves the Circumpolar Current its properties correspond to those of Antarctic Circumpolar Water (potential temperature 0.3°C, salinity 34.7; see Figure 6.13). The situation is not really much different from that in the Atlantic Ocean, where the water at the ocean floor is usually called Antarctic Bottom Water; however, most authors refer to the bottom water in the Indian Ocean as Circumpolar Water. The distribution of potential temperature below 4000 m (Figure 12.1) indicates two entry points. Entry into the Madagascar Basin has been well documented and occurs through gaps in the Southwest Indian Ridge near 30°S, 56 - 59°E. The flow gradually finds its way across to the Madagascar continental slope, where it forms a deep western boundary current (Swallow and Pollard, 1988). In a zonal temperature section (Figure 12.2) it is seen as a steep rise of the deep isotherms against the slope, consistent with northward geostrophic movement in which the speed increases with depth (rule 2a of Chapter 3). In the east, AABW enters the South Australia Basin via the Australian-Antarctic Discordance, a region in the inter-oceanic ridge system with multiple fractures near 50°S, 124°E south of Australia. Having filled the depths of the Great Australian Bight it moves west and then north into the Perth, Wharton, and North Australia Basins, forming a western boundary current along the Ninety East Ridge (Figure 12.2). Flow of AABW through the Mozambique Basin is blocked by Mozambique Strait; nevertheless, AABW recirculation in the basin must be swift, since observations of bottom currents in the 4500 m deep channel between the Agulhas Plateau and the African shelf gave average northward speeds underneath the Agulhas Current of 0.2 m s^{-1} (Camden-Smith *et al.*, 1981).

It is likely that both deep western boundary currents of Antarctic Bottom Water continue into the northern hemisphere, although direct observations exist only for the western pathway in the Somali Basin. This water eventually enters the Arabian Basin, where it must disappear through gradual upwelling into the overlying Deep Water. AABW from the eastern path proceeds into the Mid-Indian Ocean Basin, flowing over deep saddles in the Ninety East Ridge near 10°S and 5°S, and turns south; the differences in temperature and oxygen across the Ninety East Ridge below 3000 m (Figure 12.2) testify for the different ages of the waters on either side of the ridge. The inflow into the Mid-Indian Ocean Basin is a trickle of 0.5 Sv, but entrainment of Deep Water during the overflow may increase

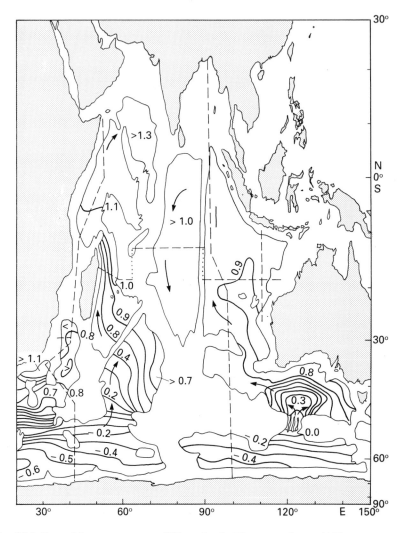

FIG. 12.1. Potential temperature at 4000 m depth. The approximate 4000 m contour is shown as a thin line. Arrows indicate the path of Antarctic Bottom Water. Thin broken lines indicate the locations of the sections shown in Figs. 12.2 - 12.4, 12.6c, and 12.10. Adapted from Wyrtki (1971), with modifications from Rodman and Gordon (1982).

transport in the western boundary current to 2 Sv. Again, the AABW must eventually disappear through upwelling.

The depth range from 3800 m upward to about 1500 - 2000 m and above is occupied by *Indian Deep Water* (IDW). Based on water mass properties the transition from Bottom to Deep Water is gradual, and some authors refuse to use the terms Bottom and Deep Water, referring to lower and upper deep water instead. This may appear logical since in the southern Indian Ocean both water masses have been observed to move northward together. A look at meridional hydrological sections (Figs. 12.3 and 12.4) proves, however, that the distinction

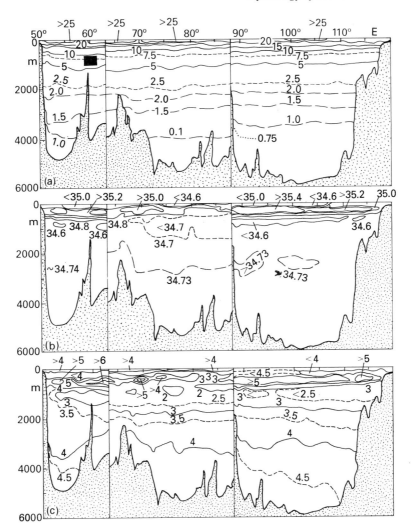

Fig. 12.2. A section across the Indian Ocean from Madagascar to Australia along 18°S, between 63°E and 88°E along 12°S. (a) Potential temperature (°C), (b) salinity, (c) oxygen. Adapted from Warren (1981b, 1982). See Fig. 12.1 for location of section.

between Deep and Bottom Water is justified. Indian Deep Water is characterized by a salinity maximum in the southern hemisphere exceeding 34.8 in the west and reaching 34.75 in the east. It occupies the depth range 2000 m - 3800 m north of 45°S and comes to within 500 m of the surface further south. Its properties in the high-salinity core near 40°E (temperature 2°C, salinity 35.85, oxygen 4.7 ml/l) match the properties of North Atlantic Deep Water in the Atlantic sector of the Southern Ocean (Figure 6.13) exactly. The obvious conclusion is that IDW is of NADW origin. Unlike Antarctic Bottom Water, Indian Deep Water is not formed in the Southern Ocean; it represents that fraction of NADW which is

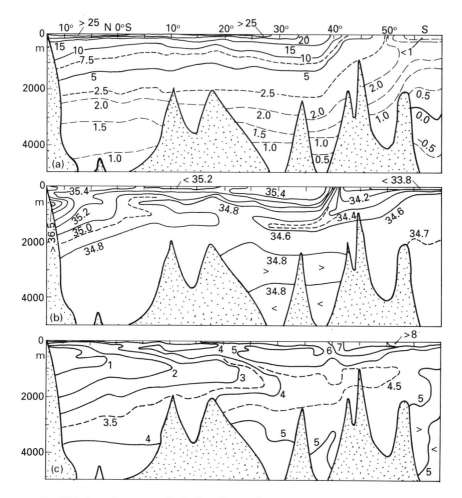

FIG. 12.3. A section across the Indian Ocean along approximately 40°E and following the African coast. (a) Potential temperature (°C), (b) salinity, (c) oxygen (ml/l). From Wyrtki (1971). See Fig. 12.1 for location of section.

not converted into Intermediate Water in the Atlantic sector but carried across into the Indian Ocean with the upper Circumpolar Current, along the path already discussed in Chapter 9 (Figure 9.6).

Like the flow of Bottom Water, the flow of Indian Deep Water is northward and concentrated in western boundary currents. In Figure 12.2 it is clearly seen in the oxygen data along the Ninety East Ridge and in the temperature and oxygen data below 2500 m along the Madagascar shelf. The transport of the combined flow of AABW and IDW in these boundary currents is estimated at 6 Sv in the eastern basin and 5 Sv in the western basin. A third western boundary current of Deep Water is indicated in Figure 12.2 along the Central Indian Ridge at 2200 m - 3200 m, since water depths south of the Mid-Indian Ocean Basin are deep enough to allow direct advection of IDW from the Southern Ocean. It flows

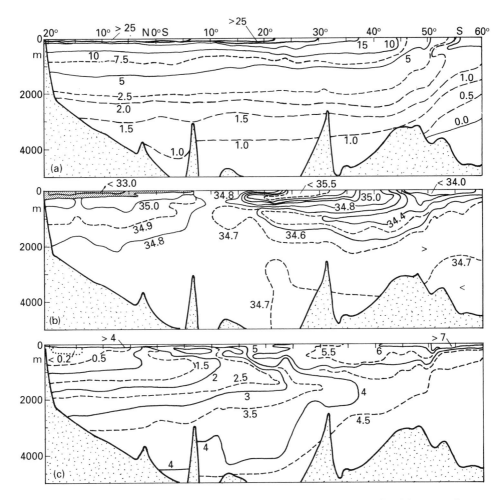

FIG. 12.4. A section across the Indian Ocean along approximately 95°E. (a) Potential temperature (°C), (b) salinity, (c) oxygen (ml/l). From Wyrtki (1971). See Fig. 12.1 for location.

against the slow southward spread of AABW underneath, carrying about 3 - 5 Sv.

Little is known at present about the sub-thermocline circulation in the northern Indian Ocean. The only northern sources capable of supplying water to depths below the permanent thermocline are the Red Sea and the Persian Gulf. *Red Sea Water* enters the Indian Ocean with a temperature close to 22°C and a salinity near 39 (Figure 12.5), giving it a density (σ_t) of about 27.25. In the Arabian Sea the same density is found at depths of 600 - 800 m and progressively deeper toward south until it reaches 1000 - 1100 m near 30°S. *Persian Gulf Water* has similar temperatures and salinities but slightly lower densities $(\sigma_t$ 26.7 and above) found at about 250 - 300 m in the Arabian Sea and at 500 - 600 m south of Madagascar. Compared with the flow of IDW from the south, supply from the two mediterranean seas is small, and given their densities, both sources are incapable

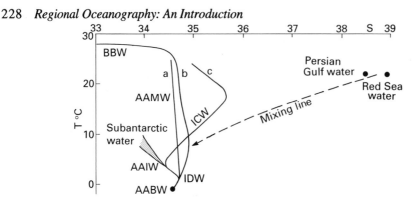

Fɪɢ. 12.5. T-S diagram showing the source water masses for the Indian Ocean and their effect on the temperature - salinity structure in different regions. Curve *a* is representative for the region between Australia and Indonesia (120°E), curve *b* for the Bay of Bengal (88°E) to 10°S, curve *c* for the subtropics south of 10°S. Intrusions of Red Sea and Persian Gulf Water can produce departures from the T-S curves near the isopycnal surface σ_t = 27.2 (the "mixing line").

of reaching much deeper than 1000 m in the northern hemisphere and 1500 m in the south. It appears therefore that Indian Deep Water penetrates northward in the western boundary current from where it spreads eastward and upward into the Arabian Sea and Bay of Bengal. Its properties are modified along the way by mixing with thermocline water above, upwelling of Antarctic Bottom Water from below, and injections of Red Sea and Persian Gulf Water at their respective densities.

To compensate for northward flow of Bottom Water below and Intermediate Water above, some southward movement must occur in the depth range of Indian Deep Water in both hemispheres. The distribution of salinity and oxygen (Figure 12.3) indicates this for the depth range 2000 m and below, i.e. the upper range of the distribution of Deep Water.

The influence of Red Sea Water and Persian Gulf Water on the water mass structure of the northern Indian Ocean is seen in the salinity maximum found in the upper kilometre in the north and closer to 1500 m some 3000 km further south (Figure 12.3). Details of their spreading are only known from the Somali Current region where both water masses are seen as density-compensated lenses and intrusions about 200 - 500 m thick and some 100 km in diameter (Figure 12.6a). Circulation outside the Somali Current at the depth of these intrusions and of the Intermediate Water must be extremely sluggish; at 1000 m the oxygen concentration falls below 1.2 ml/l everywhere north of the equator. Mixing in the boundary current must therefore be important in converting the lenses and intrusions into the smooth salinity maximum seen in the large-scale distribution of properties.

South of 10°S a conspicuous salinity minimum near 1000 m (Figs. 12.3 and 12.4) indicates the presence of *Antarctic Intermediate Water* (AAIW). Although the minimum can be followed to the surface in the Polar Front (Antarctic Convergence), this may not indicate formation in the Indian Ocean sector but be the result of advection from the Atlantic Ocean. AAIW source properties in the Indian are the same as in the other oceans, with temperatures near 2.0 - 2.5°C and salinities around 33.8. When it enters the subtropical gyre, AAIW has a temperature of 3 - 4°C and 34.3 salinity. Its depth range comes within reach of the equatorial current system, which blocks its progress into the northern hemisphere. The distribution of AAIW is therefore limited to the region south of 10°S. Cross-equatorial flow underneath the Zanzibar Current is southward, allowing the passage of Red Sea Water but no northward propagation of AAIW. Red Sea Water is therefore found embedded in AABW in the Agulhas Current (Figure 12.6b and c). It occurs as layers of 300 - 800 m thickness at about 1500 m depth; occasionally, it is contained in lenses of 50 - 100 km diameter.

The final fate of Antarctic Intermediate Water in the Indian Ocean still has to be established. It is possible that all Antarctic Intermediate Water from the Indian Ocean manages to escape into the Atlantic Ocean with Agulhas Current eddies. An alternative exit would be passage through the Great Australian Bight

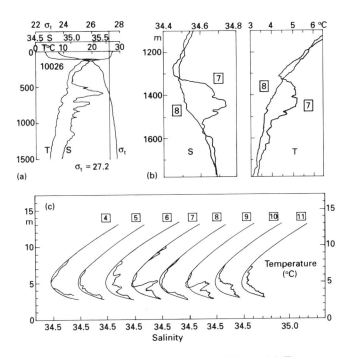

FIG 12.6. Observations of Red Sea and Persian Gulf Water. (a) Temperature (°C), salinity, and density (σ_t) at a station in the Somali Current near 3°N (the density 27.2 marks the separation between Persian Gulf Water above and Red Sea Water below. Note also the uniform salinity at 300 - 400 m and 800 - 1100 m indicative of the presence of AAMW), (b) temperature (°C) and salinity against depth at two stations in the Agulhas Current near 29°S, with little (stn 8) and strong (stn 7) presence of Red Sea Water, (c) T-S diagrams from eight stations (stns 4 - 11) across the Agulhas Current along 29°S (see Fig. 12.1 for location). The diagrams are shifted along the salinity axis by 0.3 units. The smooth curve is the mean from 20 stations without Red Sea Water presence and shows the AAIW salinity minimum. Adapted from Gründlingh (1985b).

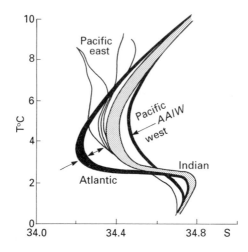

FIG. 12.7. T-S diagrams for various regions in the latitude band 40 - 45°S. The range of T-S diagrams in the Indian Ocean is shaded, while a variety of individual T-S curves is shown for the eastern Pacific Ocean (80° - 160°W). The region marked "Pacific west" is located in the southern Tasman Sea and southeast of New Zealand (150°E - 160°W); its higher salinity indicates that any exchange between the Great Australian Bight and the Pacific Ocean cannot pass through these regions but has to pass to the south. From Piola and Georgi (1982).

into the Pacific Ocean. Comparison of T-S diagrams for the latitude band 40 - 45°S shows little variation of T-S properties in the AAIW in each ocean but significant differences between the three oceans. This has been interpreted as an indication that the circulation of AAIW in the southern subtropical gyres is closed within each ocean. The data certainly exclude contact of AAIW of Indian and Atlantic origin at 40 - 45°S latitude; but Agulhas Current eddies are formed just north of 40°S and drift away toward the equator. Export of AAIW into the Atlantic Ocean thus remains a possibility. The other route, passage into the Pacific Ocean, is very unlikely despite the apparent similarity of AAIW properties in both oceans, as an AAIW variety with particularly high salinity in the southern Tasman Sea and southeast of New Zealand breaks the continuity of water mass properties (Figure 12.7). This suggests little contact between the Great Australian Bight and the Pacific Ocean at the level of the AAIW north of 45°S. South of that latitude observations of subsurface float movement obtained only very recently show westward movement at AAIW level, from the Pacific Ocean into the Great Australian Bight.

Water masses of the thermocline and surface layer

Two water masses occupy the thermocline of the Indian Ocean (Figure 12.8). *Indian Central Water* (ICW) is a subtropical water mass formed and subducted in the Subtropical Convergence (STC), as described in detail in Chapter 5. It

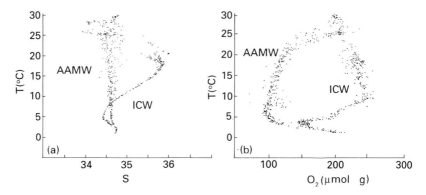

FIG. 12.8. Water mass properties of Indian Central Water (ICW) and Australasian Mediterranean Water (AAMW). (a) T-S diagram, (b) T-O$_2$ diagram (see Fig.9.4 for units). The data for ICW are from 25 - 30°S and east of 105°E, the data for AAMW from 7 - 15°S, 120 - 125°E. From Tomczak and Large (1989).

originates from the Indian Ocean sector of the STC; negative values in Figure 5.7 indicate that south of 30°S subduction occurs from the Agulhas retroflection into the Great Australian Bight. In hydrological properties ICW is identical to South Atlantic and Western South Pacific Central Water. *Australasian Mediterranean Water* (AAMW), on the other hand, is a tropical water mass derived from Pacific Ocean Central Water and formed during transit through the Australasian Mediterranean Sea, as discussed in Chapter 13. There is no established nomenclature for this water mass; Banda Sea Water, Indonesian Throughflow Water, and other names are found in the literature. The water enters the Indian Ocean between Timor and the Northwest Shelf and through the various passages between the islands east of Bali. The transport of AAMW into the Indian Ocean, a key quantity in models of the recirculation of North Atlantic Deep Water (see Chapter 7), is unknown at present. This question will be addressed in much more detail in Chapter 13 where it will be argued that a transport of 15 Sv or more may be a good estimate.

The large impact of AAMW on the hydrological structure of the Indian Ocean thermocline certainly points towards a large supply of Mediterranean Water. Outflow into the Indian Ocean occurs over the entire upper kilometre of the water column. The low salinity of the outflowing water makes salinity a good indicator for the presence of AAMW down to 600 m; at that depth the temperature is in the range 7 - 8°C, the T-S curves of ICW and AAMW intersect (Figure 12.8), and the salinity contrast between the water masses disappears. Above 600 m, maps of salinity on depth or density surfaces (Figure 12.9) show the path of AAMW as a band of low salinity centred on 10°S, from the entry point in the east to Madagascar in the west. Below 600 m the same pattern can be seen in the silicate distribution, where the presence of AAMW down to 1000 m depth is indicated by a silicate maximum. The jet-like inflow of AAMW produces one of the strongest frontal systems of the world ocean's thermocline (Figure 12.10). The front indicates that there is little meridional motion in the thermocline across

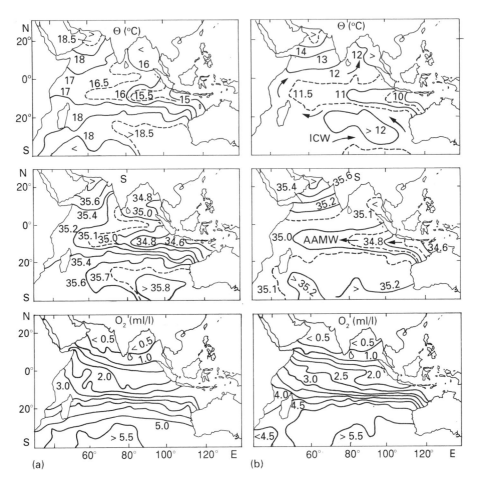

Fig. 12.9. Annual mean temperature (°C), salinity, and oxygen (ml/l) in the thermocline on isopycnal surfaces. (a) On the σ_θ surface 25.7, located in the depth range 150 - 200 m, (b) on the σ_θ surface 26.7, located in the depth range 300 - 450 m. Arrows indicate the movement of ICW and AAMW. After You and Tomczak (1993)

10 - 15°S from the point of AAMW inflow in the east to the point where the South Equatorial Current splits into two branches on approaching Madagascar. This leaves the western boundary current as the only region for advective transfer of thermocline water between the southern and northern Indian Ocean.

Being closed in the subtropics, the northern Indian Ocean does not have its own subtropical convergence; its thermocline water has to be replenished from the tropics and further south. Supply of Indian Central Water to the northern hemisphere is clearly seen on the 26.7 σ_θ surface of Figure 12.9. At that density subduction in the STC occurs at 11.5 - 12.0°C and a salinity near 35.1. This water type dominates the density surface south of the front at 10°S and enters the northern Indian Ocean with the western boundary current. Oxygen values are fairly uniform south of the front, suggesting reasonably swift recirculation of ICW

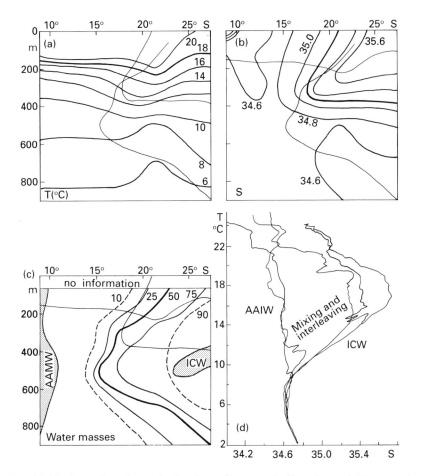

Fɪɢ. 12.10. A section through the front between Indian Central Water and Australasian Mediterranean Water along 110°E. (a) Temperature (°C), (b) salinity, and (c) water masses (% of ICW content) from bottle casts along 110°E, (d) T-S diagrams from selected CTD stations, showing evidence of interleaving in the frontal zone. The positions of the thermocline (indicated by the 18°C isotherm), halocline (the 35.2 isohaline), and water mass boundary (defined as the 50% ICW or 50% AAMW contour) are indicated in (a), (b), and (c). See Fig. 12.1 for location of the section.

in the subtropical gyre. Transition into the northern hemisphere is accompanied by a rapid fall in oxygen values, indicating rapid aging along the path. The decrease in oxygen values continues into the Bay of Bengal, which contains the oldest Central Water. The oxygen decrease in the northern Indian Ocean can be explained if it is recalled that transfer of ICW between the hemispheres is restricted to the Southwest Monsoon season. The annual net transfer rate is therefore small, and circulation of ICW in the northern Indian Ocean is slow.

In the model of NADW recirculation discussed in Chapter 7, AAMW is assumed to enter the Agulhas Current and finally the Atlantic Ocean. Closer inspection of Figure 12.9 shows that this cannot be true for *all* AAMW. Some AAMW contributes to the renewal of thermocline water in the northern Indian Ocean. This is evident from the salinity distribution of Figure 12.9 on the 25.7 σ_θ level which shows significant freshening of ICW along its path from the Subtropical Convergence to the Bay of Bengal. The salinity decrease east of Madagascar is apparently the result of mixing with AAMW at the end of the zonal jet. Further freshening is observed in the Bay of Bengal near 90°E, this time presumably resulting from AAMW advection from the southeast. Again, it has to be remem–bered that the figures show only the net result of a process with strong monsoonal variation. The final fate of ICW and AAMW in the northern Indian Ocean is not known. In the Bay of Bengal oxygen values fall below 0.5 m/l above 600 m depth and below 0.2 ml/l at the 200 m level, and in the inner Arabian Sea they are below 0.2 ml/l from 200 m to 1000 m depth. These values - the lowest in the world ocean thermocline - indicate a very low renewal rate for the thermocline waters of the northern Indian Ocean. Some water, however, must always leave the thermocline, to make room for new supply. Downward diffusion into the Deep Water would only increase the difficulties with the recirculation of Deep and Bottom Water and therefore appears unlikely. Upward diffusion into the surface layer remains as the only alternative. The observed variation of temperature and salinity on the 25.7 σ_θ surface (Figure 12.9) indicates that some mixing across isopycnals must occur in the upper thermocline. More work is definitely required to clarify these issues.

Nearly uniform salinity over the temperature range of the thermocline, the main characteristic of AAMW, is maintained along the entire length of the zonal jet. Mixing with higher salinity water on either side increases the salinity from less than 34.7 at the inflow point to 34.9 and above in the west (Figure 12.9a). Water with nearly uniform salinity near 35.0 has often been called Indian Equatorial Water. Lack of observations from the Indonesian outflow region led several authors (including Sverdrup *et al.*, 1942) to believe that it is formed in the western equatorial Indian Ocean. It is now clear that water mass formation does not occur in that region and that the so-called Equatorial Water consists of Australasian Mediterranean Water, with a good dose of Central Water to lift its salinity.

Hydrological characteristics of the water masses of the surface layer vary strongly with the seasons, more so in the Indian than in any other ocean. Low surface salinity in the tropics produces a salinity maximum at the top of the permanent thermocline. It is found below 200 m near 15°S and approaches the surface near 35°S (Figure 12.4). The corresponding water type has often been given the status of a water mass; in reality it only identifies the high salinity end of Indian Central Water. Monsoonal river input from the Indian and Indochinese subcontinent produces a low salinity water mass known as *Bay of Bengal Water* (BBW). It spreads across the Bay in a nearly 100 m thick layer, producing a strong halocline underneath (Figure 12.11). Supply of this water is sufficient to keep the surface salinity in the eastern Bay below 33.0 throughout the year. Its influence extends well into the tropics (Figure 12.11). During October - December it reaches the area along the western Indian coast with the East Indian Winter Jet. Salinities along the western Indian coast return to oceanic values for a brief period during April - June, but the Summer Monsoon season brings increased runoff from rivers and lowers the salinity again. From the point of view of water mass classification the low salinity water along the western Indian coast can be subsumed under Bay of Bengal Water, on account of its nearly identical properties (Figure 12.11).

Although the main halocline which delineates the boundary between BBW and ICW is located close to 100 m, small but significant salinity gradients occur well above that depth. They result from the fact that river water spreads across salt water in a thin film. The hydrological structure of BBW is thus somewhat reminiscent of the structure in an estuary: little or no variation of temperature with depth but important variations of salinity. In the surface mixed layer the salinity variations are erased very quickly by wind mixing; but winds in the Bay are usually light, and the mixed layer is rarely deeper than 50 m (Figure 5.6). Salinity variations below the mixed layer but above the main halocline/thermocline are maintained. A characteristic feature of Bay of Bengal Water is therefore the existence of a barrier layer throughout the year (Figure 5.7). In contrast to the western Pacific Ocean, where the barrier layer is maintained by surface layer dilution from local rainfall, the barrier layer in the Bay of Bengal owes its existence to advection of low salinity water diluted from monsoonal river runoff. The consequences for the heat budget, outlined in Chapter 5, are the same: The net heat flux into the Bay of Bengal from the atmosphere is small but positive

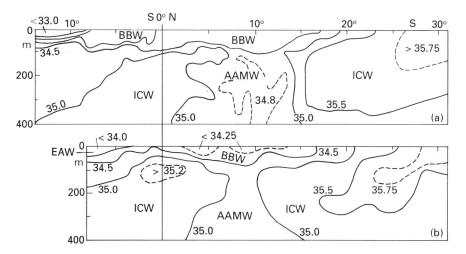

Fɪɢ. 12.11. Meridional sections of salinity showing the spreading of Bay of Bengal Water. (a) for the component originating in the inner Bay of Bengal, along 92°E, (b) for the component found along the west coast of India (for the purpose of identification labelled "East Arabian Sea Water" or EAW in the figure), along 75°E. See text for abbreviations. From Wyrtki (1971).

(Figure 1.6), so a weak heat sink is required to close the heat balance. Entrainment of cold water from below, usually the most effective heat sink, is not available as a process where a barrier layer exists. Export of heat to the Indian Ocean is a possibility consistent with the movement of surface water indicated by the surface property distributions. It is also possible that the rivers themselves contribute to the heat balance and not only to the freshwater balance.

CHAPTER 13

Adjacent seas of the Indian Ocean and the Australasian Mediterranean Sea (the Indonesian throughflow)

Being the smallest of all oceans, the Indian Ocean does not have the large number of distinct subregions found in the Pacific and Atlantic Oceans. Regions known under their own names include the Bay of Bengal and the Arabian Sea already discussed in the previous chapter, the Mozambique Strait (mentioned in the discussion of the western boundary currents), and the Great Australian Bight, clearly the least researched part of the Indian Ocean. Malacca Strait and the Andaman Sea form the transition region between the Bay of Bengal and the adjacent seas of the Pacific Ocean in Southeast Asia. The only regional seas that have some impact on the hydrography of the Indian Ocean and therefore require separate discussion are the Red Sea and the Persian Gulf. Since that discussion will not provide sufficient material for a full-length chapter, we include here the description of the Australasian Mediterranean Sea and what is often known as the Indonesian throughflow, i.e. the exchange of water between the Pacific and Indian Oceans. The Australasian Mediterranean Sea is of course a regional sea of the Pacific Ocean; but its impact on the Indian Ocean is much bigger than its influence on Pacific hydrography, and its inclusion in this chapter is justified on that ground alone.

The Red Sea

The Red Sea can be considered the prototype of a concentration basin. It is a deep mediterranean sea with a relatively shallow sill in a region where evaporation vastly exceeds precipitation (evaporation 200 cm per year, rainfall 7 cm per year, giving a net annual water loss of nearly 2 m). In such a basin water entering from the ocean in the surface layer undergoes a salinity increase and gets denser as it flows towards the inner end of the sea. This provokes vertical convection and guarantees continuous renewal of the water in the lower layer, which eventually leaves the basin in an undercurrent over the sill.

Geologically, the Red Sea is a rift valley formed during the separation of Africa and Arabia. Its topography (Figure 13.1) shows a maximum depth near 2900 m and a sill depth of about 110 m, significantly less than the average depth of 560 m.

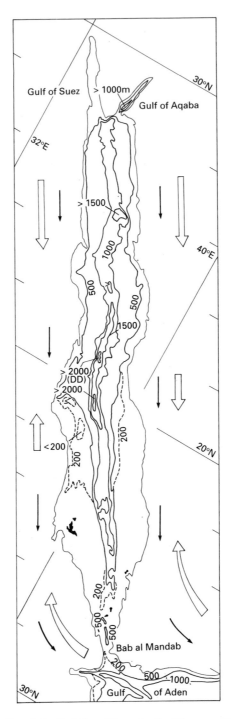

FIG. 13.1. Topography of the Red Sea. *DD* marks the Discovery Deep, the depression where the first observations of hot brines were made. Arrows indicate prevalent wind directions for summer (full arrows) and winter (open arrows). Depths are in m.

It is about 2000 km long but on average only 250 km wide. At its northern end it includes the shallow Gulf of Suez with average depths between 50 - 80 m and the Gulf of Aqaba, a smaller version of the Red Sea itself with a maximum depth near 1800 m and a sill depth close to 300 m.

Figure 13.2 gives a hydrographic section along the axis of the Red Sea and into the Gulf of Suez. The most notable feature is the extremely high salinity which makes the Red Sea the most saline region of the world ocean and gives it the character of an inverse estuary. The long and narrow shape of the basin isolates the inner part from direct exchange with the open ocean, so surface salinity increases continuously from 36 at the Strait of Bab el Mandeb to above 40 in the interior. Highest salinities above 42 are attained in the northern Red Sea and the shallow Gulf of Suez. In both regions the winter months, when the sea surface temperature in the Gulf of Suez sinks below 20°C (it ranges between 27 - 30°C during summer), are a period of active convection. Which of the two regions is responsible for the formation of deep water has been a matter of debate for many years. Recent observations of bomb radiocarbon indicate that both contribute, but in different ways. The winter water from the Gulf of Suez is very dense; it falls down the continental slope and fills the depths of the Red Sea below 1000 m. When the water starts its descent it has a temperature below 18°C and a salinity above 42; but mixing quickly modifies this, and the water soon ends up with the standard Red Sea deep water properties of 21.5°C and 40.6 salinity. Winter water from the northern Red Sea is slightly less dense; it slides down on the appropriate isopycnal surface and spreads below the surface mixed layer. The resulting circulation is indicated in Figure 13.2. A lower circulation cell with slow upward movement, gradual loss of oxygen, and a northward return flow at its upper limit is capped by southward movement of water from the second source. This is reflected in an oxygen minimum below the thermocline, indicating that the oldest water is found at the top of the lower circulation cell. It is worth noting that even at the oxygen minimum, Red Sea Water has a higher oxygen content than the very old Indian Central Water found at the same depth in the Indian Ocean thermocline; consequently, east of Bab el Mandeb Red Sea Water manifests itself through an oxygen maximum. Oxygen values in the 150 - 200 m thick surface layer of the Red Sea correspond to saturation values, which at these high temperatures are relatively low (less than 4 ml/l in summer).

Estimates for the residence time of deep water are just as controversial as identification of its sources and vary from a few years to two centuries. Recent radiocarbon data indicate a residence time slightly less than 40 years. This is the best available estimate at present.

The outflow of Red Sea water into the Arabian Sea is clearly visible in Figure 13.2, most prominently in salinity. It is seen that the dense water flows down the continental slope to a depth of 1500 m and more; but the bulk of the water is modified by mixing soon after passing Bab el Mandeb and spreads between 500 - 1000 m with a temperature of 13 - 14°C and a salinity of 36.5 or

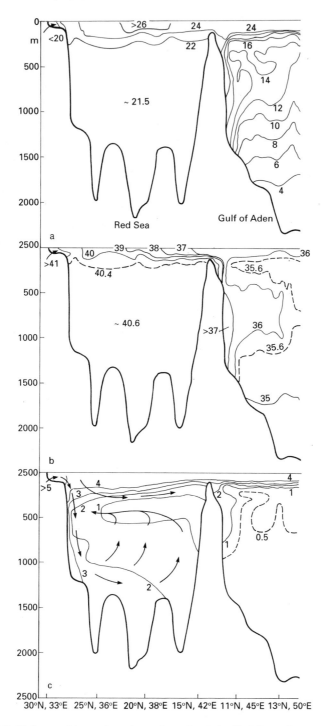

FIG. 13.2. Hydrographic section along the axis of the Red Sea, winter conditions.
(a) Potential temperature (°C), (b) salinity, (c) oxygen (ml/l). Arrows indicate the
flow of deep water as derived in Cember (1988).

less. The magnitude of the outflow at Bab el Mandeb is quite small. Measurements during the summer of 1982 indicated only 0.25 - 0.3 Sv; similar observations over a few months in 1965 showed intermittent outflow of about 0.5 Sv depending on the direction of the wind. Nevertheless, the concentration of salt in the outflow is sufficient to guarantee that Red Sea Water can be traced in the Indian Ocean thermocline even into the southern hemisphere, as we saw in the last chapter.

Seasonal variations in the outflow are related to the monsoons, which determine the surface circulation in the Red Sea in general. Winds over the Red Sea form part of the general monsoon system of the Indian Ocean but are modified by the influence of the land, which establishes a belt of low air pressure from Asia towards northern Africa during summer and a centre of high pressure over northern Africa during winter. The resulting pressure gradients against the region of constant air pressure in the tropics produce northwesterly winds throughout the year north of 20°S. South of that latitude winds are from the northwest during summer but reverse to southeasterly during winter. Throughout the entire region winds are generally stronger in winter than in summer. The resulting flow is northward in winter, south of 20°S under the direct forcing of the wind which supports surface inflow through Bab el Mandeb, further north driven by the sinking of surface water even though winds north of 20°S are predominantly from the northwest. In summer the wind opposes surface inflow, but during most of the time its strength is insufficient to suppress the surface inflow necessary to maintain the water budget. Occasionally water movement at the surface is southward and inflow occurs intermittently at intermediate depths. It is estimated that the water stays in the upper layer circulation for about 6 years before sinking.

A remarkable feature of the Red Sea is the extremely high water temperature and salinity found in various depressions of the sea floor (Figure 13.3). This is the result of geothermal heating through vents in the ocean crust, which brings minerals contained in the crust and in the sediment into solution. The resulting brine is dense enough to remain at the ocean floor even at very high temperatures. Values close to 58°C have been recorded, together with "salinities"

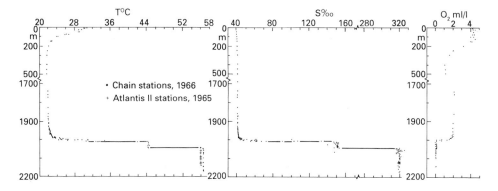

Fig. 13.3. Temperature, "salinity" (see explanation in text) and oxygen in a hydrothermal vent of the Red Sea. From Brewer *et al.* (1969).

in excess of 300. The salinity readings were obtained by diluting brine samples until their salinity came into the range of normal CTD instruments. Such determinations are invariably incorrect since the universal rule that the relative composition of sea salt remains the same throughout the world ocean does not hold in brines brought up from fissures in the earth's crust - the content in metal ions is much higher than in the normal salinity mix. Gravimetric salinity determinations (weighing the sample before and after evaporation of the water) are more accurate; they still give salinities in excess of 250.

The Red Sea was the first region where hot brines were discovered at the sea floor (Figure 13.1). Similarly high temperatures and salinities are now known to exist above fissures in mid-ocean ridges of other ocean regions; temperatures in excess of 320°C have been measured at hydrothermal vents in the Pacific Ocean. More commonly, vents are associated with seepage of hypersaline water at environmental temperatures; this is the case for the majority of vents in the Pacific Ocean and those reported from the Gulf of Mexico. Large brine deposits can only accumulate where such vents are located in topographic depressions. The deposits in the Red Sea are possibly large enough to warrant commercial metal extraction at some time in the future.

The Persian Gulf

The hydrography of the Persian Gulf is very similar to that of the Red Sea; but the much smaller volume of the Persian Gulf greatly reduces its impact on the Indian Ocean. Its length of 800 km together with an average width of 200 km gives it an area comparable to that of the Red Sea. Atmospheric conditions do not differ much between the two mediterranean seas (except that winds are from the north or northeast throughout the year over the entire area). The Persian Gulf is therefore a concentration basin, too. The rate of water loss at the surface is only slightly reduced by river runoff from the Euphrates and Tigris rivers. The main difference is that the Persian Gulf belongs entirely to the continental shelf, has a mean water depth of only 25 m and, with a sill depth at the Strait of Hormuz only marginally above its average depth, cannot hold back large quantities of salty deep water.

Figure 13.4 gives a hydrographic section through the Persian Gulf and the adjoining part of the Arabian Sea. Despite the difference in volume and residence time, the waters entering the Indian Ocean from the Persian Gulf and the Red Sea have very similar characteristics. The major difference is in oxygen content, which is markedly higher in Persian Gulf Water because the residence time is much shorter. Since Indian Central Water near the Strait of Hormuz is older and its oxygen content lower than at Bab el Mandeb, the oxygen maximum produced by the outflowing water is even more marked. Persian Gulf Water also tends to have somewhat lower density than Red Sea Water (on account of its higher temperature) and therefore tends to stay above the main thermocline rather than penetrating it. Nevertheless, at some distance from the Strait of

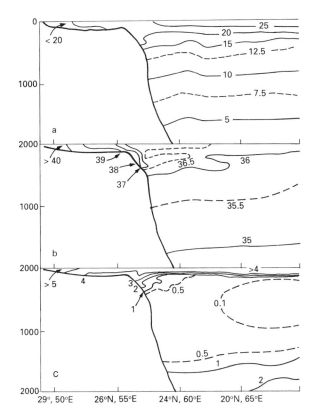

Fig. 13.4. Hydrographic section along the axis of the Persian Gulf, winter conditions. (a) Potential temperature (°C), (b) salinity, (c) oxygen (ml/l).

Hormuz it is often difficult to separate traces of Red Sea and Persian Gulf Water, and the two water masses are often regarded as one.

The Australasian Mediterranean Sea and the Indonesian throughflow

Of all the regional seas of the world ocean, the Australasian Mediterranean Sea displays without doubt the most complicated topography. It consists of a series of very deep basins with very limited interconnections, each basin being characterized by its own variety of bottom water usually of great age. The exact number of deep basins found within its borders is difficult to define; sea charts usually recognise at least eight basins under their own name (Figure 13.5). The largest and deepest is the Banda Sea which has depths in excess of 4500 m in the southeast (also known as the South Banda Sea) and in the northwest (the North Banda Sea), separated by a ridge of less than 3000 m depth; largest depths are near 7440 m in the south and 5800 m in the north. The Sulawesi Sea (formerly known as the Celebes Sea) is a single basin of similar size deeper than 5000 m over most of its area. Between these two major basins are three basins

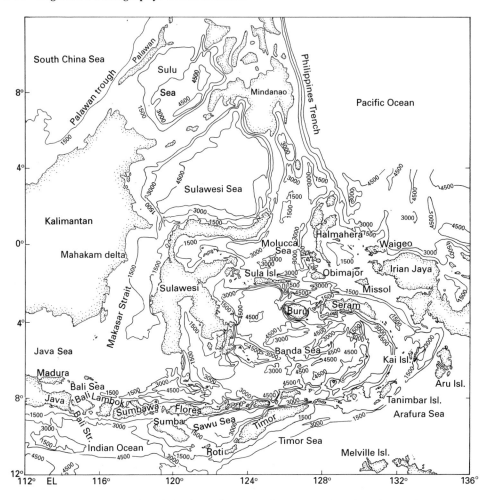

FIG. 13.5. Topography of the Australasian Mediterranean Sea. Depths are in m.
LS: Lombok Strait, MS: Makassar Strait.

deeper than 3000 m, the Molucca, Halmahera, and Seram Seas, the latter being deeper than 5300 m. North of the Sulawesi Sea and enclosed by the islands of the Philippines is the Sulu Sea, which has depths in excess of 4500 m. The Flores Sea is located in the south, connecting the Banda Sea with the shallow Java Sea and reaching nearly 6400 m in a deep depression. The Sawu Sea, which reaches nearly 3500 m depth, is the southernmost basin between Timor, Sumba, and Flores. Another important topographic feature is Makasar Strait between the Sulawesi and Java Seas; it is shallow in the west but over 2000 m deep in the east where it is connected without obstruction to the Sulawesi Sea in the north.

The climate of the Australasian Mediterranean Sea is characterized by monsoonal winds and high rainfall. Winds blow from the south during May - September, curving across the equator with a westward component in the south and an eastward component in the north, and in nearly exactly the opposite direction during November - March (Figure 1.2). Rain occurs at all times of the year and exceeds 400 cm in the annual mean near the junction of the Intertropical and South Pacific Convergences. As in the western equatorial Pacific Ocean, it is released from strong localized convection cells that are only a few kilometres in diameter, reach high into the upper atmosphere, and are surrounded by cloud-free regions of sinking air. As a consequence, solar heat input over the Australasian Mediterranean Sea is high (Figure 1.5) despite relatively high cloud-cover (Figure 8.5). Evaporative heat loss is high on account of high sea surface temperatures; but on balance the ocean receives more heat than is lost to the atmosphere. There is also more freshwater gain than evaporation, and the *P-E* balance (Figure 1.7) is strongly positive.

The atmospheric conditions leave no doubt that the Australasian Mediterranean Sea is a dilution basin. However, its circulation differs significantly from the schematic diagram of Figure 7.1. In the diagram, water that enters below the surface layer freshens as it is entrained into the surface layer and exits the basin with reduced salinity; deep water renewal through vertical convection is inhibited during all seasons by the high stability of the water column and is thus extremely slow, being determined by the rate of inflow over the sill. In the Australasian Mediterranean Sea, deep water renewal follows this scheme, but the circulation is markedly different. Nearly all inflow of high salinity water across the sill between the Sulawesi Sea and the Pacific Ocean proper occurs over the entire water depth, and nearly all outflow of low salinity water occurs into the Indian Ocean, again from the surface to the bottom of the passages between the south Indonesian islands. The modification is caused by the need for a net depth-integrated transport from the Pacific to the Indian Ocean. This requirement stems from the necessity to maintain constant pressure around islands. Constant pressure around Australia leads to a difference in depth-integrated steric height of about 70 m^2 between the western Australian coast (which shows the same values as the Australian east coast) and the east coast of the South Pacific Ocean. The net northward flow between Australia and Chile therefore has to pass through the Indonesian seas. Without this requirement, the freshened water would leave the Indonesian basins into both oceans - depending on the monsoon season - in a well defined surface layer.

Direct current measurements in the Australasian Mediterranean Sea are available for only a few locations and are often of short duration. The net transport is believed to be westward at all times, from the Pacific to the Indian Ocean. It occurs as a western boundary current (i.e. with highest velocities along Mindanao and Kalimantan) and is made up of two components, the surface current driven by the monsoons and the deeper reaching interoceanic throughflow. Although the wind-driven flow opposes the throughflow during May - September and follows it during November - March, the total westward transport reaches a maximum in August and goes through a minimum in February. The reason for this apparent contradiction is seen when the circulation of the Australasian Mediterranean

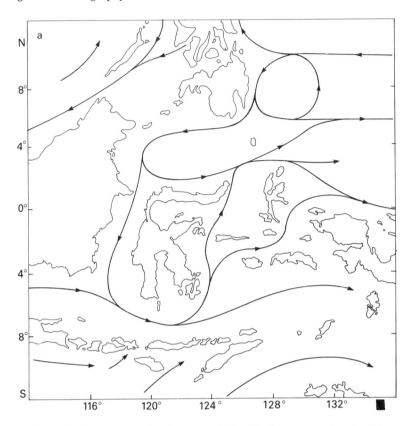

FIG. 13.6a. Surface currents in the Australasian Mediterranean Sea. In February
(north monsoon, minimum throughflow).

Sea is considered in conjunction with that of the Indian and Pacific Oceans.
During November - March the Equatorial Countercurrent of the Indian Ocean
is fully developed, supplying water to the region where the outflow from the
Australasian Mediterranean Sea occurs and raising the sea level. As a result the
pressure gradient from the Pacific into the Indian Ocean is small, and the
throughflow is at its minimum. During May - September the Equatorial
Countercurrent in the Indian Ocean is replaced by the South Equatorial Current
which expands northward under the south monsoon, drawing water away from
the eastern Indian Ocean. This lowers the sea level in comparison to the Pacific
Ocean and produces maximum throughflow.

First estimates for the throughflow based on geostrophic calculations from a
very limited data base gave annual mean values of 2 Sv or less. More recent studies
indicate that the throughflow maximum should be in the range 12 - 20 Sv, while
the minimum is estimated at 2 - 5 Sv. None of these estimates are derived from
direct current observations. Some are the result of numerical models of the world
ocean circulation with fairly coarse resolution and are derived as balances between
total Pacific and Indian Ocean transports. Others are based on Sverdrup dynamics

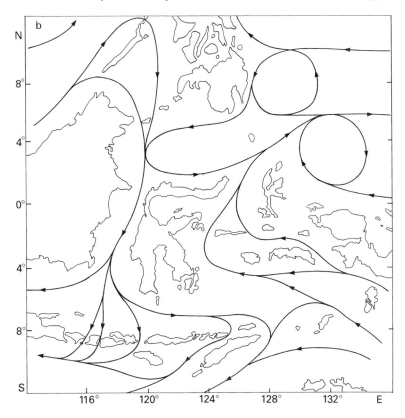

FIG. 13.6b. Surface currents in the Australasian Mediterranean Sea. August
(south monsoon, maximum throughflow).

and calculate the transport from hydrographic observations. The current is
concentrated in the upper layers and decays markedly with depth, with little
transport occurring below 500 m; this makes the assumption of a depth of no
motion below 1000 m reasonably acceptable. Recent observations from current
meters moored in the west Flores Sea and in Lombok Strait for the period January
1985 - March 1986 showed consistent flow from the Pacific to the Indian Ocean
of 0.9 m s^{-1} and more during August. During October - March the flow was
interrupted by frequent reversals of 10 - 20 day duration, but when it set
southwestward it still attained 0.6 m s^{-1}. These and similar observations indicate
that the fairly large transports inferred from numerical models and Sverdrup
calculations might well be realistic. Annual mean transport through Lombok Strait
works out at about 1.7 Sv, with virtually no flow during November - January, 1 Sv
in February - June, and maximum transport of 4 Sv in August (Murray and Arief,
1988). Given that Lombok Strait is one of the minor passages, these figures suggest
fairly large total throughflow. A recent attempt to estimate the flow through all
channels into the Indian Ocean from geostrophic calculations gave a total
transport of 24 Sv, again pointing towards a large throughflow.

Surface currents can reverse seasonally despite continuous net westward throughflow; this is known for Lifamatola Strait, the passage from the Molucca Sea to the Buru Basin which leads into the North Banda Sea, where the current sets northward during August but southward during February. A sketch of the surface circulation constructed to the best of available knowledge is given in Figure 13.6.

Currents below 500 m depth are even less well surveyed than upper layer currents. Estimates from geostrophic calculations indicate concentration of the flow in the upper few hundred metres (18 Sv or $^3/_4$ of the total transport of the estimate mentioned above were found in the layer 0 - 150 m). The few observations that are available show, however, that surprisingly large velocities do occur close to the ocean floor at some locations. Current meters moored in Lifamatola Strait during January and February of 1985 in 1940 m water depth gave mean speeds of 0.61 m s^{-1} about 100 m above the bottom and 0.40 m s^{-1} about 400 m above the bottom. At both depths the currents regularly exceeded 1 m s^{-1} during spring tides. Such large velocities have to be associated with strong mixing. This will become evident in the discussion of bottom water renewal below.

Figure 13.7 gives a hydrographic section from the Pacific Ocean north of Halmahera through the Molucca, Banda and Sawu Seas into the Indian Ocean south of Timor, obtained during one of the rare expeditions into the region. Unfortunately no salinity data were obtained on the Pacific side, so the character of the Australasian Mediterranean Sea as a dilution basin between the two oceans does not come out as clearly as it could. However, salinity in the Banda Basin is seen to vary by less than 0.06 over the entire water column, and lowest salinities are found in the Sawu Sea. The effect of freshwater input at the surface comes out more clearly in a comparison of T-S diagrams along the path of the throughflow (Figure 13.8) which shows that the vertical salinity gradient of the Pacific Central Water virtually disappears during the passage through the Indonesian seas. Mixing with Indian Central Water restores the gradient and brings the T-S diagram nearly back to its original form. This water mass conversion affects the upper 1000 m of the water column, despite the fact that adding freshwater at the surface increases the stability. Turbulent mixing must therefore occur over a large depth range and must be able to overcome the strong density gradient. Indications for strong mixing at great depth can be seen at sills such as Lifamatola Strait. Figure 13.7 shows that the water that fills the Buru and Banda basins is drawn from about 1500 m depth some 500 m above the sill depth, indicating that strong bottom currents of probably tidal character are able to mix the water over some hundreds of metres. Support for this conclusion comes from the oxygen data (Figure 13.9) which indicate an oxygen maximum above the sill as a result of downward mixing of water from above.

The mixing process in the upper 1000 m of the Australasian Mediterranean Sea is unique since it achieves nearly complete homogenization of the salinity field without destroying the temperature stratification. This excludes deep vertical

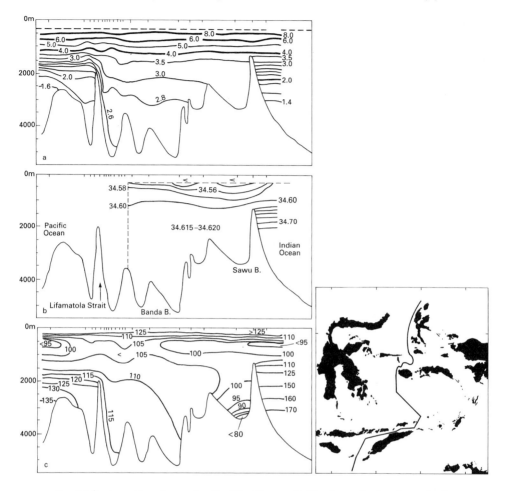

Fɪɢ 13.7. A section from the Pacific Ocean (left) through the Australasian Mediterranean Sea into the Indian Ocean. (a) Potential temperature (°C), (b) salinity, (c) oxygen (μmol/l; see Fig. 9.4 for conversion to ml/l). Tick marks along the top indicate station positions. From van Aken *et al.* (1988).

convection as the main mixing agent and requires highly turbulent flow well below the layer affected by the wind. Most likely the turbulence is concentrated near sills and related to strong tidal currents. The turbulence does, of course, affect salinity and temperature in identical fashion, so different surface boundary conditions for salinity and temperature are required to erase a salinity gradient without eliminating the temperature gradient. While a high freshwater input at the surface is responsible for homogenizing the salinity, maintaining the temperature gradient in the presence of strong mixing is impossible without a large input of heat to keep the sea surface temperature up. The atmospheric conditions found in the Australasian Mediterranean Seas can thus be deduced from its T-S properties.

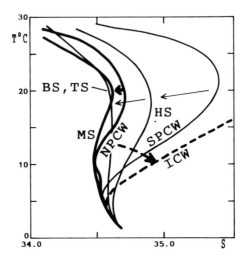

Fig. 13.8. Temperature-salinity diagrams along the path of the Indonesian through-flow, showing the transformation of Pacific Central into Australasian Mediterranean Water (demonstrating the character of the Australasian Mediterranean Sea as a dilution basin) and subsequently into Indian Central Water. South Pacific Central Water (SPCW) passes through the Halmahera Sea (HS) into the South Banda (BS) and Timor Seas (TS). North Pacific Central Water (NPCW) passes through Makassar Strait (MS) to the Timor Sea (TS). Both are then converted into Indian Central Water (ICW). Adapted from Ffield and Gordon (1992).

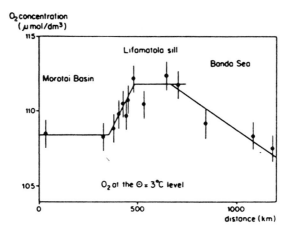

FIG. 13.9. Oxygen (units see Fig. 13.7) at the level of the 3°C potential temperature isotherm in Lifamatola Strait. From van Aken *et al.* (1988).

The turbulence of the upper layer does not reach much below the sill depths of the various basins. Nevertheless, oxygen values in the deep basins do not differ dramatically from those of the waters above, indicating reasonably short renewal times for the water below the sill depths. Figure 13.10 gives renewal paths and age estimates for bottom water derived from radiocarbon measurements. Most of the water participates in the interoceanic throughflow and has transit times

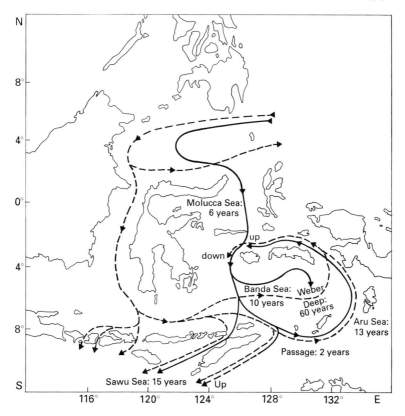

FIG. 13.10. Sketch of intermediate (broken lines) and deep water movement (full lines). Etimated transit times of bottom water are also indicated. Adapted fromvan Bennekom (1988).

of a few years. Water movement through the Seram Sea is from the east, partly in continuation of a loop from Makasar Strait through the South Banda Sea, partly as inflow from the Molucea Sea. The oldest water in the region is probably found towards the end of the loop in the deep depression of the South Banda Sea below 7000 m (the Weber Deep).

CHAPTER 14

The Atlantic Ocean

A glance at the distribution of high quality ocean data (Figure 2.3) tells us that the Atlantic Ocean is by far the best researched part of the world ocean. This is particularly true of the North Atlantic Ocean, the home ground of many oceanographic research institutions of the USA and Europe. We therefore have a wealth of information, and our task in describing the essential features of the Atlantic Ocean will not so much consist of finding reasonable estimates for missing data but finding the correct level of generalization from a bewildering and complex data set.

Bottom topography

Several outstanding topographic features distinguish the Atlantic Ocean from the Pacific and Indian Oceans. First of all, the Atlantic Ocean extends both into the Arctic and Antarctic regions, giving it a total meridional extent - if the Atlantic part of the Southern Ocean is included - of over 21,000 km from Bering Strait through the Arctic Mediterranean Sea to the Antarctic continent. In comparison, its largest zonal distance, between the Gulf of Mexico and the coast of north west Africa, spans little more than 8,300 km. Secondly, the Atlantic Ocean has the largest number of adjacent seas, including mediterranean seas which influence the characteristics of its waters. Finally, the Atlantic Ocean is divided rather equally into a series of eastern and western basins by the Mid-Atlantic Ridge, which in many parts rises to less than 1000 m depth, reaches the 2000 m depth contour nearly everywhere, and consequently has a strong impact on the circulation of the deeper layers.

When all its adjacent seas are included, the Atlantic Ocean covers an area of $106.6 \cdot 10^6$ km^2. Without the Arctic Mediterranean and the Atlantic part of the Southern Ocean, its size amounts to $74 \cdot 10^6$ km^2, slightly less than the size of the Southern Ocean. Although all its abyssal basins are deeper than 5000 m and most extend beyond 6000 m depth in their deepest parts (Figure 14.1), the average depth of the Atlantic Ocean is 3300 m, less than the mean depths of both the Pacific and Indian Oceans. This results from the fact that shelf seas (including its adjacent and mediterranean seas) account for over 13% of the surface area of the Atlantic Ocean, which is two to three times the percentage found in the other oceans.

Three of the features shown in Figure 14.1 deserve special mention. The first is the difference in depth east and west of the Mid-Atlantic Ridge near 30°S. The Rio Grande Rise comes up to about 650 m; but west of it the Rio Grande Gap allows passage of deep water near the 4400 m level. In contrast, the Walvis Ridge in the east, which does not reach 700 m depth, blocks flow at the 4000 m level. The second is the Romanche Fracture Zone (Figure 8.2) some 20 km north of the equator which allows movement of water between the western and eastern deep basins at the 4500 m level (its deepest part, the Romanche Deep, exceeds 7700 m depth but connects only to the western basins). Other fracture zones north of the equator have similar characteristics; but the Romanche Fracture Zone is the first opportunity for water coming from the south to break through the barrier posed by the Mid-Atlantic Ridge. The third feature is the Gibbs Fracture Zone near 53°N which allows passage of water at the 3000 m level; its importance for the spreading of Arctic Bottom Water was already discussed in Chapter 7.

Of interest from the point of view of oceanography are the sill characteristics of the five mediterranean seas. The Arctic Mediterranean Sea, which is by far the largest comprising 13% of the Atlantic Ocean area, was already discussed in Chapter 7; its sill is about 1700 km wide and generally less than 500 m deep with passages exceeding 600 m depth in Denmark Strait and 800 m in the Faroe Bank Channel. The Strait of Gibraltar, the point of communication between the Eurafrican Mediterranean Sea and the main Atlantic Ocean, spans a distance of 22 km with a sill depth of 320 m. The American Mediterranean Sea has several connections with the Atlantic Ocean basins, the major ones being east of Puerto Rico and between Cuba and Haiti where sill depths are in the vicinity of 1700 m and between Florida and the Bahamas with a sill depth near 750 m. Baffin Bay communicates through the 350 km wide Davis Strait where the sill depth is less than 600 m. Finally, communication with the Baltic Sea is severely restricted by the shallow and narrow system of passages of Skagerrak, Kattegat, Sund and Belt where the sill depth is only 18 m.

The wind regime

The information needed from the atmosphere is again included in Figures 1.2 - 1.4. An outstanding feature is the large seasonal variation of northern hemisphere winds in comparison to the low variability of the wind field in the subtropical zone of the southern hemisphere. This is similar to the situation in the Pacific Ocean and again caused by the impact of the Siberian and to a lesser extent North American land masses on the air pressure distribution. As a result the subtropical high pressure belt, which in the northern winter runs from the Florida - Bermuda region across the Canary Islands, the Azores, and Madeira and continues across the Sahara and the Eurafrican Mediterranean Sea into central Siberia, is reduced during summer to a cell of high pressure with its centre

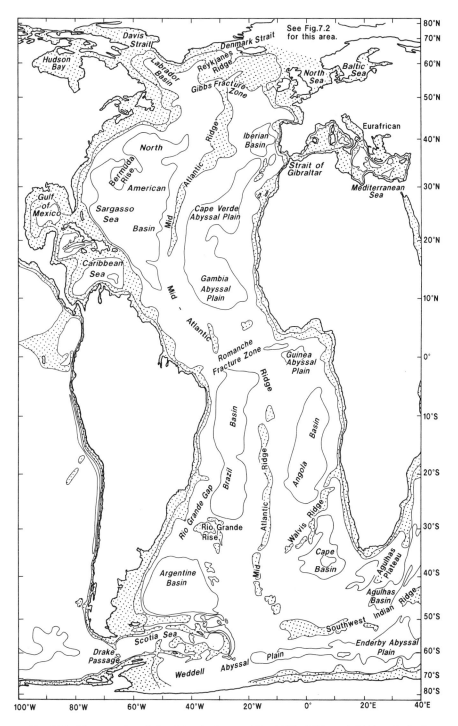

FIG. 14.1. Topography of the Atlantic Ocean. The 1000, 3000, and 5000 m isobaths are shown, and regions less than 3000 m deep are stippled.

near the Azores. This is the well-known Azores High which dominates European summer weather, bringing winds of moderate strength. During winter, the contrast between cold air over Siberia and air heated by the advection of warm water in the Norwegian Current region leads to the development of the equally well-known Icelandic Low with its strong Westerlies, which follow the isobars between the subtropical high pressure belt and the low pressure to the north. The seasonal disturbance of the subtropical high pressure belt in the southern hemisphere is much less developed, and the Westerlies show correspondingly less seasonal variation there.

The Trade Winds are somewhat stronger in winter (February north of the equator and August in the south) than in summer on both hemispheres. Seasonal wind reversals of monsoon characteristics are of minor importance in the Atlantic Ocean; their occurrence is limited to two small regions, along the African coastline from Senegal to Ivory Coast and in the Florida - Bermuda area. Important seasonal change in wind direction is observed along the east coast of North America which experiences offshore winds during most of the year but warm alongshore winds in summer.

The mean wind stress distribution of the South Atlantic Ocean shows close resemblance to that of the Indian Ocean. The maximum Westerlies do not lie quite so far north as in the Indian Ocean (at about 50°S instead of 45°S), but the maximum Trade Winds occur at very similar latitudes (about 15°S, associated with somewhat smaller wind stress curls). The Doldrum belt, or Intertropical Convergence Zone (ITCZ), is found north of the equator, rather like the North Pacific ITCZ but not as accurately zonal; its annual mean position angles from the equator off Brazil to about 7°N off Sierra Leone.

North of the ITCZ the mean wind stress distribution more closely resembles that of the North Pacific Ocean, though the Atlantic Northeast Trades are not quite as strong in comparison. Their maximum strength is at about 15°N. The North Atlantic Westerlies enter the ocean from the northwest, similar to the North Pacific Westerlies. They bring cold, dry air out over the Gulf Stream, just as the Pacific winds bring cold dry air from Siberia out over the Kuroshio. As their Pacific counterpart, the Atlantic Westerlies veer round to a definite southwesterly direction in the eastern Atlantic Ocean, and the axis of maximum westerly strength is also oriented along a line running east-north-east. The polar easterlies of the Arctic region are more vigorous in the Atlantic than in any other ocean.

The integrated flow

When the Sverdrup balance was introduced and tested in Chapter 4 we noted that the largest discrepancies between the integrated flow fields deduced from wind stress and CTD data are found in the Atlantic Ocean. We now go back to Figure 4.4 and Figures 4.5 or 4.6 for a more detailed comparison, keeping in mind that with the exception of the Southern Ocean, the CTD-derived flow pattern should describe the actual situation quite well. The largest discrepancy between the two flow fields occurs south of 34°S; it was discussed in Chapters 4 and 11. North of 34°S, the subtropical gyres of both hemispheres are well reproduced from both atmospheric and oceanic data, as in the other oceans. To be more specific, the gradient of depth-integrated steric height across the North and South Equatorial Currents is calculated fairly well from both data sets, and the gradient across the equator in the Atlantic seen in the CTD-derived pattern (one contour crosses the equator; P increases westward, as in the Pacific Ocean) also occurs in the wind-calculated pattern (even though no contour happens to cross the equator in this case). That this must be so is evident by inspection of Figure 1.2 which shows weak mean westerly winds along the equator in the Atlantic Ocean; hence the only term on the right hand side of eqn (4.7) at the equator is negative, and P must increase towards the west. However, the agreement between Figures 4.4 and 4.5 or 4.6 is not as good in the Atlantic as in the other oceans. The major reason for this is the recirculation of North Atlantic Deep Water, which was mentioned already in Chapter 7 and will be further discussed in Chapter 15. It makes the assumption of a depth of no motion less acceptable than in the other oceans. The transport of thermocline water from the Indian into the Atlantic Ocean which is part of the North Atlantic Deep Water recirculation is also not included in the flow pattern derived from wind data.

In the region where the two circulation patterns compare well, the Sverdrup relation reveals the existence of strong subtropical gyres in both hemispheres and a weaker subpolar gyre in the northern hemisphere. The gyre boundaries coincide reasonably well with the contour of zero curl(τ/f) (Figure 4.3). The northern subtropical gyre consists of (Figure 14.2) the North Equatorial Current with its centre near 15°N, the Antilles Current east of, and the Caribbean Current through the American Mediterranean Sea, the Florida Current, the Gulf Stream, the Azores Current, and the Portugal and Canary Currents. The southern gyre is made up of the South Equatorial Current which is centred in the southern hemisphere but extends just across the equator, the Brazil Current, the South Atlantic Current, and the Benguela Current. The subpolar gyre of the northern hemisphere is modified by interaction with the Arctic circulation, to the extent that it is hardly recognizable as a gyre. It involves the North Atlantic Current, the Irminger Current, the East and West Greenland Currents, and the Labrador Current, with substantial water exchange with the Arctic Mediterranean Sea through the North Atlantic Current (and its extension into the Norwegian Current) and the East Greenland Current.

The Sverdrup relation performs particularly well near the equator, where geostrophic gradients are very small. It reveals the existence of an equatorial countercurrent between the North and South Equatorial Currents. As in the Pacific Ocean, this countercurrent flows down the Doldrums; but it is broader and less intense. This results from the reduced width of the Atlantic Ocean and

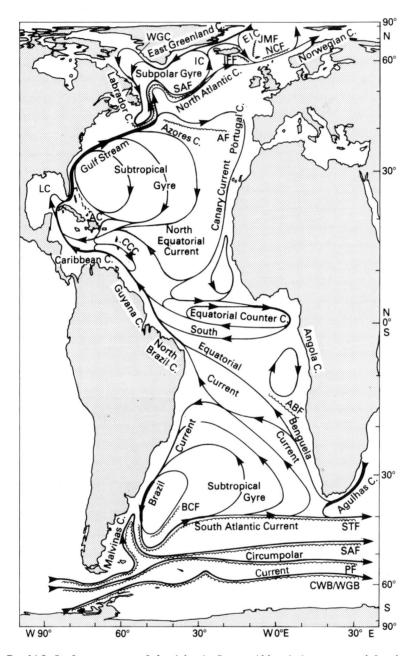

Fig. 14.2. Surface currents of the Atlantic Ocean. Abbreviations are used for the
East Iceland (EIC), Irminger (IC), West Greenland (WGC), Loop (LC) and Antilles
(AC) Currents and the Caribbean Countercurrent (CCC). Other abbreviations refer
to fronts: JMF: Jan Mayen Front, NCF: Norwegian Current Front, IFF: Iceland - Faroe
Front, SAF: Subarctic Front, AF: Azores Front, ABF: Angola - Benguela Front, BCF:
Brazil Current Front, STF: Subtropical Front, SAF: Subantarctic Front, PF: Polar
Front, CWB/WGB: Continental Water Boundary / Weddell Gyre Boundary. Adapted
from Duncan *et al.* (1982), Krauss (1986) and Peterson and Stramma (1991).

from the fact that the Doldrums (or ITCZ) are not strictly zonal but angle across from Brazil to Sierra Leone, as mentioned earlier.

A notable discrepancy between Figures 4.4 and 4.5 or 4.6 is the failure of the wind-calculated pattern to reproduce the intense crowding of the contours of depth-integrated steric height off North America near Cape Hatteras (35°N). A similar failure occurs in the north-east Pacific Ocean, but it is not as severe there; in the Atlantic Ocean, the wind-calculated flow follows the coast to Labrador (50°N) before flowing east, whereas it in fact breaks away from the coast at Cape Hatteras (as indicated in the CTD-based flow field) and takes on the character of an intense jet.

The equatorial current system

As in the Pacific Ocean, the equatorial current system displays a banded structure when investigated in detail. Figure 14.3 is a schematic summary of all its elements as they occur in mid-year. The *Equatorial Undercurrent* (EUC) is the strongest, with maximum speeds exceeding 1.2 m s^{-1} in its core at about 100 m depth and transports up to 15 Sv. It is driven and maintained by the same mechanism as in the Pacific Ocean (see Chapter 8), strongest in the west and weakening along its path as a result of frictional losses to the surrounding waters. Observations show that it swings back and forth between two extreme positions 90 km either side of the equator at a rate of once every 2 - 3 weeks, while speed and transport oscillate between the maxima given above and their respective minima of 0.6 m s^{-1} and 4 Sv. The EUC was discovered by the early oceanographer John Young Buchanan during the *Challenger* expedition of 1872 - 1876 and described in 1886, but this discovery was forgotten until the discovery of the Pacific EUC in 1952 triggered a search for an analogous current in the Atlantic Ocean. CTD data reveal the presence of the EUC through the vertical spreading of isotherms in the thermocline (Figure 14.4); in the eastern Atlantic Ocean it can be seen as a prominent subsurface salinity maximum.

The three equatorial currents known from the depth-integrated circulation dominate the surface flow (Figure 14.3) and the hydrography (Figure 14.4; see Chapter 8 for a discussion of the relationship between thermocline slope and currents) but appear more complicated in detail. The *North Equatorial Current* (NEC) is a region of broad and uniform westward flow north of 10°N with speeds of $0.1 - 0.3 \text{ m s}^{-1}$. The eastward flowing *North Equatorial Countercurrent* (NECC, the countercurrent seen in the depth-integrated flow field) has similar speeds; it is highly seasonal and nearly disappears in February when the Trades in the northern hemisphere are strongest (Figure 14.5). The *South Equatorial Current* (SEC), again a region of broad and uniform westward flow with similar speeds, extends from about 3°N to at least 15°S. Just as in the Pacific Ocean it is interspersed with eastward flow both at the surface and below the thermocline. The *South Equatorial Countercurrent* (SECC) is weak, narrow and variable and therefore not resolved by Figure 14.5, which is based on 2° averages in latitude.

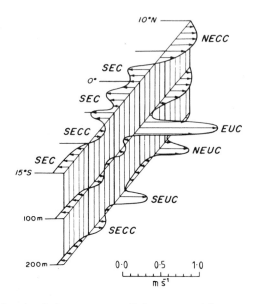

FIG. 14.3. A sketch of the structure of the equatorial current system during August. For abbreviations see text. After Peterson and Stramma (1991).

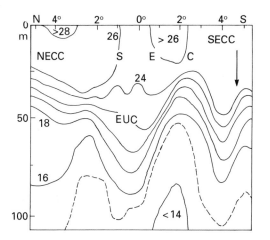

FIG. 14.4. Temperature section (°C) across the central part of the equatorial current system along 5°W. For abbreviations see text. Note the low surface temperature at the equator due to upwelling, the weakening of the thermocline in the EUC, and the poleward rise of the thermocline in the countercurrents. Adapted from Moore *et al.* (1978) .

It often shows maximum speed (around 0.1 m s⁻¹) below 100 m depth and is masked by weak westward flow at the surface. The *North Equatorial Undercurrent* (NEUC) and the *South Equatorial Undercurrent* (SEUC) are both narrow and swift, with maximum speeds of 0.4 m s⁻¹ near 200 m depth.

The most conspicuous feature of the equatorial circulation is the strong cross-equatorial transport along the South American coast in the *North Brazil Current.*

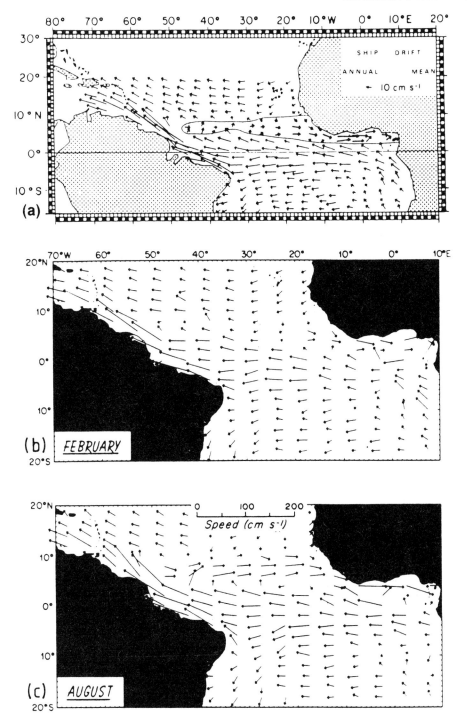

F_{IG}. 14.5. Surface currents in the equatorial region as derived from ship drift data. (a) Annual mean, (b) February, (c) August. From Arnault (1987) and Richardson and Walsh (1986).

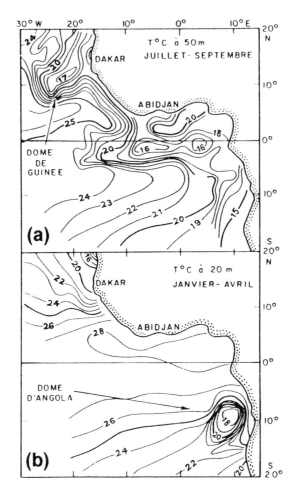

FIG. 14.6. The Angola Dome and the Guinea Dome as seen in temperature data from 20 m and 50 m depth. From Peterson and Stramma (1991)

Of the 16 Sv carried across 30°W in the South Equatorial Current during February/March, only 4 Sv are carried south into the Brazil Current while 12 Sv cross the equator (Stramma *et al.*, 1990). This is close to the estimated 15 Sv needed to feed the Deep Water source in the North Atlantic Ocean.

Little exchange between hemispheres occurs in the eastern part of the equatorial zone, the termination region of all eastward flow. The South Equatorial Countercurrent turns south, driving a cyclonic gyre centred at 13°S, 4°E which extends from just below the surface to at least 300 m depth with velocities approaching 0.5 m s⁻¹ near the African coast where this relatively strong subsurface flow is known as the *Angola Current* (Figure 14.2). By opposing the northward movement of the Benguela Current it creates the Angola - Benguela Front, a feature seen in the temperature of the upper 50 m and in the salinity distribution to at least 200 m depth.

The North Equatorial Countercurrent is prevented from flowing north by the east - west orientation of the coastline; it intensifies to an average 0.4 m s^{-1} along the Ivory Coast before its energy is dissipated in the Gulf of Guinea. However, some of its flow does escape north and combines with the North Equatorial Undercurrent to drive a small cyclonic gyre centred at 10°N, 22°W. A similar small gyre, centred near 10°S, 9°E and clearly distinct from the larger gyre which incorporates the Angola Current, is driven by the South Equatorial Undercurrent. We know from our Rules 1, 1a, and 2 of Chapter 3 that cyclonic flow is accompanied by a sea surface depression and an elevation of the thermocline in the centre of the gyre (compare Figure 2.7, or Figures 3.3 and 3.4 which show the same rules operating in an anticyclonic gyre), so in a plot of temperature at constant depth the two gyres should show up as local temperature minima. Figure 14.6 proves that this is indeed the case but only in summer when the Trades of the respective hemisphere are weakest and the Undercurrents strongest. Because of the observed doming of the thermocline in summer the gyres are known as the *Angola* and *Guinea Domes*. The associated circulation exists throughout the year, although weaker in winter, and reaches to at least 150 m depth.

Western boundary currents

The Sverdrup calculation of Chapter 4 gave integrated volume transports for the Gulf Stream and the Brazil Current of 30 Sv. These numbers are modest in comparison to the results for the Kuroshio (50 Sv) or the Agulhas Current (70 Sv). They can be explained in part as reflecting the weakness of the Atlantic annual mean wind stress and the narrowness of the basin. However, they underestimate the Gulf Stream transport by a large margin. This failure of the Sverdrup calculation is a consequence of the recirculation of North Atlantic Deep Water. The westward intensification of all ocean currents influences the flow of Deep Water, too; so both the southward transport of Deep Water at depth and the northward flow of the recirculation below and above the thermocline are concentrated on the western side of the ocean. This adds some 15 Sv to the Gulf Stream transport in the upper 1500 m and subtracts the same amount from the transport of the Brazil Current. This large difference between the two major currents in the Atlantic Ocean does not come out in the vertically integrated flow (Figure 4.7), which shows complete separation of the oceanic gyres along the American coast near 12°S, 6°N, 18°N, and 50°N and similar transports for both boundary currents. This is true for the *wind-driven* component of the flow (i.e. excluding the Deep Water recirculation which is a result of thermohaline forcing), and it is correct when the flow is integrated over all depth, but it is misleading when taken as representative of the circulation in the upper ocean.

For these reasons, the strongest of the western boundary currents is the *Gulf Stream,* so called because it was originally believed to represent a drainage flow from the Gulf of Mexico. It has now been known for many decades that this is not correct and that the flow through the Strait of Florida stems directly from Yucatan Strait and passes the Gulf to the south. Even this flow constitutes only a portion of the source waters of the Gulf Stream. It turns out that it is better to speak of the Gulf Stream System and its various components, the Florida Current, the Gulf Stream proper, the Gulf Stream Extension, and its continuation as the North Atlantic and Azores Currents.

The *Florida Current* is fed from that part of the North Equatorial Current that passes through Yucatan Strait, with a possible contribution from the North Brazil Current (see below). In Florida Strait this current carries about 30 Sv with speeds in excess of 1.8 m s^1. On average, the current is strongest in March, when it carries 11 Sv more than in November. Its transport is increased along the coast of northern Florida through input from the second path of the North Equatorial Current (the Antilles Current, see below). Recirculation of Gulf Stream water in the Sargasso Sea increases its transport further. By the time the flow separates from the shelf near Cape Hatteras - a distance of 1200 km downstream - it has reached a transport of 70 - 100 Sv (much more than the 30 Sv suggested by the integrated flow calculation of Chapter 4). For the next 2500 km the *Gulf Stream* proper flows across the open ocean as a free inertial jet. Its transport increases initially through inflow from the Sargasso Sea recirculation region to reach a maximum of 90 - 150 Sv near 65°W. The current then begins to lose water to the Sargasso Sea recirculation, its transport falling to 50 - 90 Sv near the Newfoundland Rise (50°W, also known as the Grand Banks). Throughout its path current speed remains large at the surface and decreases rapidly with depth, but the flow usually extends to the ocean floor (Figure 14.7).

In the region east of 50°W, which is sometimes referred to as the Gulf Stream Extension, the flow branches into three distinctly different regimes (Figure 14.8). The *North Atlantic Current* continues in a northeastward direction towards Scotland and withdraws about 30 Sv from the subtropical gyre, to feed the Norwegian Current and eventually contribute to Arctic Bottom Water formation. The *Azores Current* is part of the subtropical gyre; it carries some 15 Sv along 35 - 40°N to feed the Canary Current. The remaining transport does not participate in the ocean-wide subtropical gyre but is returned to the Florida Current and Gulf Stream via the much shorter path of the Sargasso Sea recirculation system.

Free inertial jets which penetrate into the open ocean become unstable along their path. They form meanders which eventually separate as eddies. Meanders which separate poleward of the jet develop into anticyclonic (warm-core) eddies, those separating equatorward produce cyclonic (cold-core) eddies (Figure 14.9). Because of their hydrographic structure - a ring of Gulf Stream water with velocities comparable to those of the Gulf Stream itself, isolating water of different properties from the surrounding ocean - these eddies are often referred to as rings. Most of the Gulf Stream rings are formed in the Gulf Stream Extension region and move slowly back against the direction of the main current

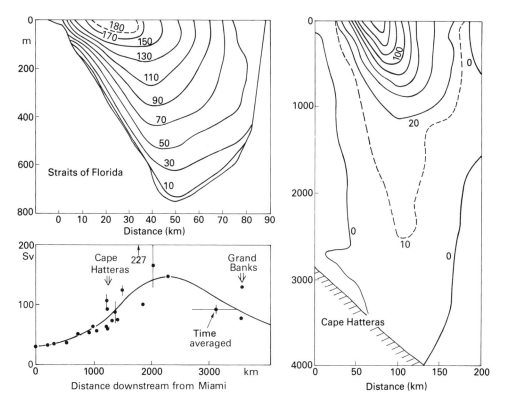

Fɪɢ. 14.7. A summary of Gulf Stream volume transports reported in the literature (based on Richardson (1985) and additional more recent data); and two sections of annual mean velocity in the Florida Current and the Gulf Stream at Cape Hatteras, based on continuous vertical profiles of velocity from cruises over a 2 - 3 year period. Note the different depth and distance scales. From Leaman *et al.* (1989).

(Figure 14.10). Rings formed north of the Gulf Stream are restricted in their movement and often merge with the main flow after a short journey eastward; but the cold-core eddies to the south dominate the Sargasso Sea recirculation region, where some 10 rings can be found at any particular time. Satellite images of sea surface temperature such as Figure 14.11 display them as isolated regions of warm water north of the Gulf Stream and regions of cold water to the south. In the world map of eddy energy (Figure 4.8) the Sargasso Sea recirculation region stands out as one of the most energetic.

Most transport estimates for the Gulf Stream are based on geostrophic calculations which, according to our Rule 2 in Chapter 3, should be accurate to within 20%. The associated pressure gradient is maintained by a drop in sea level across the current of some 0.5 m towards the coast and, according to our Rule 1a, a corresponding thermocline rise of about 500 m. This is demonstrated by Figure 14.12 which also reveals the existence of two countercurrents, one inshore - between the continental slope and the Gulf Stream - and one offshore, as part of the long-term mean situation. Actual velocities at any particular time

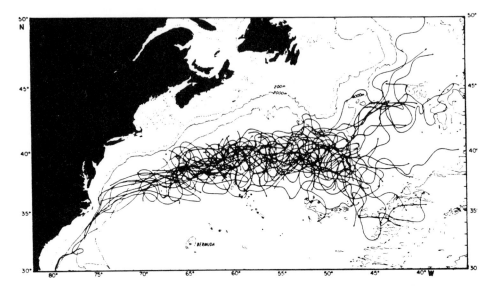

FIG. 14.8. Paths of satellite-tracked buoys in the Gulf Stream system. Most buoy tracks are from the period 1977 - 1981, some tracks going back to 1971. For clarity, only buoys with average velocity exceeding 0.5 m s⁻¹ were used and loops indicative of meanders or eddies were removed. The branching of the Gulf Stream into the North Atlantic Current, Azores Current, and Sargasso Sea recirculation is visible in the tracks east of 55°W. From Richardson (1983a).

can be much larger, since the strong currents in the rings disappear in the mean and variability in the position of the Gulf Stream acts to reduce the peak velocity in the mean as well. Observed peak velocities usually exceed 1.5 m s⁻¹. The Gulf Stream is an important heat sink for the ocean. Net annual mean heat loss, caused by advection of cold dry continental air from the west, exceeds 200 W m⁻² (Figure 1.6). A brief period of net heat gain occurs from late May to August when warm saturated air is advected from the south (Figure 1.2).

The *Labrador Current* is the western boundary current of the subpolar gyre. This gyre receives considerable input of Arctic water from the East Greenland Current. Measurements south of Cape Farewell indicate speeds of 0.3 m s⁻¹ on the shelf and above the ocean floor at depths of 2000 - 3000 m and 0.15 m s⁻¹ at the surface, for the combined flow of the East Greenland and Irminger Currents. Transport estimates for the Irminger Current amount to 8 - 11 Sv. Even if this is combined with the estimated 5 Sv for the East Greenland Current of Chapter 7, it does not explain the 34 Sv derived by Thompson *et al.* (1986) for the West Greenland and Labrador Currents from hydrographic section data. Substantial recirculation must therefore occur in the Labrador Sea if these estimates are correct. Earlier estimates of 10 Sv or less were based on geostrophic calculations with 1500 m reference depth, clearly not deep enough for western boundary currents which extend to the ocean floor. The Labrador Current is strongest in February when on average it carries 6 Sv more water than in August. It is also more variable in winter, with a standard deviation of 9 Sv in February but only 1 Sv in August.

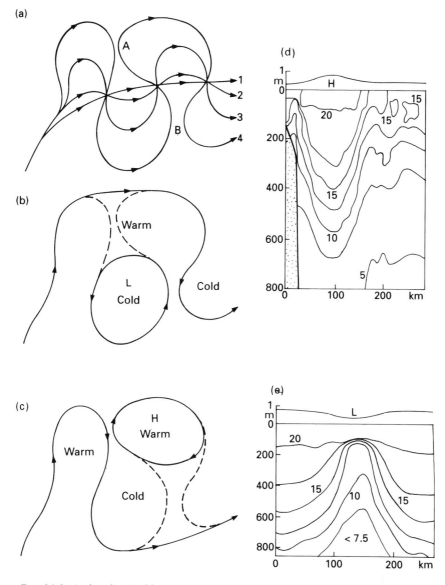

FIG. 14.9. A sketch of eddy formation in a free inertial jet and the associated hydrographic structure. (a) Path of the jet at succesive times 1 - 4. (b) A cyclonic (cold-core) ring formed after merger of the path at location A. The open line is the Gulf Stream path after ring formation, the closed line the ring, the dotted line the path just before eddy formation. (c) A similar representation of an anticyclonic (warm-core) ring formed if the jet merges at location B instead. H and L indicate high and low pressure. (d) A temperature section (°C) through an anticyclonic ring. (e) A section through a cyclonic ring. The shape of the sea surface shown in (d) and (e) was not measured but follows from Rules 1, 1a, and 2 of Chapter 3. Panels (a) - (c) show a northern hemisphere jet; the situation in the southern hemisphere is the mirror image with respect to the equator. Panels (d) and (e) apply to both hemispheres; they are adapted from Richardson (1983b).

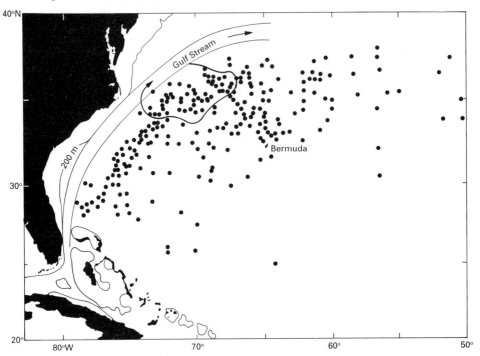

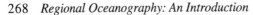

FIG. 14.10. Geographical distribution of 225 cold-core rings reported for the period 1932 - 1983. Ring movement is generally towatds the southwest until the rings decay or are absorbed ag;iin into the Gulf Stream. The arrow, the path of a ring observed in 1977, gives an example of typical ring movement. Adapted from Richardson (1983b).

FIG. 14.11. Infrared satellite image of the Gulf Stream System. The Gulf Stream is seen as a band of warm water between the colder Slope Water region and the warmer Sargasso Sea. Two rings can be seen; both contain water of Gulf Stream temperature, but the northern ring is of the warm-core type and has anti-cyclonic rotation, while the ring in the south is a cold-core ring with cyclonic rotation. The region shown covers approximately 36°N - 42°N, 65°W - 77°W. From Richardson (1983b).

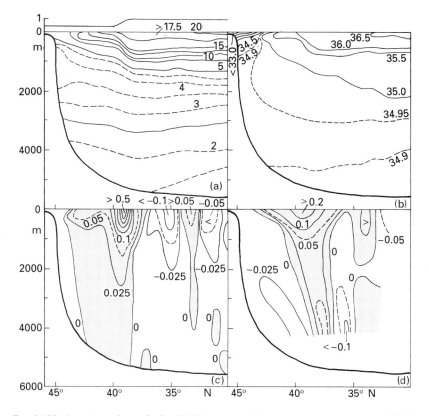

FIG. 14.12. A section through the Gulf Stream and its countercurrents across 55°W.
(a) Potential temperature (°C), (b) salinity, (c) geostrophic current (m s^{-1}) relative
to the sea floor, (d) mean current (m s^{-1}) as derived from a combination of drifters,
subsurface floats, and current meter moorings. The sections are based on data from
8 cruises between 1959 - 1983. From Richardson (1985). The shape of the sea surface
as seen in more recent satellite altimeter observations is sketched above the
temperature panel.

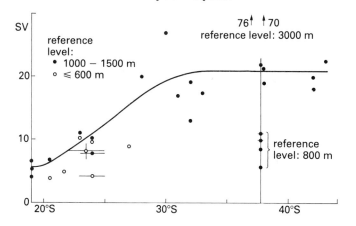

FIG. 14.13. A summary of Brazil Current transports reported in the literature. Unless
indicated otherwise, transports assume a level of no motion between 1000 m and
1500 m. After Peterson and Stramma (1991).

The western boundary current of the south Atlantic subtropical gyre, the *Brazil Current*, begins near 10°S with a trickle of 4 Sv supplied by the South Equatorial Current. Over the next 1500 km its strength increases to little more than 10 Sv through incorporation of water from the recirculation region over the Brazil Basin. The current is comparatively shallow, nearly half of the flow occurring on the shelf with the current axis above the 200 m isobath. In deeper water northward flow of Antarctic Intermediate Water is embedded in the Current at intermediate depths (below 400 m). A well-defined recirculation cell south of the Rio Grande Rise (the analogon to the Sargasso Sea recirculation regime of the Gulf Stream) leads to an increase in transport to 19 - 22 Sv near 38°S (Figure 14.13), which corresponds to a rate of increase comparable to that observed in the Gulf Stream. All these estimates are derived from geostrophic calculations with levels of no motion near or above 1500 m and therefore do not include the considerable transport of North Atlantic Deep Water below. More recent estimates which use 3000 m as level of no motion give total transports of 70 - 76 Sv near 38°S (Peterson and Stramma, 1991). The difference is larger than can be explained by the transport of Deep Water and indicates that significant recirculation must occur in the south Atlantic Ocean below 1500 m depth.

The Brazil Current separates from the shelf somewhere between 33 and 38°S, forming an intense front with the cold water of the *Malvinas Current*, a jet-like northward looping excursion of the Circumpolar Current also known as the Falkland Current (Figure 14.14). The separation point is more northerly during summer than winter, possibly as part of a general northward shift of the subtropical gyre in response to the more northern position of the atmospheric high pressure system (Figure 1.3) and northward movement of the contour of zero curl(τ/f) during summer (December - February). The southernmost extent of the warm Brazil Current after separation from the shelf varies between 38°S and 46°S on times scales of two months and is linked with the formation of eddies, the mechanism being very similar to that of the East Australian Current (Figure 14.15; see also Figure 8.19). Observed current speeds in Brazil Current eddies are near 0.8 m s^{-1}; transport estimates are in the vicinity of 20 Sv. Most eddies escape from the recirculation region and are swept eastward with the *South Atlantic Current*. This can be seen in the distribution of eddy energy of Figure 4.8; the large area of high eddy energy centred on 40°S, 52°W corresponds to the region of eddy formation, its tail along 48°S to the path of the decaying eddies. The two separate regions of high eddy energy east of South America also indicate that the South Atlantic and Circumpolar Currents are clearly different regimes. Geostrophic determinations of zonal transport east of 10°W between 30°S (the centre of the subtropical gyre) and 60°S invariably indicate a transport minimum near 45°S, indicating a separation zone between the South Atlantic and Circumpolar Currents.

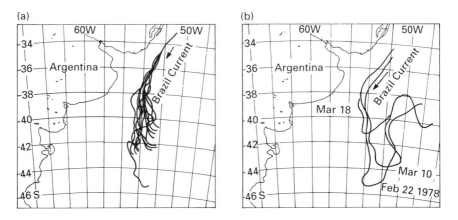

FIG. 14.14. The separation region of the Brazil Current. (a) mean position of the Brazil Current as indicated by the position of the thermal front between the Brazil and Malvinas Currents during Spetember 1975 - April 1976; (b) a succession of three positions of the thermal front, indicating northward retreat of the Brazil Current. Two eddies formed between 22 February and 18 March; they are not included here. From Legeckis and Gordon (1982).

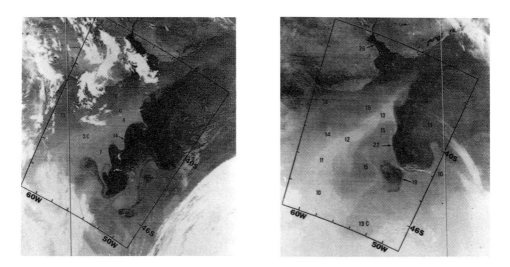

FIG. 14.15. Infrared satellite images of the Brazil Current separation obtained in October 1975 (left) and January 1976 (right). Dark is warm, light is cold; numbers show temperatures in °C. A recently formed eddy with a temperature of 18°C is seen in January 1976 south of the Brazil Current. From Legeckis and Gordon (1982).

Before concluding this section we mention the *North Brazil Current* and *Guyana Current* as another western boundary current system of the Atlantic Ocean. From the point of view of North Atlantic Deep Water recirculation it would be pleasing to see both as elements of continuous northward flow in and above the thermocline which starts at 16°S in the South Equatorial Current and continues through the American Mediterranean Sea to 27°N, eventually feeding into the

Florida Current. Although this current system has received much less attention than is warranted by its important role in the global transport of heat, it is fair to say that the continuity of northward flow at the surface is questionable. There is no doubt about the existence of the North Brazil Current; observed surface speeds in excess of 0.8 m s^{-1} testify for its character as a jet-like boundary current. The character of the Guyana Current is much more obscure; eddies related to flow instability have been reported, but some researchers doubt whether the Guyana Current exists as a permanent current. There has also been some documentation (Duncan *et al.*, 1982) that the Antilles Current is not identifiable as a permanent feature of the circulation and may indeed not exist as a continuous current. Since the flow from the North Equatorial Current has to reach the Florida Current somehow, net mean movement in both the Guyana and Antilles Currents has to be toward northwest. The topic will be taken up again in the discussion of the American Mediterranean Sea in Chapter 16.

Eastern boundary currents and coastal upwelling

South of 45°N the circulation in the eastern part of the Atlantic Ocean has many similarities with that of the eastern Pacific Ocean. In the northern hemisphere the Canary Current is a broad region of moderate flow where the temperate waters of the Azores Current are converted into the subtropical water that feeds into the North Equatorial Current. In the southern hemisphere the same process occurs in the Benguela Current. Both currents are therefore characterized, when compared with currents in the western Atlantic Ocean at the same latitudes, by relatively low temperatures. As in the Pacific Ocean, equatorward winds along the eastern edge of the ocean, from Cape of Good Hope to near the equator and from Spain to about 10°N, increase the temperature contrast by adding the effect of coastal upwelling. Although the currents associated with the upwelling and those which constitute the recirculation in the subtropical gyres further offshore are dynamically independent features, the names Canary Current and Benguela Current are usually applied to both. As in other eastern ocean basins, currents in the eastern Atlantic Ocean are dominated by geostrophic eddies (an example from the vicinity of the Canary Current region is shown in Figure 4.9). Current reversals caused by passing eddies are common.

The dynamics of coastal upwelling were discussed in Chapter 8, so it is sufficient here to concentrate on regional aspects and identify the various elements of coastal upwelling systems in the Atlantic context. The *Benguela Current* upwelling system (Figure 14.16) is the stronger of the two, lowering annual mean sea surface temperatures to 14°C and less close to the coast - two degrees and more below the values seen in Figure 2.5 near the coast which indicate the effect of equatorward flow in the subtropical gyre. It is strongest in the south during spring and summer when the Trades are steady; during winter (July - September), it extends northward but becomes more intermittent because the Trades, although

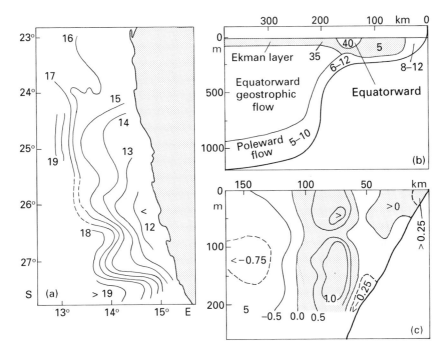

Fig. 14.16. The Benguela Current upwelling system. (a) Sea surface temperature (°C) in the northern part as observed during February 1966. (b) Sketch of the mean circulation. Transverse flow occurs in the bottom and Ekman layers; average speeds are given in cm s⁻¹, westward flow is shaded. Major alongshore (poleward or equatorward) flows are also indicated. (c) Observations of the poleward jet (m s⁻¹, northward flow is shaded) from January 1973 in the south near 34°S. Adapted from Bang (1971), Nelson (1989), and Bang and Andrews (1974).

stronger, are interrupted by the passage of eastward travelling atmospheric lows. The width of the upwelling region coincides with the width of the shelf (200 km). Velocities in the equatorward surface flow are in the range 0.05 - 0.20 m s⁻¹; in the coastal jet near the shelf break they exceed 0.5 m s⁻¹. Poleward flow occurs on the shelf above the bottom and over the shelf break with speeds of 0.05 - 0.1 m s⁻¹, advecting oxygen-poor water from the waters off Angola; the resulting oxygen minimum along the slope can be observed over a distance of 1600 km to 30°S. The interface between equatorward surface movement and poleward flow underneath often reaches the surface on the inner shelf, producing poleward flow along the coast.

Further offshore beyond the shelf break, the equatorward surface layer flow merges with the equatorward transport of thermocline water in the Benguela Current, while poleward movement above the ocean floor continues uninterrupted, feeding into the cyclonic circulation of the deeper waters discussed in the next chapter. The dynamic independence of the recirculation in the subtropical gyre and the coastal upwelling is seen in the fact that the Benguela Current gradually leaves the coast between 30°S and 25°S, while the upwelling reaches further north to Cape Frio (18°S). Geostrophic transport in the gyre circulation relative to

1500 - 2000 m depth is estimated at 20 - 25 Sv. This compares with a maximum of 7 Sv in the jet of the upwelling system (Peterson and Stramma, 1991).

Strong seasonal variability and large contrast between the waters in the north and south are the main characteristics of the *Canary Current* upwelling system. Although the width of the upwelling region is narrow (less than 100 km), it exceeds the width of the shelf in most places. Observations on the shelf, which on average is only 60 - 80 m deep, show an unusually shallow Ekman layer at the surface with offshore movement extending to about 30 m depth, an intermediate layer of equatorward geostrophic flow, and a bottom layer with onshore flow (Figure 14.17c). An equatorward surface jet occurs just inshore of the shelf edge, while the undercurrent is usually restricted to the continental

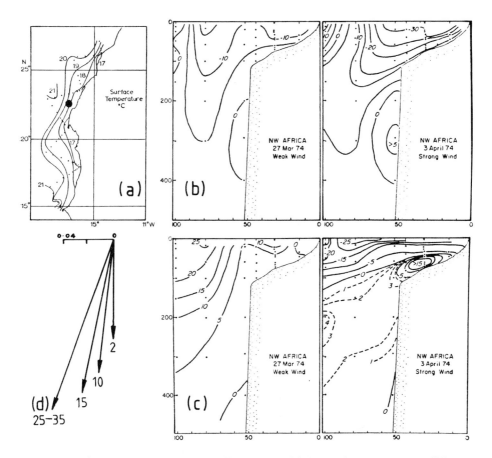

FIG. 14.17. The Canary Current upwelling system. (a) Sea surface temperature (°C) as observed in April/May 1969; (b) longshore and (c) onshore velocities (cm s⁻¹, positive is northward and eastward) during periods of weak and strong wind, (d) observed mean velocities over a 29 day period at the position indicated by the dot in panel (a), in 74 m water depth; numbers indicate distance from the bottom. Note the alignment of the current at mid-depth with the direction of the coast, and the shoreward turning of the current as the bottom is approached. From Hughes and Barton (1974), Huyer (1976), and Tomczak and Hughes (1980).

slope (Figure 14.17b). Velocities in all components of the current system are similar to those reported from the Benguela Current upwelling system.

The Canary Current upwelling reaches its southernmost extent in winter when the Trades are strongest (Figure 14.18). It then extends well past Cap Blanc, the separation point of the Canary Current from the African coast (Figure 14.2). The boundary between the westward turning Canary Current and the cyclonic circulation around the Guinea Dome marks the boundary between North Atlantic Central Water and South Atlantic Central Water, the water masses of the thermocline (which will be discussed in detail in Chapter 15). Low salinity South Atlantic Central Water is transported poleward with the surface current found along the coast of Mauritania. The undercurrent of the upwelling circulation is the continuation of this surface current. During summer when upwelling is restricted to the region north of Cap Blanc (21°N), poleward flow dominates the surface and subsurface layers south of Cap Blanc offshore and inshore; during winter it is restricted to subsurface flow along the continental slope. The depth

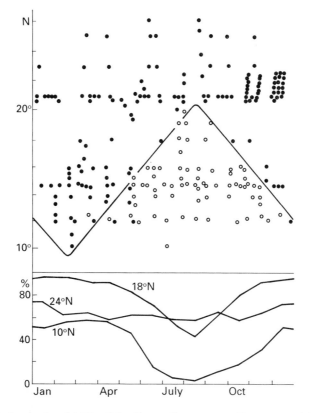

FIG. 14.18. Seasonal variability of the Canary Current upwelling system. (a) southern boundary of the upwelling region; dots mark observed upwelling, circles observed absence of upwelling. (b) frequency of occurrence of winds favourable for upwelling (wind direction is in the quarter between alongshore toward south and exactly offshore). Adapted from Schemainda et al. (1975).

of the undercurrent increases along its way to 300 - 600 m off Cape Bojador (27°N). In hydrographic observations it is evident as a salinity minimum caused by its high content of South Atlantic Central Water (Figure 14.19).

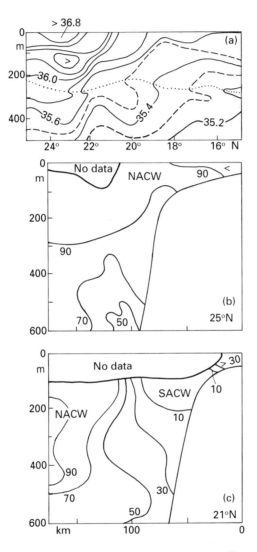

FIG. 14.19 The undercurrent of the Canary Current upwelling system as seen in hydrographic observations. (a) a salinity section along the continental slope, showing saline North Atlantic Central Water north and low salinity South Atlantic Central Water south of 20 - 22°N and the salinity anomaly on the σ_θ = 26.8 density surface (thin dotted line) caused by advection of SACW, (b) distribution of water masses in a section across the shelf and slope at 25°N, expressed as % NACW content (SACW content is 100 - %NACW), (c) a similar section at 21°N. The data for (a) were collected in April 1969, the data for (b) and (c) in February 1975. Note that the undercurrent is already well submerged at 21°N during 1969 but still close to the surface at 21°N in 1975. Adapted from Hughes and Barton (1974) and Tomczak and Hughes (1980).

A rather unique coastal upwelling region is found along the coasts of Ghana and the Ivory Coast where the African continent forms some 2000 km of zonally oriented coastline. Winds in this region are always very light and never favourable for upwelling. The sea surface temperature, however, is observed to drop regularly by several degrees, for periods of 14 days during northern summer (Figure 14.20). These temperature variations are coupled with reversals of the currents on the shelf, periodic lifting of the thermocline, and advection of nutrient-rich water towards the coast. The upwelling, which is clearly not related to local wind

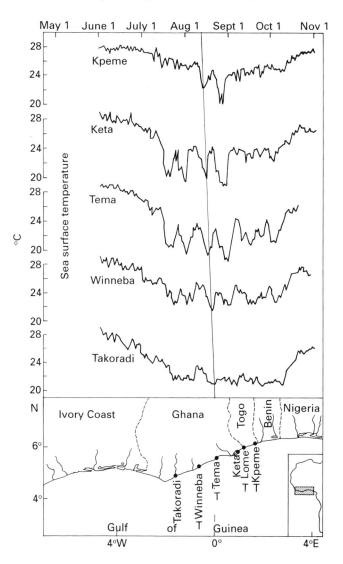

Fig. 14.20. Sea surface temperature (°C) as observed in 1974 at various locations in the Gulf of Guinea, showing periodic upwelling caused by waves of 14 day period during summer. Note the westward propagation indicated by the tilt of the line through the temperature minima. Adapted from Moore *et al.* (1978)

conditions, is believed to be caused by variations in the wind field over the *western* equatorial Atlantic Ocean which produce wave-like disturbances of the thermocline in the equatorial region known as Kelvin waves. Equatorial Kelvin waves are a major component of interannual variations in the circulation of the Pacific Ocean; a detailed discussion of their dynamics is therefore included in Chapter 19. For the purpose of the present discussion it is sufficient to note that they consist of a series of depressions and bulges of the thermocline, move eastward along the equator at about 200 km per day, and when reaching the eastern coastline continue poleward. The progression of the thermocline bulges and depressions is of course linked with significant horizontal transport of water, i.e. variations in the currents.

In the Atlantic Ocean, equatorial Kelvin waves generated off the coast of Brazil reach the Gulf of Guinea in little more than one month. They continue northward and then eastward along the African coast where they are recorded as strong regular upwelling events. For the local fishery they are of great importance, since they replenish the coastal waters with nutrients by lifting the nutrient-rich waters of the oceanic thermocline onto the shelf.

CHAPTER 15

Hydrology of the Atlantic Ocean

The hydrology of the Atlantic Ocean basins is deeply affected by the formation and recirculation of North Atlantic Deep Water, which was discussed in Chapter 7. The injection of surface water into the deeper layers is responsible for the high oxygen content of the Atlantic Ocean. It is also intricately linked with high surface salinities, as will become evident in Chapter 20. When compared with other ocean basins, the basins of the Atlantic Ocean are therefore characterized by relatively high values of salinity and dissolved oxygen.

Precipitation, evaporation, and river runoff

Precipitation over the Atlantic Ocean varies between 10 cm per year in the subtropics, with minima near St. Helena and the Cape Verde Islands, and more than 200 cm per year in the tropics. The region of highest rainfall follows the Intertropical Convergence Zone (ITCZ) in a narrow band along 5°N. A second band of high rainfall, with values of 100 - 150 cm per year, follows the path of the storm systems in the Westerlies of the North Atlantic Ocean from Florida (28 - 38°N) to Ireland, Scotland, and Norway (50 - 70°N). In contrast to the situation in the Pacific Ocean, no significant decrease in annual mean precipitation is observed from west to east; however, rainfall is not uniform across the band through the year. Most of the rain near Florida falls during summer, whereas closer to Europe it rains mainly in winter. A third band of high rainfall with similar precipitation values is associated with the Westerlies of the South Atlantic Ocean and extends along 45 - 55°S.

The precipitation-evaporation balance (P-E; Figure 1.7) reflects the rainfall distribution closely, since over most of the region evaporation varies much less than precipitation. The influence of the ITCZ is seen as a region of positive P-E values north of the equator. In the vicinity of South America the region extends to 30°S and along the coast of Panama, a result of extreme annual mean rainfall conditions over land. A similar southward extension is found near the African continent. The band of high rainfall in the northern hemisphere Westerlies is also evident as a region of positive P-E balance; its counterpart in the southern hemisphere does not come out very well due to lack of data.

Compared to the Pacific and Indian Oceans, the total downward freshwater flux (i.e. the P-E balance averaged over the ocean area) is obviously smaller in the Atlantic than in the other two oceans. Maximum $P - E$ values are considerably less, and the areas with less than 100 cm per year cover a proportionately much larger area. The effect on the sea surface salinity is somewhat alleviated by the fact that the land drainage area of the Atlantic Ocean is much larger; it includes nearly all of the American continent, Europe, large parts of Africa, and northern Asia (Siberia). Many of the world's largest rivers — including the Amazon, Orinoco, Mississippi, St. Lawrence, Rhine, Niger, and Congo Rivers — empty into the Atlantic Ocean, others — the Nile, Ob, Jenisej, Lena, and Kolyma Rivers — into its mediterranean seas. In these adjacent seas river runoff plays an important role in the salinity balance and consequently influences their circulation. Overall, however, the contribution from rivers to the freshwater flux of the Atlantic Ocean cannot compensate for the low level of rainfall over the sea surface.

Sea surface temperature and salinity

As noted earlier, the map of sea surface salinity (SSS; Figure 2.5b) resembles the *P-E* distribution (Figure 1.7) outside the polar and subpolar regions. Poleward of the Westerlies, the SSS values decrease further, despite the decrease in rainfall and *P-E* values, as a result of freshwater supply from glaciers and icebergs. In the north Atlantic Ocean this effect is concentrated in the west and linked with advection by the East and West Greenland Currents and the Labrador Current. This produces a sharp salinity increase across the boundary between the Labrador Current and the Gulf Stream (the Polar Front). A similar effect is seen in the southern hemisphere along the boundary between the Malvinas and Brazil Currents. The low SSS values along South Africa and Namibia, on the other hand, are the result of Indian Ocean water extrusions from the Agulhas Current, which were discussed in Chapters 11 and 12.

In the subtropics, the water with high salinities flows westward with the North and South Equatorial Currents. Continuous evaporation along its way increases the surface salinity further. The SSS maxima are therefore shifted westward relative to the *P-E* maxima. In the southern hemisphere this process continues into the South American shelf, where the Brazil Current advects the high salinity water southward. In the northern hemisphere the region of high surface salinity does not reach the American shelf because the North Brazil, Guyana, and Antilles Currents carry tropical water of low salinity across the equator into the northern hemisphere. This water is needed to feed the Gulf Stream and its extensions and is therefore not available to dilute the waters to the east. (This contrasts with the situation in the Pacific Ocean, where the low salinity water of the western region is recirculated in the Equatorial Countercurrent.) The highest sea surface salinities of the world ocean(not including mediterranean seas) are therefore found in the region of the Canary and North Equatorial Currents.

The distribution of sea surface temperature (SST; Figure 2.5a) shows similarities with the Pacific Ocean, particularly in the southern hemisphere where surface temperatures are much the same in the central parts of both oceans. Advection by the Brazil Current and upwelling along the Namibian coast are responsible for the marked SST differences between west and east in the subtropics. The thermal equator is at about 5°N and coincides with the Doldrums or ITCZ. In the west it extends northward into the Gulf of Mexico. The region of weak and variable winds is limited to the narrow band of the ITCZ; there is no analogue to the large region of extremely light winds found in the region north of Papua New Guinea. This is probably the reason why maximum SST values in the Atlantic Ocean are 2°C lower than in the Pacific Ocean. The contouring interval of Figure 2.5a shows the highest temperatures as above 26.0°C; actual annual mean SST values are in fact above 27.0°C over most of the region.

The major feature of the SST distribution is the marked departure of the isotherms from a zonal distribution and the associated crowding along the Polar Front in the northern hemisphere. The temperature difference between the east and west coasts north of 40°N and its consequences for the local climate have often been noted. In fact, the SST difference between the shelves off northern Japan and Oregon is only marginally smaller than the SST difference between the shelves off Newfoundland and France (about 6°C and 8°C, respectively). But in the Pacific Ocean this difference develops over more than twice the zonal distance available in the Atlantic Ocean, and the isotherms cross the latitude circles at a much smaller angle. The departure from zonal isotherm orientation in the North Atlantic Ocean is enhanced by water exchange with the Arctic Mediterranean Sea; the 5°C isotherm (and the 35 isohaline) angles across the ocean basin from 45°N near Newfoundland to 72°N off Spitsbergen.

Abyssal water masses

When compared with the other two oceans, the abyssal layers of the Atlantic Ocean display a hydrographic structure full of texture and variety. This results mainly from water exchange with mediterranean basins, particularly the Arctic and Eurafrican Mediterranean Seas.

Below 4000 m depth, all Atlantic Ocean basins are occupied by *Antarctic Bottom Water* (AABW). This water mass spreads northward from the Circumpolar Current and penetrates the basins east and west of the Mid-Atlantic Ridge. On the eastern side its progress comes to a halt at the Walvis Ridge; but on the western side it penetrates well into the northern hemisphere past 50°N. A map of potential temperature below 4000 m depth (Figure 15.1) shows a gradual temperature increase from the Southern Ocean to the Labrador Basin through mixing with the overlying waters. It also indicates how Antarctic Bottom Water enters the eastern basins north of the Walvis Ridge from the equator by passing through the Romanche Fracture Zone (Figure 8.2). As a result, potential temperature increases slowly both northward and southward from the equator in the eastern basins, and potential temperatures north and south of the Walvis Ridge differ by more than 1°C.

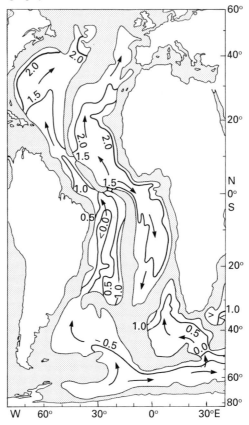

Fɪɢ. 15.1. Potential temperature below 4000 m depth and inferred movement of
Antarctic Bottom Water. Adapted from Wüst (1936).

The occurrence of *Arctic Bottom Water* (ABW) in undiluted form is restricted
to the immediate vicinity of the Greenland-Iceland-Scotland Ridge. As already
discussed in Chapter 7 its main impact is its contribution to the formation of
North Atlantic Deep Water (NADW) which fills the depth range between 1000 m
and 4000 m. In vertical sections (Figure 15.2) it is seen as a layer of relatively
high salinity (above 34.9) and oxygen (above 5.5 ml/l) extending southward from
the Labrador Sea to the Antarctic Divergence. More detailed inspection reveals
two oxygen maxima in the subtropics, at 2000 - 3000 m and 3500 - 4000 m depth,
indicating the existence of two distinct Deep Water varieties. The upper maximum
can be traced back to the surface near 55°N and reflects the spreading of NADW
formed by mixing Arctic Bottom Water with the product of deep winter
convection in the Labrador Sea. The lower maximum has its origin in the
Greenland-Iceland-Scotland overflow region and indicates that some Deep Water
is formed before the Arctic Bottom Water reaches the Labrador Sea, through
mixing of overflow water with the surrounding waters. East of the Mid-Atlantic
Ridge this is the only mechanism for the formation of Deep Water, and this
NADW variety, variably known as lower or eastern NADW, is therefore particularly

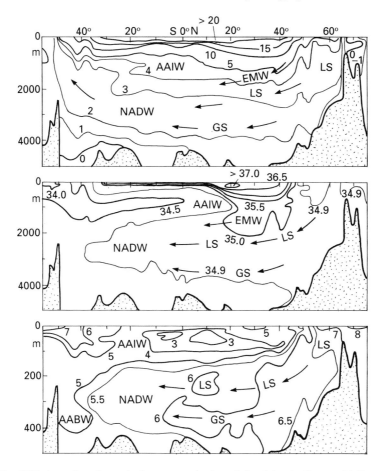

FIG. 15.2. A section through the western basins of the Atlantic Ocean. (a) Potential temperature (°C), (b) salinity, (c) oxygen (ml/l). See Fig. 15.7 for position of section. AABW: Antarctic Bottom Water, AAIW: Antarctic Intermediate Water, NADW: North Atlantic Deep Water originating from the Labrador Sea (LS) or the Greenland Sea (GS) or containing a contribution of Eurafrican Mediterranean Water (EMW). Adapted from Bainbridge (1980).

prominent in the eastern basins. Deep Water of Labrador Sea origin, referred to as middle or western NADW, is less dense than the eastern variety, and the two varieties remain vertically layered along their southward paths.

The long-term stability of NADW properties depends on the degree of atmospheric and oceanic variability during its formation period. A section through the Labrador Sea (Figure 15.3) shows a huge volume of nearly homogeneous water, with temperatures of 3.0 - 3.6°C and salinities of 34.86 - 34.96 and consistently high oxygen content, surrounded by strong cyclonic circulation. This *Labrador Sea Water* is the product of deep convection during the winter months. Observations show that deep winter convection is not an annual event; Clarke and Gascard (1983) report the formation of 10^5 km^3 of water with 2.9°C and 34.84 salinity in 1976 but virtually no formation of new water in 1978. Present

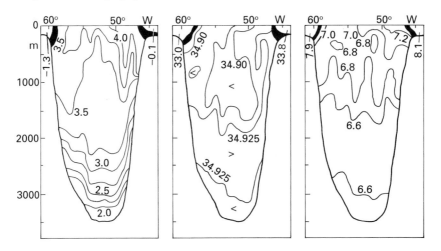

FIG. 15.3. A section through the southern Labrador Sea along approximately 60°N.
(a) temperature (°C), (b) salinity, (c) oxygen (ml/l). Data from Osborne
et al. (1991).

estimates are that convection occurs in 6 out of 10 years. This produces significant variation in the properties of Labrador Sea Water; for the years 1937, 1966, and 1976 Clarke and Gascard report respective T-S combinations of 3.17°C and 34.88, 3.4°C and 34.9, and 2.9°C and 34.84. However, not all of this variability is passed on to the NADW since mixing of Labrador Sea Water with Arctic Bottom Water does not occur during a single passage through the Labrador Sea. Variations in the rate of production of Labrador Sea Water are therefore averaged over the number of loops the Arctic Bottom Water performs around the area. A rough calculation based on the volume transports and velocities of the last chapter gives 2 - 3 loops, performed over 12 - 18 months. This suggests that interannual variations of Labrador Sea Water properties are transmitted into NADW at about half the original magnitude.

A third variety of North Atlantic Deep Water sometimes found in the literature as upper NADW is really NADW from the Labrador Sea (middle or western NADW) with traces of *Eurafrican Mediterranean Water* (EMW). This water mass leaves the Strait of Gibraltar with a temperature of about 13.5°C and a salinity of 37.8; but within less than 250 km its temperature and salinity are reduced by mixing to 11 - 12°C and 36.0 - 36.2. Starting from these characteristics EMW spreads isopycnally across the ocean, mixing gradually with the Deep Water above and below. Relative to NADW of the same density it has anomalously high salinities and temperatures. Figure 15.4 shows it as a salinity and temperature maximum at 1000 m depth near the upper distribution limit of NADW. The Mediterranean Water is carried northward along the Portuguese shelf under the influence of the Coriolis force and mixes into the subtropical gyre circulation, eventually spreading southward and westward. The core of the salinity and temperature anomaly sinks as the water spreads, and at the 2000 m level (Figures 2.5e and f) traces of Mediterranean Water can be seen all across the North Atlantic

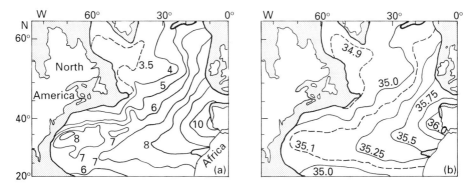

FIG. 15.4. Temperature (°C) (a) and salinity (b) in the North Atlantic Ocean at 1000 m depth.

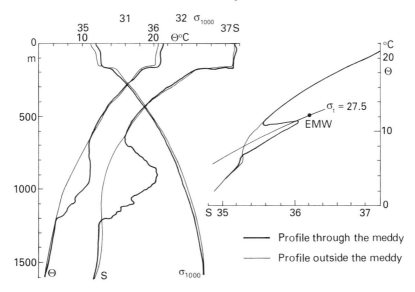

Profile through the meddy

Profile outside the meddy

FIG. 15.5. An example of a "meddy", a rotating lense of Mediterranean Water found some 2500 km south-west of the Strait of Gibraltar near 26°N, 29°W. Note that the density profiles inside and outside are nearly identical. (σ_t gives density at atmospheric pressure, σ_{1000} at a pressure equivalent to 1000 m depth.) Adapted from Armi and Stommel (1983).

Ocean, with some high salinity water crossing the equator in the west and proceeding southward.

Although the influence of the Mediterranean Water is strong enough to put its mark on the long-term mean distribution of oceanic properties, it is wrong to imagine the spreading of EMW as a process of smooth isopycnal movement with equally smooth diapycnal diffusion. It has to be remembered that the eastern basins of all oceans are characterized by slow mean motion but high eddy activity. Mediterranean Water is therefore injected into the NADW in the form of subsurface eddies, rotating lenses which contain a high proportion of Mediterranean Water in its core. The rotation shelters EMW from the surrounding

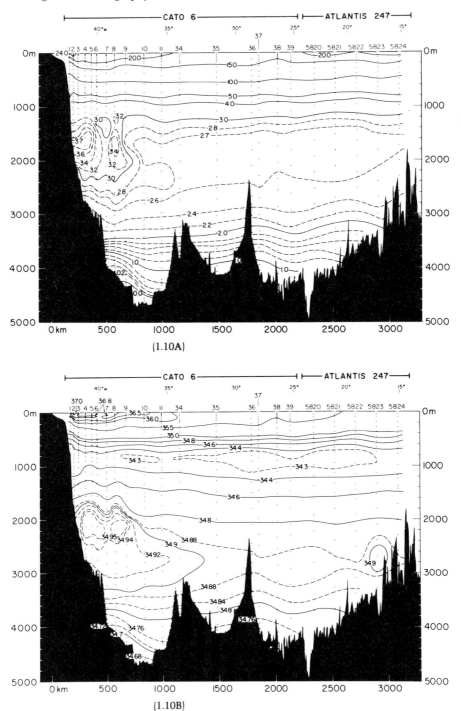

FIG. 15.6. A section through the western Atlantic Ocean along 30°S. (a) Potential temperature (°C), (b) salinity. From Warren (1981a).

NADW; it prevents mixing and keeps the lense together over large distances. Lenses of Mediterranean Water, often referred to as "meddies" (Figure 15.5), have been found as far afield as in the Sargasso Sea. Direct observations in eddies from the Canary Current region showed rotational velocities of 0.2 m s^1 in general southward movement of about 0.05 m s^1. The salinity and temperature anomalies found in the long-term average distribution have to be seen as the result of a process in which many such meddies travel through the upper NADW range at any particular moment in time, slowly releasing their load of extra salt and heat into the surrounding Deep Water.

Along the western boundary of the ocean the mean flow of Deep and Bottom Water becomes stronger than eddy-related movement and can therefore be seen in hydrographic sections. Figure 15.6 shows the Deep Water as a salinity maximum at 2000 - 3000 m and a temperature maximum at 1400 - 2000 m, concentrated against the South American shelf. Both features are nearly 1000 km wide, probably wider than the associated boundary current as a result of mixing. Intensification of Antarctic Bottom Water flow is indicated by the shape of the isotherms and isohalines below 4000 m; upward slope towards the coast is consistent with a northward "thermal wind" increasing in speed with depth (Rule 2a of Chapter 3). A similar intensification occurs on the western side of the Cape Basin along the Walvis Ridge. Because the basin is closed in the north below the 3000 m level, the flow follows the depth contours in cyclonic motion, and the Bottom Water leaves the basin on the eastern side towards the Indian Ocean (Nelson, 1989). The current is swift enough to remove sediment along the base of the continental rise and produce a band of exposed rockface at 5000 m depth.

Antarctic Circumpolar Water has the same density as North Atlantic Deep Water but is colder and fresher (Figure 6.13). In the absence of NADW it would take its place in the Atlantic Ocean; however, the southward advance of the Deep Water reduces its influence. Detailed analysis (Reid, 1989) shows northward propagation of some Circumpolar Water both below and above the Deep Water.

Above the Deep and Circumpolar Waters is the Intermediate Water, characterized as in the other oceans by its low salinity. Figure 15.7 gives the depth of the salinity minimum produced by the spreading of this water mass and the salinity at that depth. The outstanding feature is the pronounced lack of symmetry relative to the equator. The dominant water mass is the *Antarctic Intermediate Water* (AAIW). Formed mostly in the eastern south Pacific and entering into the Atlantic Ocean through Drake Passage and with the Malvinas Current, it spreads isopycnally into the northern hemisphere. Concentration of its flow along the western boundary is indicated by the northward extension of the isohalines with the Guyana and Antilles Currents and in Figure 15.6 by the widening of the isohalines around the salinity minimum at 1000 m depth. In the eastern basins its movement is masked by eddies, particularly Agulhas Current eddies propagating northward (Chapter 11). Observations along the south African continental rise near 1000 m depth (Nelson, 1989) indicate that AAIW participates in the cyclonic motion

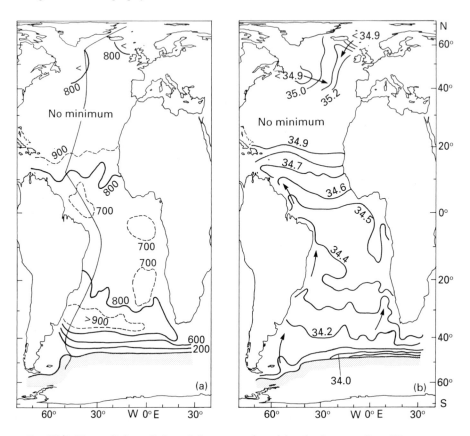

FIG. 15.7. Maps of the salinity minimum produced by the Intermediate Waters.
(a) Depth of the minimum, (b) salinity at the depth of the minimum. The location
of the section shown in Fig. 15.2 is also indicated in (a). After Wüst (1936) and
Dietrich *et al.* (1980).

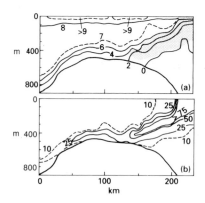

FIG. 15.8. A section across the Iceland-Scotland Ridge 150 km northwest of the Faroe
Islands showing the spreading of Arctic Intermediate Water. (a) Temperature (°C),
with the region of low salinity (<34.9) shaded; (b) presence of Arctic Intermediate
Water (percentage of volume). Adapted from Meincke (1978).

of abyssal water masses in the Cape Basin, even though the Walvis Ridge does not pose a barrier to flow at AAIW level.

As discussed in Chapter 6, formation of AAIW in the Atlantic Ocean itself occurs through water mass conversion in the Southern Ocean with limited direct atmospheric contact. The only region where winter convection contributes to AAIW formation is in the Scotia Sea. Most of the AAIW enters from a formation region in the eastern Pacific Ocean (England *et al.*, in press). As a consequence, Atlantic AAIW differs little from AAIW in the other two oceans. Close to the formation region it has a temperature near 2.2°C and a salinity of about 33.8. Mixing with water from above and below erodes the salinity minimum; by the time AAIW reaches the Subtropical Convergence it has properties closer to 3°C in temperature and about 34.3 in salinity. The gradual weakening and eventual disappearance of the minimum towards north can also be seen in the T-S diagrams of Figure 15.9.

The occurrence of *Arctic Intermediate Water* (AIW) is restricted to two small regions in the north, a western variety formed in the southern Labrador Sea at temperatures near 3°C and 34.5 salinity, and an eastern variety which originates in the Iceland Sea at temperatures below 2°C and near 34.6 salinity. Both are subducted in locations along the Polar Front of the north Atlantic Ocean, in the west at the boundary between the Gulf Stream and the Labrador Current and in the east along a frontal region between the North Atlantic and East Iceland Currents. Their influence on the hydrography is limited by their proximity to the formation regions for NADW, which absorbs their low salinities over short distances. The eastern variety in particular cannot be recognized much beyond the sill where it sinks; its salinity minimum does not extend past 60°N (Figure 15.7). The influence of the western variety is felt most strongly in the North Atlantic Current, but its salinity minimum can be traced into the Bay of Biscay and southward to 40°N.

Because of the rapid absorption of Arctic Intermediate Water into the Deep and Bottom Water complex the existence of Intermediate Water in the north Atlantic Ocean is often ignored. A hydrographic section across the Iceland-Scotland Ridge (Figure 15.8), however, gives clear evidence that this water fits our definition of a water mass as a body of water with a common formation history: The Arctic Bottom Water, with salinities near 35.0 and temperatures below 3°C, is retained behind the sill and enters the Atlantic Ocean episodically, while the Intermediate Water sinks from the surface and is continuously subducted.

Water masses of the thermocline and surface layer

Two well-defined water masses occupy the Atlantic thermocline. Both are characterized by nearly straight T-S relationships. A south to north succession of T-S curves (Figure 15.9) shows a sudden shift to lower salinities some

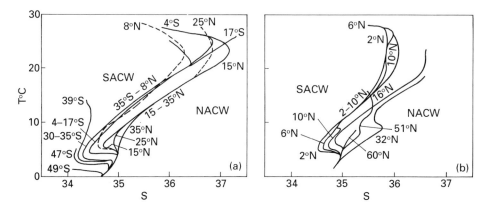

FIG. 15.9. T-S diagrams for stations along two meridional sections. (a) Western basins,
(b) eastern basin (northern hemisphere). Note the northward weakening of the
AAIW salinity minimum, the deep salinity maximum produced by the inflow of
Eurafrican Mediterranean Water (most prominently at 32°N in the east), and   the
sudden transition from SACW to NACW south of 15°N. Data from
Osborne *et al.* (1991).

1500 km north of the equator, indicating different hydrographic properties north
and south of about 15°N.

South Atlantic Central Water (SACW), the water mass south of 15°N, shows rather
uniform properties throughout its range. Its T-S curve is well described by a
straight line between the T-S points 5°C, 34.3 and 20°C, 36.0 and is virtually the
same as the T-S curves of Indian and Western South Pacific Central Water. This
reflects the common formation history of all Central Waters in the southern
hemisphere, which are subducted in the Subtropical Convergence (STC).
Although the STC is well defined and continuous across the south Atlantic Ocean,
detailed comparison between the T-S relationship along a meridional track across
the STC with the T-S curve of SACW (in the manner described with Figure 5.4)
reveals that the T-S properties of SACW in the tropics are closer to those typical
for the Subtropical Convergence of the western Indian Ocean near 60 - 70°E,
than those found along the STC in the Atlantic Ocean (Sprintall and Tomczak,
1993). This indicates that much of the SACW is not subducted at the Atlantic
portion of the STC but is in fact Indian Central Water (ICW) brought into the
Atlantic Ocean by Agulhas Current eddies (see Chapter 11), in agreement with
the ideas of North Atlantic Deep Water recirculation discussed in Chapter 7.
Mixing in the eddy separation region and possibly in the Agulhas Current itself
does not change the T-S characteristics of the inflowing ICW but redistributes
the contributions of the water types which make up the T-S curve, enhancing
in particular the volume of water near 13°C. This water type, also known as
13° Water (Tsuchiya, 1986), thus turns into a variety of Subtropical Mode Water;
the associated thermostad can be traced from Namibia to the coast of Brazil near
10°S, along the North Brazil Current and into the eastward flowing components

of the equatorial current system. It is worth noting that, unlike other Subtropical Mode Water varieties, 13° Water is not formed in contact with the atmosphere.

Some Central Water formation does occur in the western south Atlantic Ocean, in the confluence zone of the Brazil and Malvinas Currents (Gordon, 1981). It is responsible for a high salinity variety of SACW (36 salinity is reached at 17°C instead of 20°C; see Figure 15.9). This SACW variety is recirculated within the southern subtropical gyre and therefore restricted to the western south Pacific Ocean.

North Atlantic Central Water (NACW) can again be characterized by a nearly straight line in the T-S diagram, with some variation of the T-S relationship within the water mass. Typically, the T-S curve connects the T-S point 7°C, 35.0 with the points 18°C, 36.7 in the east and 20°C, 36.7 in the west. The regional differences stem from property variations in the formation region. Although curl(τ/f) is negative over most of the north Atlantic Ocean (Figure 4.3), a subtropical convergence as a region of more or less uniform subduction of Central Water from the surface cannot be identified. This is apparently because the subtropical gyre extends further north in the eastern north Atlantic than in the north Pacific Ocean, allowing the surface waters of the North Atlantic Current to cool much more during their northeastward passage than those of the North Pacific Current. Winter convection in the north Atlantic Ocean therefore reaches much deeper. Figure 15.10 shows that at the end of winter it reaches the bottom

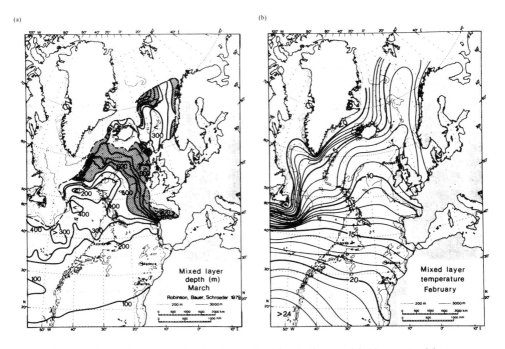

FIG. 15.10. Winter convection in the north Atlantic Ocean. (a) Mean mixed layer depth (m) in March, (b) mean mixed layer temperature (°C) inFebruary. Adapted from Robinson *et al.* (1979).

of the thermocline along the European shelf as far south as northern Spain. This affects all temperatures below 12°C. As a consequence, NACW enters the thermocline at those temperatures by a process of horizontal "injection" rather than isopycnal subduction, and its properties are influenced by water masses from the Arctic Mediterranean and Labrador Seas which participate in the convection. The deepest convection occurs in the northeast, and the corresponding vertical transfer of salt from the surface layer reaches its greatest depth there, raising the salt content of the eastern NACW variety.

Above 12°C NACW is formed through the usual process, i.e. surface subduction of winter water. Again, this process does not occur uniformly across the region but involves Mode Water formation. Large volumes of Central Water are formed every winter in the Sargasso Sea at temperatures around 18°C. They appear in vertical temperature profiles as a permanent thermostad at 250 - 400 m depth (Figure 15.11) and represent a variety of Subtropical Mode Water known as the *18° Water*. A third variety, the *Madeira Mode Water*, is formed north of Madeira and indicated by a summer thermostad at 70 - 150 m depth (Figure 15.11). Both mode waters together contribute more than half the volume of NACW.

The transition from South to North Atlantic Central Water occurs as a front along approximately 15°N which extends from below the mixed layer to the bottom of the thermocline. In the east it bends northward past 20°N, following the southern limit of the Canary Current and sharpened by the confluence with the circulation around the Guinea Dome. In general, SACW penetrates northward underneath NACW (see Chapter 14 for details), giving the front a downward slope from south to north. With SACW and NACW occupying the same density range, the front is density-compensated, i.e. the effect of the temperature change across the front is compensated by the effect of the salinity change and the front

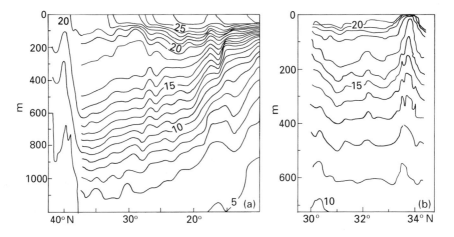

FIG. 15.11. Temperature sections indicating the presence of Subtropical Mode Water. (a) Through the Sargasso Sea along 50°W, (b) near Madeira along approximately 18°W. Note the increased isotherm spacing (thermostad) in the 17 - 18°C range. Adapted from McCartney (1982) and Siedler *et al.* (1987).

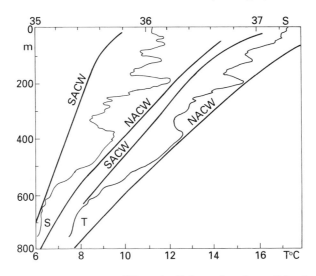

FIG. 15.12. (Left) Temperature (°C) and salinity as functions of depth at a station in the eastern part of the water mass boundary between SACW and NACW, showing strong interleaving. Reference curves for undisturbed SACW and NACW conditions are indicated. The station is located some 25 km west of the position marked in Fig. 14.17. From Tomczak and Hughes (1980).

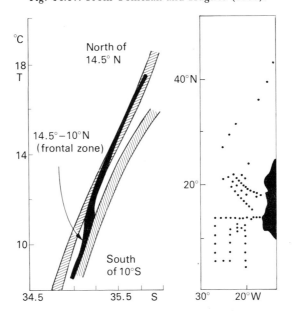

FIG. 15.13. (Right) T-S relationships in the South Atlantic Central Water of the eastern tropical Atlantic Ocean, showing two SACW varieties separated by a frontal zone between the North Equatorial Current and the Equatorial Countercurrent. The stations used for the construction of the T-S diagram are shown on the right.

is not noticed in the density field. Parcels of water from either side of the front can therefore be moved easily across the front on isopycnal surfaces. The resulting multitude of intrusions, filaments, and lenses (Figure 15.12) makes the structure of the front quite complicated. Further west the front loses its identity, as mixing between SACW and NACW in the North Equatorial Current erodes the horizontal gradients. Eventually the mixture is carried north in the Guyana and Antilles Currents to complete the route of SACW from the Agulhas Current eddies to the formation region of North Atlantic Deep Water.

Because the separation zone between both Central Waters is located more than 1500 km north of the equator and the SACW/NACW mixture produced in the west is not returned into the equatorial current system but transported northward, there is no opportunity to form a special equatorial water mass in the manner seen in the Pacific Ocean. The opposing eastward and westward equatorial flows leave, however, their mark in the hydrographic properties of SACW. Observations from the region of the Guinea Dome show a small but well-defined salinity increase, from south to north, across the boundary between the North Equatorial Countercurrent and the North Equatorial Current near 10°N (Figure 15.13). The front between the two SACW varieties slopes downward toward the north, the low salinity variety moving eastward above westward movement of the high salinity variety.

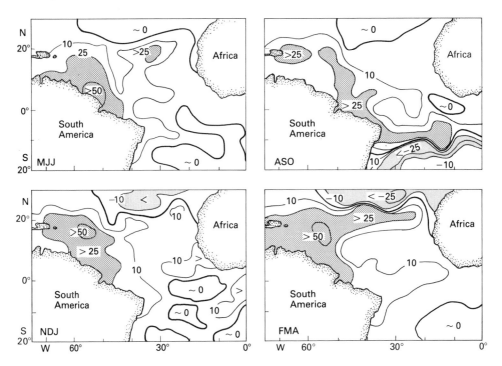

Fɪɢ. 15.14. Seasonal mean barrier layer thickness (m) in the tropical Atlantic Ocean. (a) May - July, (b) August - October, (c) November - January, (d) February - April. The barrier layer is located below the mixed layer (see Fig. 5.6 for mixed layer depth). Contours are given for layer thickness of 50 m, 25 m, 10 m, 0 m, -10 m, and -25 m. Subduction regions, indicated by values less than -10 m, are lightly shaded, regions with a barrier layer thickness >25 m are shown with dark shading. Adapted from Sprintall and Tomczak (1992).

Main aspects of the hydrographic structure above the permanent thermocline were already discussed in an earlier section of this chapter. A major feature of the tropical Atlantic Ocean is the existence of a barrier layer in the region of the Guyana Current (Figure 15.14). This region is characterized by net water loss to the atmosphere (the major region where rainfall exceeds evaporation being east of 40°W; Figure 1.7). Local freshening of the surface layer, the mechanism that produces the barrier layer in the Pacific Ocean, can therefore not be responsible here. It appears that the high salinities found at the surface in the subtropics (Figure 2.5b) are subducted towards the equator at the upper end of the temperature/salinity range of the Central Water. This creates a salinity maximum above the Central Water in the tropics, which is then advected westwards towards the equator into regions of uniform temperature in the equatorial current system. The result is salinity stratification in the isothermal surface layer. Figure 15.14 shows active subduction (indicated by negative barrier layer thickness; see Chapter 5 for an explanation of the mechanism) for August - October south of 12°S, coupled with the formation of a barrier layer to the north. During February - April the same process occurs in the vicinity of 20°N. The two sources alternate in renewing the barrier layer structure in the west, where the barrier layer is found during all seasons. In this region an accurate heat and mass budget of the surface layer cannot be achieved without taking advection into account.

A description of the hydrographic conditions in the shelf regions of the Atlantic Ocean is beyond the scope of this book, but one region deserves mention. The large volume of water between the Gulf Stream and the continental shelf is isolated from direct contact with the oceanic water masses of its depth range by the western boundary current. Its properties are formed through a complex process of interaction between water on the shelf, from the Labrador Current,

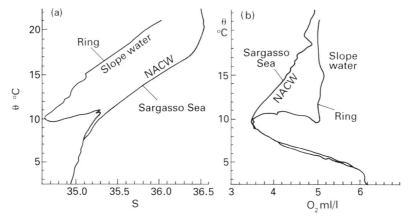

Fɪɢ. 15.15. An example of Slope Water advection into the Sargasso Sea in a Gulf Stream ring. (a) Θ-S diagram, (b) Θ-oxygen diagram, for a station outside ("Sargasso Sea") and inside ("Ring") a cyclonic Gulf Stream ring. Higher oxygen concentration in the Slope Water indicates more recent contact with the atmosphere. From Richardson (1983b).

and from the Gulf Stream. Water on the shelf has very low salinity (below 34, a result of freshwater inflow from the St. Lawrence River). The Labrador Current also carries low salinity water. Mixing of the various components produces a water mass known as Slope Water, which extends over the upper 1000 m along the north American continental rise north of Cape Hatteras (35°N) and is characterized by a nearly linear T-S relationship similar to that of NACW but with much lower salinity. This water is frequently trapped in cyclonic Gulf Stream Rings and transported across the Gulf Stream into the Sargasso Sea, as seen in the example of Figure 15.15. As a result, variations of hydrographic properties in the permanent thermocline of the Atlantic Ocean are largest in the northern Sargasso Sea.

CHAPTER 16

Adjacent seas of the Atlantic Ocean

The hydrography of the Atlantic Ocean is strongly influenced by its adjacent seas, to the degree that a discussion of this ocean would be incomplete without a description of the hydrographic conditions of its adjacent seas. This is particularly true for the Arctic Mediterranean Sea which plays a crucial role in the formation of Deep Water not only for the Atlantic but for the entire world ocean. Its importance for the global oceanic circulation justified treatment of this adjacent sea in a separate chapter (see Chapter 7).

The remaining adjacent seas can be divided on geographical arguments into four groups. The first group contains the waters connected to the Atlantic Ocean proper through the Labrador Sea and consists of the Davis Strait, Baffin Bay, the Northwest Passage, and Hudson Bay. The second group is located between Europe, Africa, and Asia and contains the Eurafrican Mediterranean Sea (which includes the Black Sea). The third group is found near the junction of North and South America and contains the American Mediterranean Sea with its subdivisions the Caribbean Sea and the Gulf of Mexico. The shallow European seas make up the fourth and last group, which contains the Irish and North Seas and the Baltic Sea with its approaches.

Davis Strait, Baffin Bay, the Northwest Passage, and Hudson Bay

The passages between the Arctic Mediterranean and the Labrador Seas, variably known as the Northwest Passage, the North Water, or the Canadian Arctic Archipelago, consist of a maze of islands, channels, straits, and basins of widely different character. Baffin Bay is a deep basin with maximum depths in excess of 2300 m. It is separated from the Labrador Sea by Davis Strait, which has a sill depth of less than 600 m. The two major connections to the Arctic basins are through Nares Strait, the Kane Basin and Smith Sound to the north and through M'Clure Strait, Viscount Melville Sound, Barrow Strait and Lancaster Sound to the west (Figure 16.1). A third passage bifurcates from Lancaster Sound towards south and connects with the Labrador Sea through Prince Regent Inlet, the Gulf of Boothia, the Foxe Basin and Foxe Channel, and Hudson Strait. Sill depths in all three passages are quite shallow; in the northern passage the sill depth is just over 200 m in Smith Sound, while in the western passages it is less than 150 m in Barrow Strait and less than 100 m in the Foxe Basin. The various deep passages between the Queen Elizabeth Islands are blocked to the south by similarly shallow water and do not play a significant role in the water exchange between the Arctic and Labrador Seas.

The circulation in the region is determined by the West Greenland Current and the throughflow from the Arctic Mediterranean Sea. Both influences combine to drive the Baffin Current, a southward flow of about 0.2 - 0.4 m s^{-1} along the western side of Baffin Bay which supplies water to the Labrador Current (Figure 16.1). Flow through the passages is generally southward and westward, but eastward countercurrents are found on the northern sides of Lancaster Sound and Hudson Strait. The countercurrents appear to be more variable in time than the eastward flow along the southern coasts but of comparable strength (0.3 - 0.5 m s^{-1}).

The hydrography of *Baffin Bay* and *Northwest Passage* shows a distinct layering of water masses. Unfortunately a generally accepted nomenclature of water masses in the region does not exist, and some names found in the literature can be quite misleading. Two conspicuous features of the vertical temperature distribution in Baffin Bay are a temperature maximum at about 500 m depth below a minimum in the range 50 - 200 m (Figure 16.2). The maximum is the result of Atlantic water inflow with the West Greenland Current. This water is sometimes called Atlantic Intermediate Water on account of its position in the water column and the fact that it is brought into the region from the south. However, its origin is in the East Greenland Current and thus ultimately in the Arctic Mediterranean Sea, and there is no relation with what is normally called Intermediate Water (water subducted near the Polar Front and characterized by a salinity minimum). Another, more appropriate name for this water is Polar Atlantic Water.

The low temperature water above the Polar Atlantic Water is also of Arctic origin but advected from the north. It is drawn from the sub-surface layer of Arctic Surface Water (see Chapter 7) and modified by some injection of brine from the surface during winter freezing. Its salinity is therefore somewhat higher than the salinity of Arctic Surface Water at the same temperatures. This water is sometimes referred to as Arctic Intermediate Water but again not related to the Intermediate Water of the temperate zone.

Above the Arctic Intermediate Water is a thin surface layer of not more than 50 m depth where water properties change with the seasons. Summer temperatures vary between -0.1 - 5.0°C, and salinities are in the range 30.0 - 35.5.

Below the Atlantic Polar Water temperature decreases and salinity increases slowly until both become virtually constant in the Baffin Bay Bottom Water below 1800 m depth (-0.4°C and 34.49 salinity). The low oxygen content (3.6 ml/l) of this water testifies for the character of Baffin Bay as a small mediterranean dilution basin and correspondingly slow renewal of its Bottom Water. The details of the formation process are not entirely clear. Irregular discharge of Arctic water through Smith Sound is believed to contribute. There is also evidence that formation of cold, saline water in shallow regions of Smith Sound and subsequent

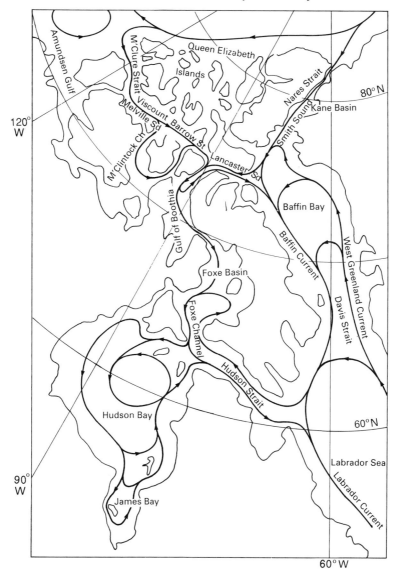

Fɪɢ. 16.1. Geography and circulation of the Northwest Passage and connected seas.

sinking along the continental slope, similar to the formation process of Arctic Bottom Water (Chapter 7), plays a major role (Bourke *et al.* 1989).

The hydrography of *Hudson Bay* is determined by its shallow water depth of on average only 250 m and by seasonally varying river discharge, which gives it the character of a large estuary. Since estuarine dynamics are not discussed in this book, a few remarks have to suffice. Foxe Basin and Hudson Bay are completely ice-covered during several months. The ice starts to break up from James Bay in June, and Hudson Bay and Foxe Basin are clear of ice by mid-August. Large amounts of freshwater are poured into the bay during this period.

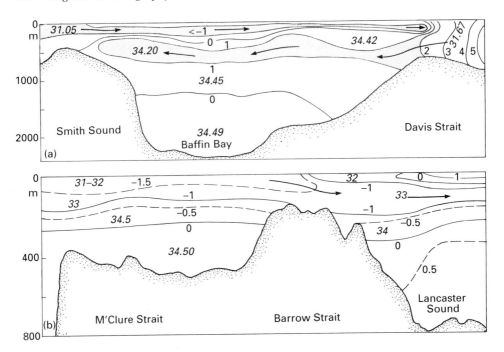

FIG. 16.2. Temperature (°C) in two sections through the Northwest Passage and Baffin Bay. (a) From Smith Sound to Davis Strait, (b) from M'Clure Strait to Lancaster Sound. Arrows indicate water movement. Typical salinities are indicated in *italics*. Adapted from Collin (1966).

As a consequence, temperature and salinity undergo large seasonal changes with a period of strong stratification during late spring and summer. This affects the circulation as well, so Figure 16.1 can only show the water movement as it prevails during two thirds of the year. Lack of ice movement during spring breakup indicates that there is little if any water movement before the stratification is established by the influx of meltwater. The effect of the runoff is felt most strongly towards October when flow across the entire northern bay is northward.

Ice coverage in other parts of Northwest Passage is similarly heavy, but satellite observations show that up to 10% of the region between Lancaster Sound, Smith Sound, and Greenland are ice-free at any time even through winter. The polynya are produced by offshore movement of ice in response to the prevailing wind. The dynamics are essentially those of coastal upwelling, but the situation is unique because in the upper few hundred metres temperature increases with depth. Upwelling thus brings warm Atlantic Polar Water to the surface and keeps the surface layer ice-free.

The Eurafrican Mediterranean Sea

The mediterranean basin between Africa, Europe, and Asia has always been regarded the prototype of a concentration basin, to the extent that it is mostly referred to not as the Eurafrican but simply *the* Mediterranean Sea. It consists of a series of deep basins mostly well connected with each other, the major exception being the Black Sea which has very limited communication with the other subdivisions (Figure 16.3). The mean depth of the Mediterranean Sea is near 1500 m; maximum depths in the various basins are between 2500 m and 5100 m (in a narrow trench off southwestern Greece). A sill between Sicily and Tunisia with a maximum depth near 400 m divides the region into the western and eastern Mediterranean Sea. Maximum depths in the two subdivisions are about 3400 m in the west and about 4200 m in the east (if the deep trench is excluded). The second connection between the western and eastern basins, the narrow Strait of Messina between Sicily and mainland Italy, has a sill depth of only 120 m and is of no significance for the general circulation; its reputation as a treacherous passage for ships stems from its strong tidal currents of 2 - 3 m s^{-1} ascribed to the two monsters Scylla and Charybdis in Homer's masterful account of the adventures of the ancient Greek navigator Odysseus.

Communication with the Atlantic Ocean proper is through the Strait of Gibraltar which is 22 km wide and has a sill depth of 320 m. This poses a severe limitation on the water exchange and in combination with the atmospheric conditions creates distinctive hydrographic conditions. During most of the year winds over the Mediterranean Sea are from the northwest and carry warm dry air, causing large evaporation. During winter the winds are often northeasterly, bringing dry but cold air into contact with the sea.

Over most of the Mediterranean Sea annual evaporation exceeds rainfall and river runoff by about 1 m, so on average the Mediterranean Sea is a concentration basin. The two exceptions are the Black and the Adriatic Seas which receive large amounts of freshwater from the Danube and Po rivers and therefore are dilution basins. Water exchange between the Atlantic Ocean proper, the western and

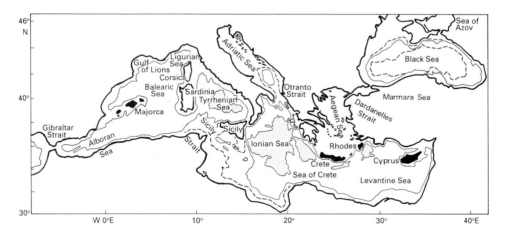

FIG. 16.3. Topography and subdivisions of the Eurafrican Mediterranean Sea. The 1000 m contour is shown, and regions deeper than 3000 m are shaded. The 200 m contour is shown as a broken line where it departs significantly from the 1000 m contour. In addition, the 2000 m contour is shown as a broken line in the Black Sea.

eastern Mediterranean Sea, and the Black Sea, which follows the principles discussed with Figure 7.1, therefore consists of inflow of Atlantic water through the Strait of Gibraltar in an upper layer, outflow of dense Mediterranean water below, inflow of relatively fresh Black Sea water through the Dardanelles in an upper layer, and outflow of Mediterranean water below (exchange with the Adriatic Sea is similar but not associated with strong currents because of the lack of topographic restriction between the Adriatic and Ionian Seas). An annual mean budget gives the following transports (where the actual numbers have been adjusted to give zero balance and do not imply accuracy to three digits):

inflow (Sv)		outflow (Sv)	
from the Atlantic	1.107	to the Atlantic	1.041
from the Black Sea	0.006		
precipitation	0.027	evaporation	0.111
river runoff	0.011		
total	1.152	total	1.152

Note that the total water exchange through Gibraltar Strait is about thirty times what is required to replace the water lost by evaporation. This is because flow between mediterranean seas and the open ocean is mainly driven by the density difference between the water masses on either side of the strait (which is, of course, the result of the freshwater balance in the mediterranean sea). Inflow velocities in excess of 1 m s^{-1} in combination with a rapidly shoaling bottom in a constricted passage result in a situation where normal ocean dynamics give way to hydraulic control of the flow; in other words, in the vicinity of the strait the flow axis and the depth of the interface between the layers are not controlled by geostrophy and deflection by the Coriolis force but by the same processes which govern the flow of water over a weir. The inflowing Atlantic water initially continues eastward as a free jet and breaks into one or two large eddies of 150 km diameter before the Coriolis force can deflect it to continue along the African coast (Figure 16.4). The changeover from the Spanish to the Algerian coast occurs in a narrow current associated with a front, known as the Almeria - Oran Front, which separates the relatively fresh Atlantic water from the salty Mediterranean water (Figure 16.5). The Atlantic inflow then continues as the Algerian Current (Figure 16.6), which for at least 300 km maintains the character of a narrow jet of less than 30 km width with average velocities of 0.4 m s^{-1}, maximum velocities of 0.8 m s^{-1}, and a total transport of about 0.5 Sv. Having advanced past 4°E the flow becomes more diffuse. In the eastern Mediterranean basins it is dominated by eddies, some large and stationary as indicated in Figure 16.6, others of only 40 - 60 km diameter but reaching to great depth. Observations have shown them to extend to at least 1000 m, with velocities exceeding 0.2 m s^{-1} above the 300 m level.

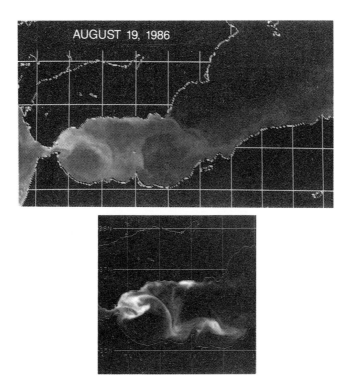

Fig. 16.4. Satellite thermal image of the Alboran Sea showing the inflow of cool Atlantic Water (white is cold, dark is warm). (a) With two eddies and the Almeria - Oran Front at 1° - 2°W, (b) with a single eddy and the front at 4°W. The two-eddy situation is far more frequent. The transition period from one to two eddies is about one month. From Tintore *et al.* (1988)

On its way east the Atlantic water encounters saltier but warmer and consequently less dense Mediterranean water. Outside the Algerian Current it therefore continues submerged under a shallow surface layer of high salinity and can be followed as a salinity minimum at 20 - 30 m depth. Being formed outside the Mediterranean Sea and identifiable throughout the basin, it can rightly be given the status of a water mass and is usually referred to as *Atlantic Water*.

The character of the Mediterranean Sea as a concentration basin requires that Atlantic Water is converted into denser water that eventually leaves the sea over the sill of the Strait of Gibraltar. This conversion process involves deep vertical convection during winter. It does not act uniformly in the entire mediterranean basin but occurs in three small regions which more than the remainder of the basin are affected by the cold northeasterly winter winds. Very cold air from Siberia is channelled through the valleys of the Alps and descends in bursts of strong winds known as mistral on the Ligurian Sea and the northern Balearic Basin; this is the region of origin for *Mediterranean Deep Water* (MDW), often also called Western Mediterranean Deep Water. Similarly cold winter winds descend on the region between Rhodes and Cyprus and on the northern and central

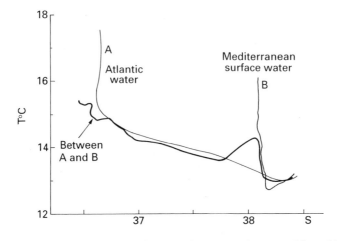

FIG. 16.5. Temperature - salinity diagrams for two stations on either side of the Almeria - Oran front (locations A and B in Fig. 16.6) and for a station inbetween. Note the layering and associated temperature inversion in the front. After Arnone *et al.* (1990).

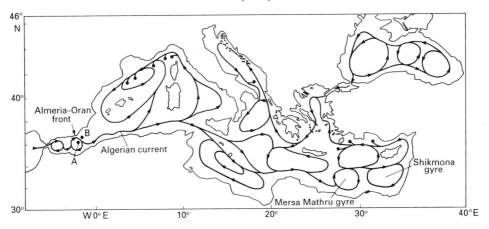

FIG. 16.6. Upper ocean currents in the Eurafrican Mediterranean Sea. Locations marked A and B in the Alboran Sea refer to the data shown in Fig. 16.5. Dots indicate areas of winter convection.

Adriatic Sea and are responsible for the formation of *Levantine Intermediate Water* (LIW). Both are characterized by high salinities, but the details of the formation mechanisms for the two water masses of the lower layer are very different and require some discussion.

The region where MDW is formed is generally under the influence of a cyclonic (anti-clockwise) wind system. This results in a cyclonic ocean circulation with Ekman transports directed outwards and upwelling in the centre. As a consequence the LIW, which is found at about 400 - 500 m depth to the south of the region, here rises to 150 - 200 m, increasing the salinity of the upper 200 m and reducing the salinity contrast over the water column. The result is a geographically well defined region of reduced vertical stability. Winter cooling

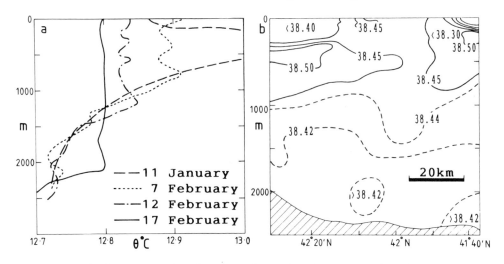

FIG. 16.7. Observations of Mediterranean Deep Water formation in the Balearic Sea. (a) Potential temperature (°C) over a one-month period in 1969 near 42°N, 5°E showing the deepening of the mixed layer to 2200 m, (b) a salinity section along 5°20'E showing a column of newly mixed water near 42°N. Adapted from Sankey (1973).

reduces the surface temperatures throughout the Mediterranean Sea and reduces the stability further; in the Ligurian Sea and the northern Balearic Basin it leaves the upper 200 m of the water column with only marginal stability. When a mistral event reaches the region, rapid additional cooling produces instability and vigorous sinking of the surface water. The sinking occurs in funnels not larger than a few tens of kilometres in diameter and is accompanied by a compensating rise of water from great depth on all sides (Figure 16.7). The water can sink some 800 m within a matter of hours and reach the 2500 m level within days (the maximum depth in the region is near 2900 m). Short and violent as these episodes of MDW formation are, the result of one such episode of a few days' duration supplies enough water to feed the lower layer outflow through the Strait of Gibraltar for several weeks. Newly formed MDW is characterized by a potential temperature of 12.6 - 12.7°C, a salinity of 38.4, and an oxygen content of 4.6 ml/l, much warmer, saltier, and better oxygenated than the North Atlantic Deep Water found at the same depth in the Atlantic Ocean. The residence time of MDW in the Ligurian Sea has been estimated at 11 ± 2 years.

Similar outbreaks of strong cold winter winds occur in the Adriatic Sea, where they are known as bora. However, the stability of the water column is also affected by river runoff which keeps the surface density low. Currents through the Strait of Otranto are therefore as expected for a dilution basin: outflow of low salinity water at the surface and inflow of high salinity Levantine Intermediate Water at its usual depth of 200 - 500 m. The modification to this simple scheme, in the form of an additional outflow layer below the inflow, comes as a result of the shallowness of the Adriatic Sea north of 42.5°N (less than 200 m decreasing to less than 100 m north of 43.5°N). This allows the water in the northern and central regions to cool very fast during a bora event and to attain a density higher than the density of LIW in the south. The dense surface water flows southward on the Italian shelf, bypassing the deep southern basin until it encounters a series of canyons near 41.5°N. It falls down the canyons, mixing vigorously with LIW on the way, and leaves Otranto Strait as an outflow below the LIW inflow. Its characteristic properties are 13°C and 38.6 salinity, slightly less than the salinity of 38.7 found in the LIW above (which is warmer - 14°C - and thus less dense). The slight reduction in salinity indicates a contribution from the Po river.

The cold but relatively fresh bottom water from the Adriatic Sea does not maintain its identity very long. It turns eastward and enters the Levantine Basin where it encounters water freshly formed in the region between Rhodes and Cyprus. The two sources mix, and together they form the Levantine Intermediate Water. The Levantine surface source is much warmer but significantly more saline (15°C and 39.1 salinity). The resulting mixture gives a potential temperature of 13.3°C, a salinity of 38.67, and a very high oxygen content of 5.0 ml/l as the characteristics of LIW.

LIW is saltier but warmer than MDW and has a slightly lower density. As a consequence, a hydrographic station in the central parts of the Mediterranean Sea shows a layering of four water masses (Figure 16.8). LIW is easily identifiable as a salinity maximum underneath the Atlantic Water minimum. A salinity section

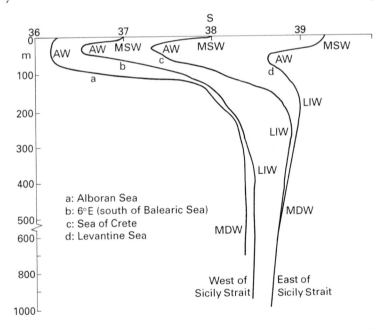

FIG. 16.8. Salinity against depth in the eastern and western Mediterranean Sea. MSW: Mediterranean surface water, AW: Atlantic Water, LIW: Levantine Intermediate Water, MDW: Mediterranean Deep Water. Note the difference in the salinity of MDW east and west of Sicily Strait; evaporation is higher in the eastern Mediterranean Sea, and the sill between Sicily and Tunisia prevents horizontal mixing between the basins.

along the axis of all major basins (Figure 16.9) shows its movement towards the Strait of Gibraltar. In general, MDW is not found above 600 m depth, so the bulk of the outflow through Gibraltar Strait must be provided by LIW. However, in the vicinity of the sill hydraulic control causes an uplift of the thermocline of several hundred metres on the African side, and MDW is able to leave the Mediterranean Sea (Figure 16.10) at an estimated rate of 0.2 Sv. Once over the sill, MDW and LIW are not much longer recognizable as separate water masses; they sink and spread as *(Eurafrican) Mediterranean Water* as described in Chapter 15.

The Black Sea has not been mentioned throughout this discussion since its impact on the Mediterranean Sea is very small and it is more appropriate to describe it as a separate mediterranean basin. The world's largest inland water basin (area 461,000 km^2, volume 537,000 km^3), the Black Sea is connected with the world ocean by a narrow passage with three subregions. Bosphorus Strait is on average 60 m deep, 31 km long and at its narrowest point only 760 m wide

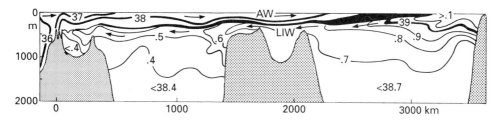

FIG. 16.9. Salinity section along the axis of the Eurafrican Mediterranean Sea. The black region indicates the salinity range 38.4 - 38.5 in the west and 38.4 - 39.1 in the east. Arrows indicate water mass movement; AW: Atlantic Water, LIW: Levantine Intermediate Water. After Wüst (1961).

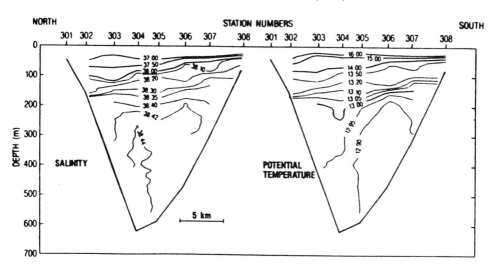

FIG. 16.10. Outflow of Levantine Intermediate Water (LIW) and Mediterranean Deep Water (MDW) through Gibraltar Strait as seen in salinity and potential temperature (°C). The section is some 15 km west of the narrowest point. MDW is identified by Θ < 12.90°C, LIW by a salinity > 38.44. From Kinder and Parrilla (1987).

and has a sill depth near 35 m. In the Marmara Sea the connection broadens to some 75 km width with depths in excess of 1000 m, but further passage to the Eurafrican Mediterranean Sea is again constricted by the Dardanelles, a more than 100 km long narrow waterway between Europe and Asia. Depths in the Black Sea exceed 2000 m throughout (Figure 16.3). The shallow Sea of Azov connects to the Black Sea through Kerch Strait (sill depth 5 m) in the north; it makes up 9% of the area but only 0.5% of the volume of the Black Sea.

Large freshwater input from the Danube, Dniester, Dnieper, Severskiy Donets, and Don rivers produces a positive freshwater balance and gives the Black Sea its character as a dilution basin. Luigi Marsigli argued in 1681 already and verified with laboratory experiments that underneath the well known surface flow that carries low-salinity water from the Black Sea through the Bosphorus, Marmara Sea, and Dardanelles, there should be a flow of salty Mediterranean water in the opposite direction, produced by the salinity (and thus density) difference between the Mediterranean and Black Seas. Modern observations confirm his ideas and give the following freshwater budget, which is believed to have existed since Bosphorus Strait opened 9000 years ago:

input (km^3/year)		output (km^3/year)	
from the Mediterranean Sea	120	to the Mediterranean Sea	260
precipitation	140	evaporation	350
river runoff	350		
total	610	total	610

If this is compared to the total volume of the Black Sea it becomes obvious that water renewal in the basin is extremely slow. This is also evident in the distribution of hydrographic properties (Figure 16.11) which show that the inflow is insufficient to keep the salinity at normal oceanic values and all oxygen is depleted below 200 m depth, resulting in the formation of hydrogen sulfide which makes the Black Sea uninhabitable below 200 m.

The apparent uniformity of T-S properties below 200 m gave rise to the idea that the inflowing Mediterranean Water sinks to the bottom and that water renewal at depth is achieved by very slow upwelling. CTD data obtained over the last 30 years (Tolmazin, 1985b) indicate that the structure is not as uniform as previously believed. They show a ribbon, about 7 - 8 m thick, of Mediterranean water above the shelf floor defying the Coriolis force by bending northward for the first 50 km or so after entering from Bosphorus Strait, apparently under hydraulic control from the shallow sills. The velocities in the inflowing water are quite modest, usually less than 0.1 m s^{-1}; however, given the weak density difference between inflow and outflow, they are sufficient to generate hydraulic control. There is evidence to suggest that the inflow is not continuous but

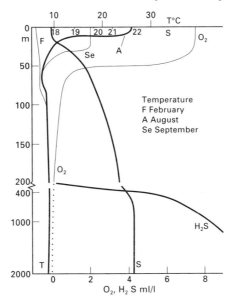

FIG. 16.11. Vertical profiles of temperature (°C), salinity, oxygen (O_2, ml/l), and hydrogen sulfide (H_2S, ml/l) typical for the Black Sea. From Tolmazin (1985a).

occasionally blocked by weather systems which depress the interface in Bosphorus Strait below the sill depth. Eventually the inflow merges with the general anti-clockwise circulation of the western Black Sea, and the Mediterranean water progresses along the Turkish shelf. All along its path it is exposed to intense mixing through bottom-induced turbulence, which causes parcels of water to separate and float away into the interior. As a result, lenses of warm saline oxygenated water are found at various depths in an environment usually devoid of oxygen. This is reflected in age estimates derived from radiocarbon measurements, which for the water in the 300 - 2160 m depth range vary from 600 to 2200 years and do not show a systematic variation with depth. Nevertheless, all water below 200 m depth must ultimately be renewed from Bosphorus Strait.

Away from the entry point for Mediterranean water the circulation in the entire water column is dominated by anti-clockwise motion along the continental slope with three anti-clockwise gyres of about identical size filling the western, central, and eastern basin (Figure 16.6). The western gyre brings water from the northwest shelf region into contact with the open Black Sea. In winter the water on the shelf is colder than that in the open sea (through a process explained for Bass Strait water in Chapter 17) and sinks as it leaves the region of shallow water depth. By spreading through the Black Sea at intermediate depth it produces the temperature minimum regularly observed near 75 m (Figure 16.11). A recent series of research cruises (Murray, 1991) will allow a much more detailed description of the circulation and water masses in the Black Sea.

The American Mediterranean Sea

The topography of the Caribbean Sea and the Gulf of Mexico (Figure 16.12) shows a succession of five basins, separated by sills of less than 2000 m depth and set apart from the main Atlantic basins by an island-studded enclosure less than 1000 m deep but containing several passages with sill depths of 740 - 2200 m. This alone identifies the region as a mediterranean sea, similar in structure to the Australasian Mediterranean Sea, a dilution basin in the tropics (see Chapter 13). The similarity extends to the fact that both seas have more than one connection with the main ocean basins and are therefore dominated in their upper layers by throughflow. The difference is that the Australasian Mediterranean Sea is located in a region of large freshwater gain and is a true dilution basin, while over the Caribbean Sea and the Gulf of Mexico evaporation exceeds precipitation by over 1 m per year, as much as in the case of the Eurafrican Mediterranean Sea and too much to be balanced by freshwater input from rivers; so the American Mediterranean Sea should really be a concentration basin. This is indeed correct: The annual mean salinity, averaged over the upper 200 metres, increases from 36.09 at the inflow through the Lesser Antilles to 36.19 in Yucatan Strait and further to 36.39 in the Strait of Florida (Etter *et al.,* 1987). However, for north Atlantic standards these salinities are quite low (Figure 2.5b); they do not reach the salinity values below the surface layer (compare Figure 16.15). The density increase associated with the concentration process is therefore insufficient to cause deep vertical convection. As a result, deep water renewal does not occur through the sinking of surface water (as observed in other concentration basins)

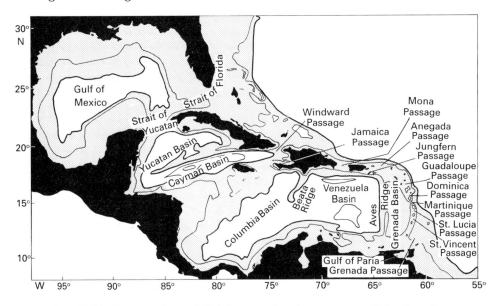

FIG. 16.12. Topography, subdivisions, and major passages of the American Mediterranean Sea. The 1000, 3000, and 5000 m isobaths are shown, and regions less than 3000 m deep are shaded.

but follows the pattern typical for dilution basins, i.e. sporadic inflow of oceanic water from outside. This makes the American Mediterranean Sea rather unique among all adjacent seas of the world ocean.

The reason for the low salinities in the surface layer is advection of Amazon River water with the Guayana and Caribbean Currents. Estimates based on radium measurements indicate that on average, 15 - 20% of the surface water that enters the Caribbean Sea is derived from the brackish waters of the Orinoco and Amazon River estuaries (Moore *et al.*, 1986). The influence of river runoff is strongly seasonal, with strongest flow occurring between May and November; surface salinities in the eastern Caribbean Sea can then drop to 33 and lower. Even when distributed over the upper 150 - 200 m, the associated density decrease is sufficient to decouple the deep circulation from that at the surface. The water in the deep basins therefore enters from the southern Sargasso Sea, the Venezuela and Columbia Basins being filled from the Jungfern Passage and the Cayman and Yucatan Basins from the Windward Passage (Figure 16.13c). A section following the path through one of the passages (Figure 16.14) shows the remarkable uniformity of potential temperatures below the sill depths and indicates how the water is drawn from a narrow layer at the depth of the sills. Direct observations show the inflow as being confined to a layer less than 200 m thick, as well as being highly variable, often modulated by strong tidal currents, and sometimes suppressed for extensive periods. The large scale distribution of potential temperature averages over many inflow episodes and gives a good indication of the water renewal. The variability of the inflow makes estimation of deep water residence times a difficult task. Numbers found in the literature range from 55 to 800 years.

Movement above the sill depths of the passages is dominated by throughflow from the Antilles to Yucatan Strait and into the Gulf of Mexico. The details of the circulation are determined by the topography and the location of the source. North Atlantic Deep Water is advected from the north and consequently enters mainly through Windward Passage, with additional inflow through Jungfern Passage (Figure 16.13b). Antarctic Intermediate Water is advected from the southern hemisphere and therefore enters the Caribbean Sea nearly exclusively through the eastern passages (Figure 16.13a). Central Water is advected with the North Equatorial Current from the east and finds its way into the Carribean Sea through both the eastern and northern passages (Figure 16.16).

The throughflow through the American Mediterranean Sea is part of the system of western boundary currents of the north Atlantic Ocean and therefore associated with large transports. The total transport through the Caribbean Sea (close to 30 Sv) is well known from detailed measurements in the Strait of Florida, through which all water from Yucatan Strait must leave. How much of this flow enters through the passages of the Lesser Antilles and how much through Windward Passage is less well established. Early geostrophic calculations put the transport through the eastern passages at 26 Sv, leaving 4 Sv for Windward Passage. More

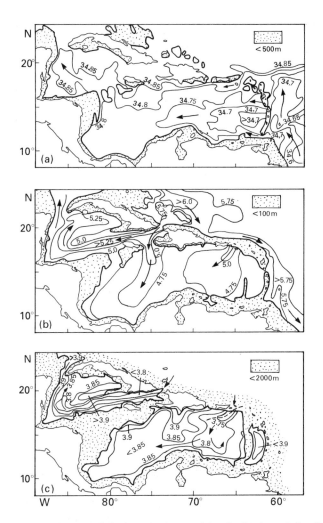

FIG. 16.13. Intermediate and deep water renewal in the basins of the Caribbean Sea. (a) Salinity at the level of the salinity minimum near 700 - 850 m, indicating the path of Antarctic Intermediate Water; (b) oxygen (ml/l) at the level of the oxygen minimum near 2000 m, indicating movement in the upper range of North Atlantic Deep Water; (c) bottom potential temperature (°C), indicating renewal paths for the water in the deep basins. The broken line indicates the location of the section of Fig. 16.14. After Wüst (1963). (The data for (b) were obtained during 1954 - 1958. Data obtained during 1932 - 1937 show oxygen levels higher by 0.3 ml/l in the east and 0.7 ml/l in the west. The indicated flow pattern is the same in both cases.)

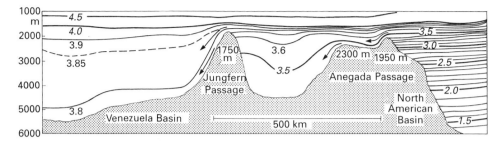

FIG. 16.14. Potential temperature (°C) along the section indicated by the broken line in Fig. 16.13c. After Wüst (1963).

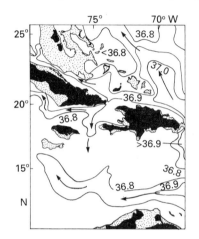

FIG. 16.15. Salinity at the salinity maximum at the top of the Central Water (approximately 150 - 200 m depth). Depths less than 200 m are shaded. After Wüst (1964).

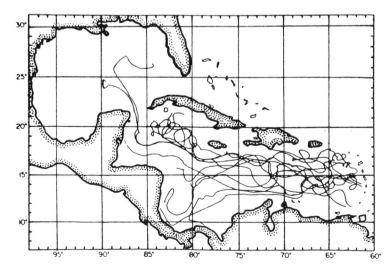

FIG. 16.16. Tracks of 19 satellite-tracked buoys for the period October 1975 - June 1976. From Kinder *et al.* (1985).

recent estimates based on a combination of direct current measurements and circulation models reduce the role of the eastern passages significantly, allocating 15 Sv to the Grenada, St. Vincent, and St. Lucia Passages and 5 Sv to the remaining passages in the Lesser Antilles, leaving 10 Sv for Windward Passage (Kinder *et al.*, 1985).

The surface flow through the Caribbean Sea has been documented by drifting buoys tracked by satellites. The tracks (Figure 16.16) show the Caribbean Current with speeds around 0.2 m s^{-1} in the Grenada Basin, 0.5 m s^{-1} in the Venezuela, Columbia, and Cayman Basins, and 0.8 m s^{-1} near Yucatan Strait. These velocities are lower than in other western boundary currents and particularly in the western Carribean Sea lower than the velocities associated with the eddies produced by the current. This makes the currents highly variable and causes occasional flow reversals from westward to eastward in the Yucatan Basin. Systematic eastward flow embedded in generally westward movement is found in the Grenada and possibly also in the Venezuela Basin. Observations above the Aves Ridge (Figure 16.17) show a banded flow structure with eastward flow of variable strength near 13°N. This flow eventually leaves the Caribbean Sea and enters the Atlantic Ocean as the Carribean Countercurrent (Figure 14.2). Near 16°N the current appears to be mostly eastward; but occasionally the flow turns westward with minimum speeds near 16°N.

The continuation of the Caribbean Current through the Gulf of Mexico is known as the Loop Current. This is a true western boundary current which separates from the continental shelf north of Yucatan Strait. It is therefore characterized by instability of its path and periodic eddy shedding. Figure 16.18 gives a summary of the circulation features associated with the Loop Current. The main path followed by the current penetrates the Gulf to about 27°N. The

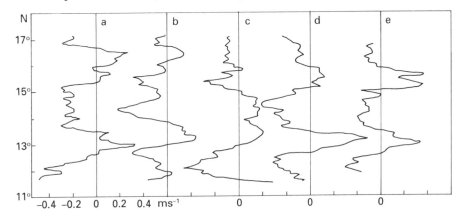

Fig. 16.17. East-west component of velocity (m s^{-1}, positive is eastward), averaged over 0 - 200 m from measurements with an acoustic doppler current meter, as a function of latitude along 63.55°W in (a) August 1985, (b) January, (c) March, (d) July, (e) October 1986. The velocity scale is correct for (a), other curves are shifted as indicated by the zero velocity line. Adapted from Smith and Morrison (1989).

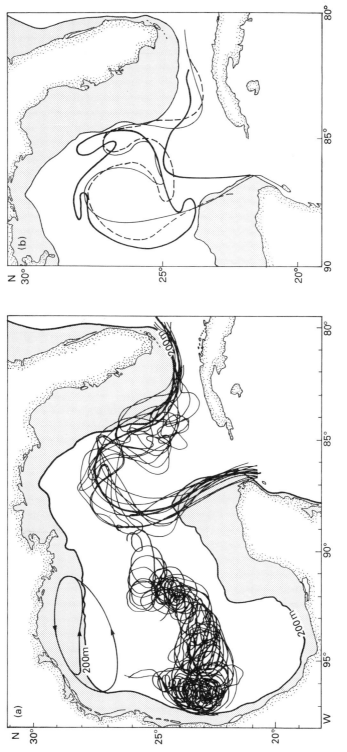

Fig 16.18. Circulation in the Gulf of Mexico. (a) Mean position of the Loop Current during 1980 - 1984 (heavy line), current positions inferred from satellite observations of sea surface temperature, tracks of satellite tracked drifting buoys indicating eddy movement in the west, and schematic circulation on the northwestern shelf; (b) observed positions of the Loop Current just before (thin and dotted lines) and after eddy shedding (heavy lines). Areas shallower than 200m are shaded.

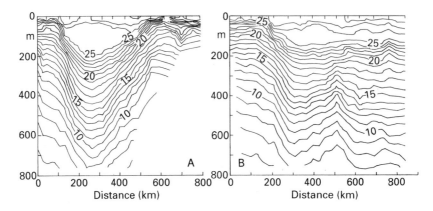

FIG. 16.19. Temperature along sections *A* and *B* shown in Fig. 16.20. Yucatan Strait
is on the right. From Lewis and Kirwan (1987).

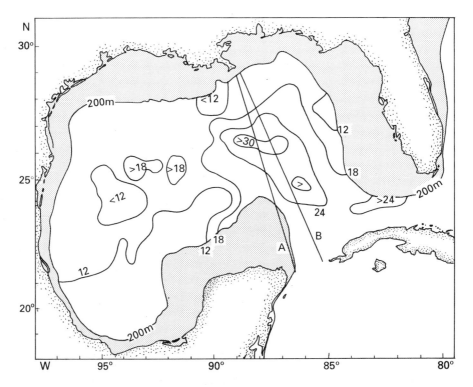

FIG. 16.20. Long term mean variability of sea level (cm). After Maul and Hermann
(1985). The lines indicate the positions of sections *A* and *B* of Fig. 16.19.

speed of the current in Yucatan Strait usually exceeds 1 m s^{-1} at the surface but
falls off with depth, reaching 0.4 m s^{-1} at about 1000 m depth. A southward
undercurrent is often found in the last 200 m above the sill depth of 1895 m.
It is highly variable, can sustain velocities of 0.05 m s^{-1} over several months and
0.15 m s^{-1} in bursts, but a mean over three years gave a net southward flow of
just under 0.02 m s^{-1}.

Occasionally the northern part of the Loop Current separates into a ring and the main current does not reach beyond 25°N for some time (Figure 16.18b). Eddy separation occurs on average every 11 months but can vary between 6 and 17 months (Vukovich, 1988). The eddies or rings are of the anticyclonic, warm core type described in detail in the discussion of the East Australian Current in Chapter 8. Figure 16.19 shows the isothermal core and the depression of the isotherms for one such eddy as it was crossed by two sections along tracks only about 100 km apart. Note the difference in the depth of the 15 - 25°C isotherms across Yucatan Strait (the southern end of the sections); the associated thermocline slope indicates the Loop Current as it leaves the Strait between the two sections. The steep slope of the isotherms further north in section B indicates where the Loop Current crosses the section to continue towards Florida Strait.

Once formed the eddies drift away from the Loop Current in a general southwestward direction at 3 - 4 km per day. Like other eddies of western boundary currents they have typical diameters of 200 - 300 km and surface speeds of 1 - 2 m s^{-1} depending on age. When they reach the western continental shelf they still induce shelf currents as high as 0.7 m s^{-1}. Figure 16.18 shows typical eddy paths as reflected in the movement of drifting buoys. It is obvious that the direction of water movement in the western Gulf of Mexico at any particular moment is determined by the eddy field. Given their rate of formation and mean drift speed, between one and three eddies are usually present at any one time. The net circulation, determined geostrophically from smoothed data, indicates anti-cyclonic (clockwise) movement of about 5 Sv around the Gulf and a smaller cyclonic feature in the north with a transport of 8 Sv and linked with the circulation on the northern shelf (Figure 16.18a).

The westward passage of the eddies is accompanied by large variations in sea level (as explained with Figures 2.7 and 3.2 - 3.4), and a map of the long-term mean variability of sea level is a useful indicator of eddy movement. Figure 16.20 indicates mean sea level variability of 0.3 m in the eddy formation region and variability levels near 0.2 m along the major eddy drift path. This amounts to about 60% of the variability level in the centre of the Gulf Stream eddy region and indicates the high level of kinetic energy that is dispersed in the western Gulf of Mexico.

The Irish Sea, the North Sea, and the Baltic Sea

The three seas in the last geographical subdivision are all part of the European continental shelf and therefore mostly shallow, the only notable exception being the Norwegian trench in the North Sea. The Irish and North Seas both have long open connections with the Atlantic Ocean proper and are dominated by strong tidal currents and frequent strong winds; they are therefore similar in character. In contrast, communication between the Baltic Sea and the Atlantic Ocean is only indirect (through the North Sea) and severely restricted, which makes the Baltic Sea the fourth mediterranean sea of the Atlantic Ocean. Our discussion of the basics of ocean dynamics in Chapters 1 - 5 did not include the modifications that occur in coastal, shelf, and estuarine areas. With these limitations in our understanding of the processes responsible for water movement and renewal in shallow seas we have to restrict our description of the circulation and hydrology on the European shelf to a few general remarks.

An important element of the dynamics of shallow seas with a good connection to the deep ocean is tidal movement. As the tide enters from the ocean the tidal current increases in magnitude as the water depth decreases. The increase in current speed is not restricted to the surface layer but occurs at all depths. The associated turbulence acts like a giant stirring mechanism trying to break down the stratification. There is therefore a competition between solar heat input at the surface, which acts to stabilize the water column, and mixing from the tides, which attempts to homogenize the water column. In deep water tidal currents are weak and the water is stratified. In shallow water the currents are strong and the water is well-mixed. The transition between the mixed region and the stratified region occurs in a frontal zone known as a shallow sea front. Figure 16.21 gives

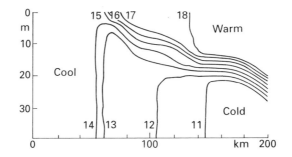

FIG. 16.21. Sketch of a temperature section through a shallow sea front, based on observations from the eastern entry to the English Channel. Note that the temperature on the shallow side (on the left) equals the temperature in the core of the thermocline.

a sketch of such a front. Shallow sea fronts occur at well defined locations which are determined by a combination of water depth h and tidal current u. Energy arguments show that the parameter which measures the competition between thermal stabilization and tidal stirring is h/u^3.

The topography of the *Irish Sea* resembles that of a channel with gentle slopes on either side. Maximum depths along the axis are near 110 - 140 m in the south and exceed 250 m between northern Ireland and Scotland. East of 4.5°W the depth rarely exceeds 50 m. If this topography is combined with the magnitude of the tidal current, the resulting distribution of h/u^3 gives the contours shown in Figure 16.22. By comparing the contours with the sea surface temperature during May 1980 it is seen that most of the Irish Sea is well-mixed throughout the year and separated from the stratified regions by shallow sea fronts in the north and south.

Mean flow through the Irish Sea is weak and difficult to measure directly, due to the dominance of strong tidal flow. It can be deduced from hydrographic

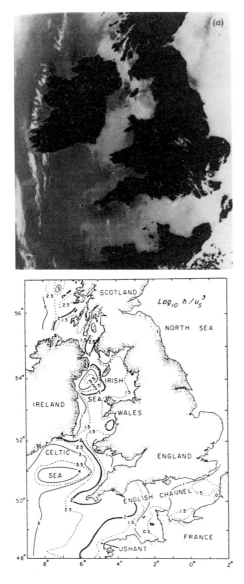

FIG. 16.22. Shallow sea fronts in the Irish Sea. (a) Sea surface temperature derived from satellite observations during May 1980; dark is warm, grey and white is cold (some small scale white features are clouds). The dark regions are stratified and therefore show warm surface temperature, while the grey regions are mixed. Fronts are found along the edges of the grey regions. (b) Contours of $\log_{10} h/u^3$; the contour closest to the position of the fronts is highlighted. Adapted from Simpson (1981).

properties, which indicate northward movement with inflow from the Celtic Sea and outflow to the Hebrides (Figure 16.23). The circulation in the eastern Irish Sea is modified by freshwater input from several rivers which lowers the sea surface salinity (Figure 16.24) and produces upward entrainment of salty water. This is

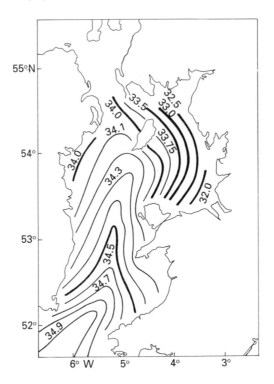

FIG. 16.23. Mean surface salinity in the Irish Sea.

associated with inflow towards the river mouths in the lower layer and outflow near the surface. The situation is typical for estuarine circulation systems, and the reader is referred to textbooks on estuaries and coastal regions for further detail.

The introduction of nuclear power stations resulted in contamination of the European shelf with radioactive caesium nucleids from two nuclear fuel reprocessing plants. Sellafield (formerly Windscale), the larger of the two, is located on the coast of the northern Irish Sea. The spreading of ^{137}Cs confirms the concept of mean northward flow and indicates the passage of Irish Sea water into the North Sea within 3 - 4 years (Figure 16.25). Some ^{137}Cs spreads into the Celtic Sea to the south by tidal dispersion and during occasional wind-driven reversals of the mean flow. The same process disperses some ^{137}Cs from La Hague, the second reprocessing plant which is located on the French coast of the English Channel, westward allowing it to enter the southern Irish Sea with the mean flow.

The inferences made from the distribution of ^{137}Cs can be extended to give information on the mean water movement of the *North Sea* as well. Effluent from La Hague spreads mainly eastward, indicating a net flow of water through the English Channel from the Bay of Biscay into the North Sea. The North Sea itself displays anti-clockwise circulation, evidenced by the movement of Sellafield effluent

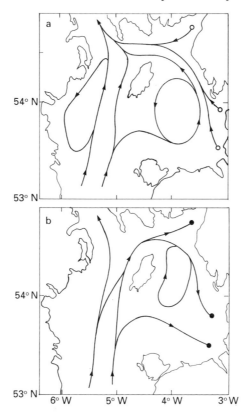

FIG. 16.24. Mean circulation in the northern Irish Sea. (a) Near-surface, (b) near-bottom. Dots and circles indicate entrainment from the near-bottom into the near-surface layer.

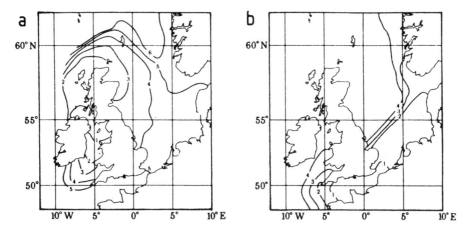

FIG. 16.25. Spreading of ^{137}Cs from Sellafield (a) and La Hague (b), expressed as average time in years required for a water particle to travel to the positions indicated by the contours. The distribution is the result of dispersion and advection with the mean flow, the latter being indicated by the asymmetry of the contours with respect to the release point. From Zimmerman (1984).

along the British coast and the spreading of La Hague effluent along the Scandinavian coast.

The North Sea is generally shallow, with depths around 120 - 150 m in the north decreasing to 100 m between Aberdeen and Stavanger, to 50 m between Hull and Skagen, and to 30 - 40 m further to the southeast. An exception is the Doggerbank, a shallow region that stretches from 2°E to 6°E just south of the 50 m isobath. At 54.3°N, 2°E it rises to 13 m and is feared by fishermen for its dangerous waves which can break in severe weather and have caused the loss of more than one vessel. The other exception is the Norwegian trench which stretches along the Norwegian coastline with depths around 300 m and a width of 35 - 80 km at the 250 m depth contour; between 7°E and 19°E it ends in a depression with a maximum depth of 700 m.

The mean circulation can be seen in the salinity distribution which changes little over the year. Figure 16.26 compares the salinity near the sea surface with the distribution near the bottom. It is seen that the salinity difference between the two surfaces is small over most of the North Sea; but significant differences are observed in the north and in particular the northeast. They are related to the water movement and indicate the presence of a two-layer circulation system in the Skagerrak and the Norwegian trench. Low-salinity water from the Baltic Sea enters the North Sea at the surface and joins the northern branch of the anti-clockwise circulation which derives most of its water from the central North Sea. The inflow is intermittent and controlled by the local wind. High salinity

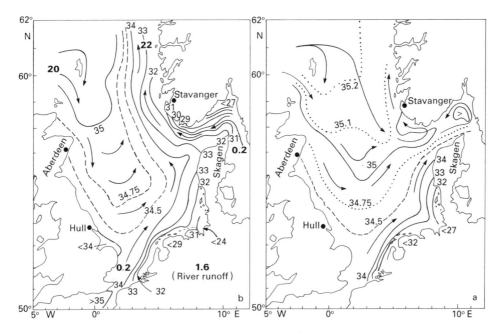

Fig. 16.26. Mean salinity and circulation in the North Sea for June. (a) At 7.5 m depth, (b) above the bottom. Bold numbers give average transport in 1000 km³ per year (1000 km³ per year = 0.03 Sv). Adapted from Goedecke *et al.* (1967).

water moves eastward from the open Atlantic Ocean along the southern slope of the Norwegian trench and returns along its northern side; some of it fills the depression in the east. Mean currents in the central and southern North Sea do not vary much with depth.

The temperature distribution shows more seasonal variation and regional structure. Sea surface temperatures in the central North Sea vary between 2 - 4°C in winter and 18 - 20°C in summer, the total range increasing monotonically from northwest to southeast. Close to the Dutch and German coast the range increases to -1 - 22°C, bringing occasional ice formation. The vertical structure of the temperature field (Figure 16.27) shows the imprint of shallow sea front dynamics (discussed above with the Irish Sea), a clear indication for strong tidal currents - tides dominate the flow field and sea level in the North Sea at any particular time -, although these fronts have not received much attention in the North Sea. Observations of currents near the sea floor show that they are associated with geostrophic flow along the fronts of up to 0.15 m s^{-1}. The development of a summer thermocline is restricted to the region deeper than 50 m and east of 0°W. Regions shallower than this depth and the waters along England and Scotland remain unstratified throughout the year due to strong tidal currents. A strong seasonal thermocline develops in the eastern and southern parts of the Norwegian trench where the salinity stratification produces high vertical stability of the water column and inhibits vertical mixing. The entire North

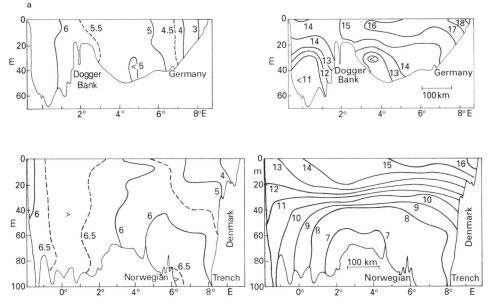

Fig. 16.27. Mean temperature (°C) for February (left) and August (right) based on data for the period 1902 - 1954. (a) Across the Doggerbank along 54.5°N, (b) from Aberdeen to Skagen along 57.5°N. Note the similarity of the isotherm distribution near the Doggerbank and the Scottish coast with the sketch of Fig. 16.21 during summer. The sharpness of the fronts is greatly reduced by averaging over 52 years. From Tomczak and Goedecke (1964).

Sea is well mixed vertically during winter; horizontal temperature differences result from the larger seasonal range in shallow coastal water.

The *Baltic Sea*, a mediterranean dilution basin, is in many aspects similar to the Black Sea. It has very restricted exchange with the open ocean and a significant freshwater surplus; as a result its salinity is also well below normal oceanic values, and oxygen is regularly depleted in the deep basins and replaced by hydrogen sulfide. Its area is comparable ($350,000$ km^2), and so is its freshwater surplus (between 300 km^3 per year in February and 770 km^3 per year in May,

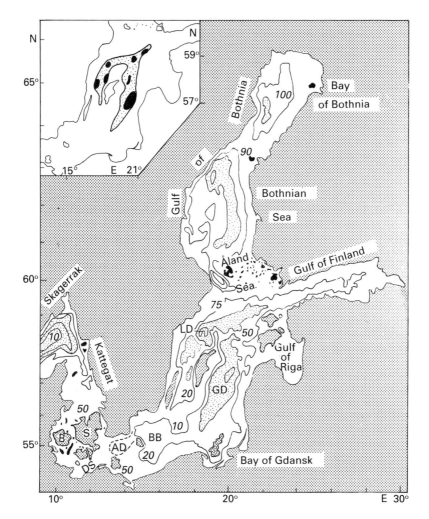

Fig. 16.28. Topography of the Baltic Sea. Depths > 100 m are stippled, with contours drawn for 50, 100, 300, and 500 m depth. AD: Arkona Deep, BB: Bornholm Basin, BS: Belt Sea, DS: Darß Sill, GD: Gotland Deep, LD: Landsort Deep. Numbers in italics give the probability of total ice coverage (*100%* = total ice cover every year). The inset shows the minimum (black) and maximum (stippled) extent of anoxic conditions in the central basins for the period 1979 - 1988; adapted from Nehring (1990).

annual mean 470 km^3 per year). Its system of passages to the North Sea and the open ocean - Belt Sea, Kattegat, and Skagerrak - is also similarly shallow and complicated (Figure 16.28; the Kattegat has a mean depth of 23 m, and the sill depth in the Belt Sea is 18 m). The all important difference lies in the volume of the two mediterranean basins: The mean depth of the Baltic Sea is about 60 m, giving it a total volume of about 20,000 km^3 or 4% of the volume of the Black Sea. Thus, while the exchange of water between the Baltic and North Seas carries

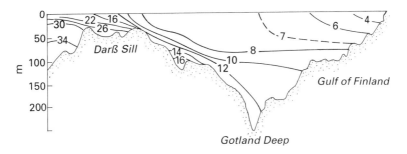

FIG. 16.29. Salinity in a section along the axis of the Baltic Sea. Note the change of contouring interval across the Darß Sill.

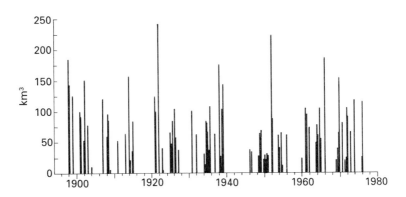

FIG. 16.30. Volumes of high salinity inflow into the Baltic Sea. The gaps during 1914 - 1918 and 1939 - 1945 are from lack of observations. From Matthäus and Franck (1990).

transports similar to those between the Black and Mediterranean Seas, the amount of basin water requiring renewal through inflow from the open ocean is much less, and de-oxygenation in the deep basins is not permanent. The average salinity (Figure 16.29), on the other hand, is even lower than in the Black Sea, since the river runoff constitutes 2% of the Baltic Sea volume.

The strong salinity stratification provides the water column with ample stability even in winter, and convective mixing during the cooling period is restricted to the upper 70 m or so. This results in a large seasonal temperature variation at the surface and in combination with the low surface salinity leads to regular ice formation (Figure 16.28). Below 70 m depth the temperature is rather stable

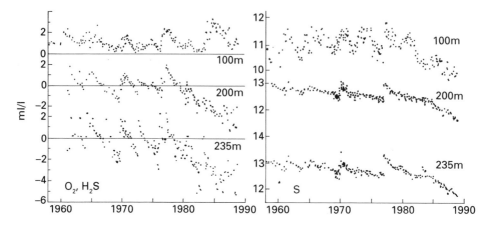

FIG. 16.31. Oxygen O_2 (ml/l), hydrogen sulfide H_2S (ml/l, plotted as negative oxygen), and salinity S at a station in the Gotland Deep. Strong inflow events occurred in 1965, 1970, 1973, and 1976. Note that after each inflow event salinity and oxygen are higher at 235 m than at 200 m, indicating that renewal takes place near the bottom and progresses upward by turbulent diffusion. After Nehring and Matthäus (1990).

and quickly approaches the 5.5°C found throughout the year in the deep basins. The thermocline at 70 m is also the upper limit for the occurrence of hydrogen sulfide during extended periods of stagnation; during most years hydrogen sulfide does not occur above 120 - 150 m depth.

Water renewal in the Baltic basins is a highly intermittent process. Individual inflow events last between ten days and three months (average duration 32 days) and are associated with a rise of sea level in the Baltic Sea of some 0.6 m. They are restricted to the period August - April; most of them occur in December. Mean inflow velocities are 0.1 m s^{-1} near the Darß Sill and 0.08 m s^{-1} on approaching the Bornholm Basin. On average, the amount of water transported into the Baltic Sea is 200 ± 100 km^3, which is 35 - 100% of the volume of the Belt Sea. The amount of high salinity water brought in by each event is, however, considerably less, because at the beginning of the inflow the lower layer of the Belt Sea contains a mixture of low salinity Baltic and high salinity North Sea water and has to be emptied before less diluted high salinity water can flow in. Figure 16.30 shows the volumes of high salinity water transported by the 90 major events which were identified for the period 1897 - 1976. It is seen that the events occur in clusters, on average every three years, and alternate with periods of 1 - 4 years without major inflow. The same variability has been found in the meridional wind component over the North Atlantic Ocean (Börngen *et al.,* 1990). This suggests that inflow into the Baltic Sea is controlled by interannual variability of the global climate.

Systematic measurements in the deep basins indicate that long term climate change is also likely to affect the hydrography of the Baltic Sea. Data collected since 1958 (Figure 16.31) indicate that several major renewal events reached the

Gotland Deep; however, it is also clear from the data that over the observation period deep water renewal is not frequent enough to prevent a general increase of hydrogen sulfide concentration and that no major inflow event occurred after 1976[*]. This is accompanied by progressive freshening, a result of diffusion through the halocline. Whether this process will continue is impossible to predict at present. What is clear from the data is that the hydrographic state of the Baltic Sea is finely tuned and susceptible to changes in the world climate. It also reacts strongly to small changes induced by human activity. Present plans to connect Denmark and Sweden by a system of bridges or tunnels can have a major impact on deep water renewal if the sill depths or cross-sections of the Belt Sea passages are changed.

[*]A major inflow event, comparable in volume to earlier major events, occurred during the winter of 1992/1993. A full assessment of its impact on water renewal in the Baltic deep basins was not possible at the time of printing.

CHAPTER 17

Aspects of advanced regional oceanography

The discussion of the Atlantic Ocean's adjacent seas completes our tour of the circulation and hydrographic structure of the world ocean. The material presented in sixteen chapters covered a lot of ground and might leave the impression that not much remains to be told. To dispel any doubt in that respect and to demonstrate that the word "Introduction" in the title of this book describes its contents correctly, we devote one chapter to some examples of what can be considered advanced regional oceanography. Compared to the introductory level of this book, advanced regional oceanography is an extension of the material in two respects. It addresses the oceanography of a given region in much greater detail than is possible in a introduction, and it explains it in terms of oceanographic concepts and processes which are beyond the scope of an introductory text. Advanced regional oceanography requires an advanced understanding of ocean dynamics (e.g. an understanding of internal waves or double diffusion). However, in this chapter we will restrict our discussion to examples that can be discussed on the basis of concepts already introduced in earlier chapters.

Example I: Modification of Central Water in the Tasman Sea

Our first example brings us back to the discussion of Central Water of the South Pacific Ocean. It was pointed out in Chapter 9 that Western South Pacific Central Water (WSPCW) which is found in the Tasman and Coral Seas and just east of New Zealand has properties virtually identical to Indian and South Atlantic Central Water. This was taken as an indication that conditions in the Subtropical Convergence (STC) of the southern hemisphere from where these Central Waters originate do not vary much around the globe. Figure 17.1 compares again the T-S diagrams of Indian Central Water (ICW) with those of WSPCW at various locations in the western Pacific Ocean. It is noted that in the Tasman Sea WSPCW is significantly more saline than ICW and that this difference increases from north to south. In contrast, WSPCW in the Coral Sea has properties very similar to those of ICW. While this indicates that the water found in the thermocline of

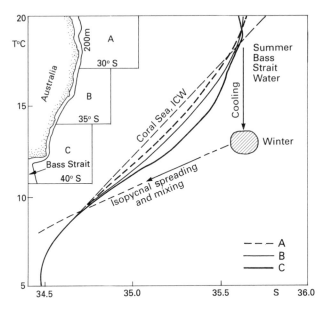

Fig. 17.1. Temperature-salinity diagrams for Indian Central Water (ICW) and for Western South Pacific Central Water (WSPCW) in the Coral Sea and for three Tasman Sea areas as identified in the inset. Also shown are the summer and winter T-S characteristics for Bass Strait Water. See text for details. Adapted from Tomczak (1981d).

the Coral Sea originates from the southern STC (i.e. from south of the Tasman Sea), it also shows that it could not have taken the most direct route through the Tasman Sea where the salinity is higher than both in the north and in the south. The regional salinity maximum in the Central Water of the Tasman Sea requires a regional source of high salinity water. This source is located at the western edge and is known as the Bass Strait Water Cascade.

Bass Strait is a shallow sea between Tasmania and mainland Australia connecting the Tasman Sea with the Great Australian Bight. It is about 80 m deep in the centre and has sills slightly deeper than 50 m on either side. During summer its hydrographic properties are the same as those of Tasman Sea surface water, but in winter it cools faster than the Tasman Sea, and its T-S characteristics differ (Figure 17.1). This comes from the fact that cooling of Tasman Sea surface water during autumn and winter deepens the surface mixed layer through convection. As long as the mixed layer is shallower than the depth of Bass Strait the same process operates in that region as well, and the mixed layer temperature in both regions is the same. When that depth is reached, Bass Strait Water cools faster than the surface water in the Tasman Sea, where continued entrainment of water from below keeps the cooling rate low. By the middle of winter the mixed layer in the Tasman Sea has reached a thickness of 100 m and more and Bass Strait Water is some 5°C colder and therefore denser than Tasman Sea surface water. A density front is established between the two regions and maintained by geostrophic adjustment, i.e. a northward current of Bass Strait Water on the

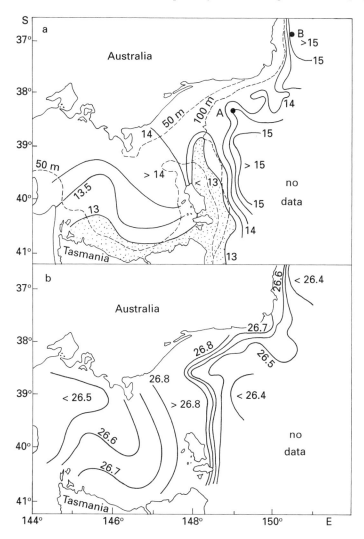

Fɪɢ. 17.2. Hydrographic conditions in Bass Strait and the adjacent Tasman Sea during winter 1981, showing the front between the two regions. (a) Surface temperature (°C), (b) surface density (σ_t). The broken lines are isobaths; the large dots indicate the positions of the stations used in Fig. 17.3. From Tomczak (1985).

western side (Figure 17.2). When this water approaches the southern coast of the Australian mainland it sinks down along the continental slope and continues as an undercurrent at the depth where it finds water of its own density. The sinking process, known as the Bas Strait Water Cascade, occurs in a well defined region and is apparently associated with a canyon. As is seen from the T-S diagram of Fig 17.1 cascade water is more saline and slightly warmer than Tasman Sea water of the same density and can therefore be identified along its path by its high salinity. As it flows eastward into the Tasman Sea the Coriolis force keeps it close to the Australian shelf, forcing it to turn northward at the southeastern corner.

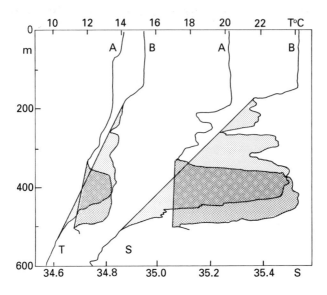

FIG. 17.3. The Bass Strait Water undercurrent in the Tasman Sea as seen in vertical profiles of temperature (T) and salinity (S). Temperature and salinity anomalies produced by the undercurrent are shaded. See Fig. 17.2 for station locations. From Tomczak (1985).

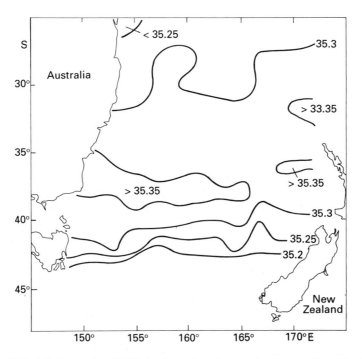

FIG. 17.4. Salinity on the 14.5°C isothermal surface in the Tasman Sea. The depth of this surface varies between less than 100 m in the south to 350 m in the north. High salinity along 37°S indicates Bass Strait Water influence. From Villanoy and Tomczak (1991).

The warm and saline undercurrent from the Bass Strait Water Cascade (Figure 17.3) proceeds northward until it encounters the East Australian Current, which sweeps it out into the open Tasman Sea. It quickly loses its identity as a water mass, through mixing in eddies and turbulence in the strong current shear associated with the East Australian Current, but its contribution to the properties of the Tasman Sea thermocline is seen in the anomalously high salinities of WSPCW at the temperature characteristic for Bass Strait Water during winter (Figure 17.4).

The impact of Bass Strait water on WSPCW properties is an example of modification of oceanic water masses by coastal, shelf, and estuarine processes. Similar situations are found in many parts of the world ocean. Another undercurrent of high salinity and temperature has been observed in the Great Australian Bight east of Spencer Gulf. As in the case of Bass Strait, the undercurrent results from an increase of the water density on the shelf (i.e. in Spencer Gulf) above the density of the adjacent oceanic surface water. In this case the density increase is the result of very high evaporation in the arid desert environment of South Australia. It is therefore caused by a salinity increase and requires only a minor drop of sea surface temperature in early autumn to start the sinking motion. The resulting undercurrent, however, shows very similar behaviour and properties. Another example of shelf influence on thermocline waters is described below.

Example 2: Mixing in the Canary Current upwelling region

In our second example we return to the discussion of coastal upwelling in the North Atlantic Ocean. The Canary Current upwelling system is probably the most complex of all coastal upwelling systems, particularly near Cape Blanc (20°S) where the frontal zone between North Atlantic Central Water (NADW) and South Atlantic Central Water (SACW), which was discussed in Chapter 15, reaches the African coast. Mixing between the two water masses, which have very different hydrographic properties, competes with the effect of upwelling, to the degree that it is often difficult to decide whether an observed increase in primary productivity is the result of upwelling, mixing, or advection. Compared with NACW, its southern counterpart SACW has similar temperatures but somewhat lower salinities and much higher nutrient concentrations. An intrusion of SACW into a region originally occupied by NACW can double or treble the nutrient levels in the euphotic zone within hours, bringing phosphate concentrations up from less then 0.5 to 1.5 μg-at/l and raising silicate concentrations from below 5 to 15 μg-at/l. Upwelling of NACW could not lift those nutrients levels above concentrations of 1.0 and 7 μg-at/l, respectively, even if it came from 500 m depth which is very unlikely. It is seen that in the vicinity of the frontal zone horizontal advection of water masses can be at least as important for primary productivity as the upwelling process itself. Because the water masses on either side have

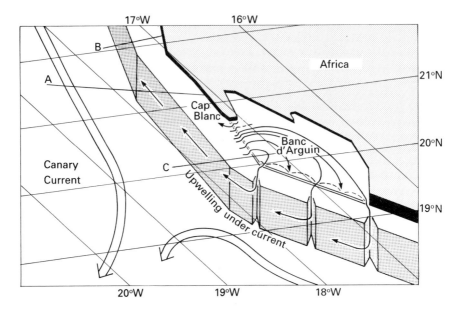

FIG. 17.5. Sketch of the circulation on Arguin Bank. Lines *A*, *B*, and *C* indicate the locations of the sections shown in Figs 17.6 and 17.8. Adapted from Peters (1976).

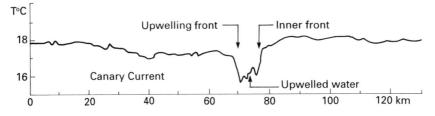

FIG. 17.6. Sea surface temperature along 20°N (section *C* in Fig. 17.5) from the Canary Current to Arguin Bank. The upwelling front, produced by the surfacing of the thermocline, separates the coastal upwelling zone to the right from the region of cold water advection in the Canary Current. The inner front separates Arguin Bank Water from oceanic water. From Tomczak and Miosga (1976).

different temperatures and salinities but identical densities, the front is density compensated and thus not associated with geostrophic flow but dynamically passive. It is therefore quite common to see parcels of water drift across the front and continue their movement as intrusions or lenses embedded in different water (A typical example of such a situation was seen in Figure 15.12). As these parcels are lifted into the euphotic zone with the upwelling, the nutrient content in the surface layer is constantly varying depending on the source water masses of the layers. The result is an extremely patchy nutrient distribution in the surface layer and a corresponding patchiness in primary productivity.

The situation is complicated further by the presence of a coastal water mass originating from Arguin Bank, a large expanse of water less than 20 m deep off

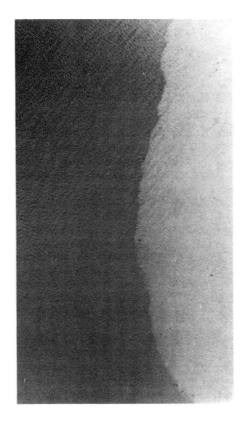

FIG. 17.7. A photograph of the sea surface in the region of the Arguin Bank front. Arguin Bank Water is on the right, the cold band of upwelling water to the left. The north-south extent of the area shown is about 18 km. From Tomczak and Miosga (1976).

the coast of Mauritania. Situated in an arid desert climate typical for coastal upwelling regions, this body of water is exposed to strong solar heating and evaporation. The effect on its hydrological properties is determined by the circulation which in turn is driven by loose coupling with the oceanic circulation. The Canary Current veers southwestward at Cape Blanc, leaving the region seaward of Arguin Bank to a northward flowing countercurrent (Figure 17.5). Flow on the bank is therefore anticyclonic; water leaves the bank in the south and is replaced from the north. There is thus little difference between oceanic water properties and those of Arguin Bank Water in the north; but in a bay with average depths around 10 m not much heat is required to warm the water and increase its salinity through evaporation, and by the time the water reaches the southern part of the bank it leaves it with a salinity above 39 and a temperature above 24°C. In the coastal upwelling region, offshore water with an equivalent density is found between 100 - 200 m depth, so Arguin Bank Water sinks to that level where it joins the general northward movement and the upwelling process.

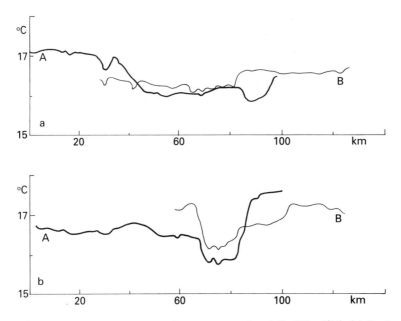

FIG. 17.8. Sea surface temperature along transects *A* and *B* of Fig. 17.5. (a) During weak upwelling, (b) during an event of strong upwelling. Note the absence of fronts and coastal warm water when the upwelling is weak. Note also how the inner front is eroded by mixing as the coastal water moves northward from *A* to *B*. The observations for (a) and (b) were taken ten days apart. From Tomczak (1981e).

This sinking motion occurs against the background of coastal upwelling and is therefore not continuous but concentrated in canyons where it occurs in bursts, alternating with vigorous upwelling events.

Similar to the situation encountered with the Bass Strait Water Cascade, the coastal water is retained at the surface by a front situated roughly inside the shelf break. This produces the unique situation that the coastal upwelling zone is bounded by two fronts (Figure 17.6) and appears in the sea surface temperature as a narrow band of cold water between warmer water on either side. The temperature contrast from upwelled water to coastal water is typically 2 - 3°C and occurs over a 10 km distance. This corresponds to a gradient of about 0.2°C km^{-1}. The maximum observed gradient, which occurs only over a narrow strip of 1 - 2 km in the centre of the front, is usually in the range 0.5 - 1.0°C km^{-1} but can reach up to 2°C km^{-1}. This is more than the gradients encountered in the upwelling front (but still an order of magnitude smaller than upwelling fronts observed in the Benguela Current upwelling region). Despite the relatively small thermal contrast the front is easily located even visually (Figure 17.7), since Arguin Bank Water is even less transparent than upwelled water, which on behalf of its high productivity is already low in transparency. Sand dust blown across the water from the Sahara desert contributes significantly to the discoloration of Arguin Bank Water.

North of Cape Blanc the upwelling system follows the dynamics sketched in Figure 8.25 which shows lowest temperatures at its inshore edge. However, during periods of intensified upwelling - so-called upwelling events, which are related to the variability of the synoptic weather systems and thus last for about 5 days - the Arguin Bank front has been observed to extend northward past Cape Blanc, indicating that the mass deficit produced by the increase in Ekman transport towards the sea is balanced mainly by northward advection of Arguin Bank Water (Figure 17.8). The main effect of the increase in wind speed is offshore movement of the upwelled water, leaving the inshore region free to be filled by warm and saline shelf water. During these events the upwelling region north of Cape Blanc shows the same structure usually found further south, i.e. a narrow band of cold upwelled water between the oceanic water of the Canary Current and northward moving water from Arguin Bank.

Ocean variability and mixing

The two examples discussed above have to suffice as evidence for the important role coastal and shelf processes can play in shaping the details of the oceanic circulation and hydrography. There are many other places along the oceanic rim where similar situations can be found. As already said at the beginning of this chapter, more insight into physical processes is required to fully understand and describe the impact of the vigorous mixing and dynamic interactions between the ocean and its shelf waters. We leave this difficult topic here and conclude the examples of advanced regional oceanography with a brief discussion of ocean variability, i.e. the role of processes occurring on time scales from days to months and space scales from centimetres to hundreds of kilometres, leaving the longer time scales and larger space scales to a detailed discussion in the last three chapters.

Ocean variability occurs in many forms. Not all forms are present everywhere in the world ocean, and identifying their regional distribution is part of advanced regional oceanography. Instability of western boundary currents and associated ring shedding is one form of ocean variability; it is part of the dynamics of these currents and restricted to well defined regions of the world ocean. Other forms (such as the eddies seen in Figure 4.9) are related to the dynamics of the vast geostrophic interior of the ocean and therefore more ubiquitous; but their intensity varies in space and time, and a task of advanced regional oceanography would be to quantify their regional occurrence (an attempt to achieve this is presented in Figure 4.8). Interleaving of water masses (as seen in Figure 15.12) is yet another form of variability; it occurs preferably in the vicinity of fronts but can be found in regions of quite uniform property distributions as well, as we saw in our discussion of the outflow of Mediterranean Water into the Atlantic

Ocean and its "meddies" (Figure 15.5).

   More detailed investigation reveals a sequence of variability scales spanning several orders of magnitude that tend to co-exist (Figure 17.9). The largest scale is set by features of the mean circulation; Figure 17.9 uses the Antarctic Polar Front as an example (compare the top panel of the figure with Figure 6.8a). The deviation of isotherms and isohalines from horizontal is an indication for geostrophic flow normal to the section. Eddies have similar scales, they differ from oceanic fronts mainly by the fact that their dimensions are comparable in all horizontal directions; in comparison, frontal dimensions are small across the front but large in a direction parallel to the front. The top panel of Figure 17.9 can therefore also be seen as representing half an eddy. The important point to observe here is that at these scales ocean variability is geostrophic, i.e. water movement in its elements is in geostrophic balance. The eddies cause departures from the long-term mean flow, and a time series of currents looks very much like a slowed-down version of turbulence as it is observed for example in a water pipe. The difference is that turbulence in a pipe is too fast to be affected by

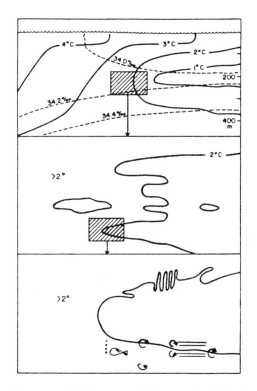

FIG. 17.9. Scales of ocean variability: oceanic fronts and eddies (top), interleaving (middle), double diffusion (bottom). Approximate overall horizontal and vertical sizes for the three sketches are 500 - 1000 km and 500 m (top), 100 - 150 km and 100 m (middle), and 20 km and 25 m (bottom), each panel being an enlargement of the shaded region in the panel above. See text for more explanation.
From Joyce (1977).

the Coriolis force. In the ocean, flow variations caused by eddies are always accompanied by geostrophic adjustment of the density field. This type of variability is therefore also called *geostrophic turbulence*. The intensity of geostrophic turbulence varies in space and time. To give an example, Figure 17.10 shows the regional distribution of eddy velocities associated with geostrophic turbulence for the North Atlantic Ocean as deduced from satellite-tracked surface drifters.

Around the edges of the eddies are found the *intrusions* (the middle panel of Figure 17.9). They cause deformations of isotherms and isohalines which have to be density-compensated to keep the stratification stable. It follows that on this scale ocean variability is not constrained by geostrophy. It would appear that this should make interleaving the dominant turbulence mechanism in the ocean, since so much less energy is required to disturb the oceanic mean state if there is no need for geostrophic adjustment. However, there is plenty of energy available from the atmosphere at the scales of geostrophic turbulence, while not much energy goes into the generation of turbulence at the interleaving scale. Advanced study would show that other processes, such as the stirring of water above sills or the injection of water from the shelf, are required to trigger significant interleaving. As a consequence, interleaving is less common than geostrophic turbulence and shows a much more uneven regional distribution.

The basis of both geostrophic turbulence and interleaving is that water particles are physically moved from one area of the mean circulation to another. If many

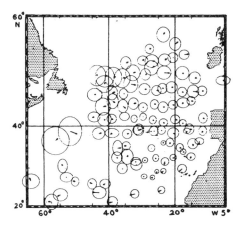

FIG. 17.10. Regional distribution of geostrophic turbulence in the North Atlantic Ocean. The arrows show the mean surface current, the axes of the ellipses give the mean north-south and east-west eddy velocities. Mean eddy velocities increase threefold from Ireland to Newfoundland. In the centre of the subtropical gyre (35°N, 30°W) mean eddy velocities are below 10% of the values reached east of Newfoundland. Note also that with few exceptions, mean eddy velocities are larger than the mean current and therefore cause reversals of the mean flow. From Krauss and Käse (1984).

such moves take place the result will be the mixing of water properties. Because all water properties, and in particular heat and salt, are exchanged with the turbulent movement of the particles, the rate of mixing is the same for all of them. This is not the case on the third and smallest turbulence scale (the bottom panel of Figure 17.9) which is linked with molecular diffusion processes. The molecular diffusivity of salt is about two orders of magnitude smaller than the molecular diffusivity of heat; in other words, it is much more difficult to exchange salt on the molecular scale than it is to exchange heat. This gives rise to striking instabilities in the stratification. As an example, consider the situation in the oceanic thermocline. Temperature decreases with depth, and so does salinity in most parts of the world ocean. On its own, the vertical salinity gradient would result in an unstable density stratification; but this instability is more than compensated by the vertical temperature gradient. Molecular diffusion tends to reduce the temperature and salinity gradients; but the temperature gradient is reduced much faster than the salinity gradient, and if this continues long enough there comes a moment when the temperature gradient is no longer sufficient to compensate for the salinity-induced instability. Convection sets in and occurs in narrow vertical tubes of rising low salinity water between narrow tubes of sinking salty water ("salt fingers"). The process is known as *double diffusion* and has been well documented in the laboratory. Observation in the ocean is much more difficult, and it is generally only possible to infer its existence from its impact on the larger scales. The physics of double diffusion are beyond the scope of this book, so two observations must do. It can be shown that a stratification characterized by a stable salinity but unstable temperature gradient is modified by double diffusion in such a way that the continuous gradients are replaced by a series of layers of uniform temperature and salinity. This type of stratification

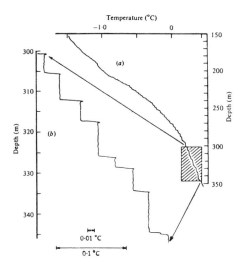

Fig. 17.11. Evidence for double diffusion in the form of homothermal layers observed under an Arctic ice island. From Neal *et al.* (1969).

is sometimes found in the Arctic and Antarctic oceans when fresher melt water overlies warm salty ocean water. The predicted layers have indeed been found under drifting ice islands where the water column is sheltered from atmospheric disturbances and molecular diffusion is the dominant mixing process (Figure 17.11). It is also known that the intensity of double diffusion depends on the ratio of the rate of change of temperature and salinity with depth (the slope of the temperature-salinity (T-S) curve in the T-S diagram). This would indicate that double diffusion plays a more prominent role in the thermocline of the Atlantic and Indian Oceans than in the Pacific Ocean which has a much steeper T-S relationship (less salinity change for the same temperature change). Detailed description of the regional differences has to be left for an advanced textbook.

CHAPTER 18

The oceans and the world's mean climate

The ultimate aim of regional oceanography is not a description of the oceans in their steady state but an understanding of the changes that occur in them over seasons, years and decades. A full understanding of these aspects, to a level where forecasting the oceanic and atmospheric circulation for months or years ahead becomes a realistic possibility, requires close cooperation between regional oceanography and geophysical fluid dynamics. In fact, it inevitably leads to an approach quite different to the approach taken in the first seventeen chapters of this book, which largely ignored the interaction processes between ocean and atmosphere and treated the ocean as a self-contained system. The last three chapters in this book now look at the nature of the interaction and its consequences. It will be seen that the steady state discussed up to this point provides a logical reference for the discussion; so the seventeen chapters were not wasted. It will also become clear that much more work needs to be done before we can claim to fully understand the interplay between ocean and atmosphere.

In earlier chapters we identified momentum transfer (wind stress), heating and cooling, and evaporation and precipitation as the key mechanisms through which the atmosphere exerts an influence on the ocean. We now adopt the opposite point of view and ask: What are the mechanisms through which the ocean exerts an influence on the atmosphere, and hence on the world's climate?

The ocean, evidently, is the dominant source of atmospheric moisture; and the latent heat released when this moisture condenses into rain or snow is the primary driving force for the atmospheric circulation (the global wind systems). The winds in turn affect the sea surface temperature in several different ways; and the sea surface temperature largely controls the magnitude and spatial distribution of the moisture flux to the atmosphere. This shows that the most important oceanic parameter for the atmosphere which provides the link between two components of a tightly coupled system is the sea surface temperature (SST). A discussion of the ocean and the world's climate therefore has to begin with a detailed understanding of the SST distribution.

A first and very elementary solution to the problem of understanding SST can be obtained by treating the ocean as a "swamp", i.e. a layer of water so thin that it can store no heat, so the heat budget is balanced locally without assistance from currents. In a "swamp" ocean, the net heat flux into the ocean at any locality is exactly zero at all times. This assumption allows an easy estimate of SST from which evaporation into the atmosphere may be estimated. The resulting "swamp temperature" (known as the equilibrium temperature, Figure 18.1) matches reality to the extent that the ocean is warm at the equator and in summer, and cold at the poles and in winter. It is quite a useful first guess against which to measure the effects of ocean currents on SST; but it cannot explain the observation that seasonal temperature variations are much larger over land than over sea. SST varies by little more than 30°C, from -2°C near the poles to 30°C in the equatorial Pacific Ocean; at any particular location its daily variation rarely exceeds 1°C, and its difference between winter and summer usually falls within ±5°C. In contrast, surface temperatures over land can vary by as much as 100°C over the earth's surface; the daily temperature range exceeds 10°C in many places, and the difference between summer and winter temperatures comes close to 100°C in extreme continental climates.

A somewhat better approximation to this situation is to treat the ocean as a passive "slab", perhaps 100 m deep, i.e. to allow it to absorb heat during summer and release it again in winter, through the formation of the seasonal mixed layer (Chapter 5). Relative to the atmosphere, the storage capacity of the ocean for heat is huge (about 1000 kcal are released by every 3100 m^3 of dry air or 1 m^3 of sea water if their temperature is lowered by 1°C). This results in a seasonal SST cycle almost three months out of phase with the solar heating and much reduced in amplitude compared to that found in places far from the sea. Thus heat storage in the mixed layer results in milder climates for coastal and island locations, and a slab model can reproduce these effects quite well.

However, both the swamp and slab models are very deficient for representing the earth's mean climate, particularly if one wishes to understand its year-to-year

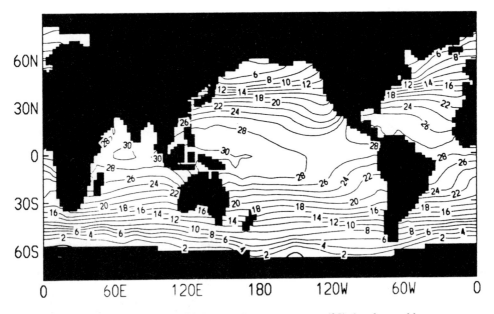

FIG. 18.1. Annual mean equilibrium surface temperature (°C) for the world ocean. This is the temperature the sea surface would have if the heat budget were locally balanced at all times. Note the similarity with the observed annual mean surface temperature (Fig. 2.5a) but also the differences in regions of strong currents and upwelling. From Hirst and Godfrey (1992).

variations. The ocean can and does absorb heat in one region, carry the heated water below the surface by subduction (Figure 5.3) or deep convection, and return the heat to the atmosphere many thousands of kilometres away and years, decades or even centuries later. In the mean, this results in the transport of heat from the equator towards the poles, tending to cool the tropics and heat the polar regions; the efficiency of this process is comparable to that of the atmosphere (Figure 18.2). The difference is that this process is carried out by ocean currents with a cycle time of many years. The strength of these currents varies, so we can expect year-to-year variations of the heat exchange with the atmosphere. To give two examples, the big region of heat gain in the eastern equatorial Pacific seen in Figure 1.6 largely disappears in some years (known as El Niño years, discussed in Chapter 19), with drastic effects on the world's climate; and there are reasons to suspect that the region of heat loss in the far north Atlantic Ocean may vary substantially from decade to decade (see Chapter 20). Climatologists believe that these year-to-year variations of the heat exchange with the atmosphere are a major contributor to the observed natural variations of climate.

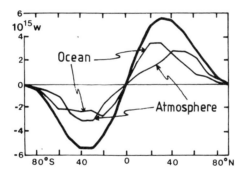

Fig. 18.2 Annual mean meridional heat flux in the ocean and the atmosphere (thin lines), and total annual mean meridional heat flux (heavy line). Adapted from Carissimo *et al.* (1985).

Similarly, long-term mean evaporation in one part of the ocean produces a continuous supply of salty water, which must be carried to some other part where there is an excess of rainfall or runoff. Once again, the transport can take many years to accomplish and is subject to interannual changes. These, too, may have effects on climate, as will be discussed in Chapter 20.

In this chapter we shall review what we know about the long-term mean surface fluxes of heat and moisture and about how the ocean currents are organized to transport heat and freshwater from where it is absorbed to where it is released. We will rely strongly on descriptions given in earlier chapters. However, our data base does not yet allow us to draw solid inferences about future climate trends, and some recourse to theoretical modelling of the climate system is necessary even in a text on regional - i.e. observational - oceanography. Numerical modellers have made considerable progress in creating realistic models of the ocean circulation in recent years, and these models allow us to explore the potential of the ocean as an active agent in the world's climate. One of their main results is that the atmosphere is particularly sensitive to small changes in sea surface temperature in the tropics; the reasons for this are discussed in the final section of this chapter, in the context of the mean seasonal *atmospheric* circulation. This information is needed in the last two chapters, where we discuss how the coupled ocean-atmosphere system operates to generate year-to-year and longer term climate variations.

Observed heat fluxes into the ocean

As mentioned in Chapter 1, it is quite difficult to estimate the net heat flux into the ocean accurately, because it is typically a small residual of four terms, two of which are relatively large and not very well determined by observations. *Solar heating* is nearly always the largest contributor, tending to warm the ocean surface. Simple rules have been devised for estimating the flux of solar radiation into the ocean from data on cloud cover, which have been collected by merchant ships' officers; world maps of solar (or "shortwave") radiation entering the ocean (such as Figure 1.5) are essentially maps of merchant ship cloud cover estimates, modulated by the geographic variation of clear-sky radiation (which is simply a function of latitude and season). Even with "perfect" cloud data, different algorithms (known as bulk formulae) differ in their estimates of the net solar radiation by about 10 - 20% (e.g. Hanawa and Kizu, 1990). Evaporation cools the ocean surface and is responsible for the second most important term in the heat budget, *latent heat loss* (Figure 18.3), particularly in the tropics. It can be estimated from other simple rules, using data on wind speed, the "wet bulb" and "dry bulb" temperatures, and sea surface temperature, all of which are also

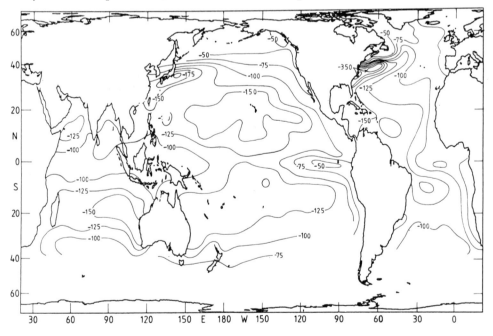

FIG. 18.3. Annual mean latent heat flux (W m^{-2}). Oceanic heat loss to the atmosphere is shown as negative numbers. From Oberhuber (1988).

routinely collected by merchant ship officers. Again, different algorithms can yield results that differ by as much as 30 W m^{-2} (Godfrey *et al.*, 1991; 30 W m^{-2} is enough to warm a layer of water 50 m thick by 0.5°C per month).

Two other types of heat flux - *sensible heat transfer* (the direct, mechanical transfer of heat between air and water when the two media have different temperatures) and *net longwave radiation* (the net escape of thermal radiation from the ocean surface) - contribute to the total heat flux; they are generally smaller than the first two but not small enough to be neglected. These also can be estimated from simple rules, using the same merchant ship data. The net heat flux is found by adding all four contributions. Figure 1.6 shows one such estimate, which was calculated using a particular set of bulk formulae. Other global or regional climatological maps use different bulk formulae or evaluate the fluxes over different time intervals and therefore come to somewhat different conclusions.

The sum of the four contributions is generally only a few tens of W m^{-2} over most parts of the ocean, whereas the rules used for estimating the two largest terms - the solar radiation and latent heat loss - have errors of this order of magnitude; the merchant ship data also contain errors. Furthermore, different techniques are used in averaging the data, which can lead to different smoothing of maxima. This (rather than the different choice of sampling times) is probably the main reason why different climatologies show fairly large differences from region to region. Nevertheless, all climatologies are qualitatively quite similar; and as has already been foreshadowed in earlier chapters, a quite plausible explanation of their general pattern can be given in terms of observed ocean currents. The fact that this is so is a tribute to the care and diligence which merchant ship officers have put into the accurate recording of data over the last hundred years or so, without direct reward to them during their lifetime. Oceanographers and climatologists owe a great debt to them.

Figure 1.6 is an annual mean picture of the ocean heat flux. However, the seasonal cycle in the ocean currents and temperatures both have large amplitudes in many parts of the world. One might therefore expect that it would be necessary to examine the heat budget season by season and then average the result, in order to understand how the ocean achieves the annual mean horizontal transports of heat implied by Figure 1.6. Surprisingly, this does not seem to be so, at least to the limited accuracy of Figure 1.6; one can understand many features of the annual mean heat flux maps quite well in terms of the annual mean currents (provided care is exercised in interpreting the results), and the mean currents are primarily driven by the mean wind stress field. Therefore we will mainly interpret the annual mean heat flux maps, region by region, with annual mean wind stresses and currents. We will discuss reasons why this approach is reasonable at the end of this chapter.

East Pacific and Atlantic upwelling regions

Along the eastern boundaries of the Pacific and Atlantic Oceans, Figure 1.6 shows narrow regions of oceanic heat gain extending from about 50°N to 50°S in the Pacific and from about 50°N to the Cape of Good Hope at 32°S in the Atlantic Ocean. These latitude limits are not very well defined, because of data limitations and the uncertainty in net heat flux estimates discussed above, but they are fairly similar to the points at which the annual mean westerly wind stresses bifurcate at the coast (Figure 1.4a). For example, the westerlies bifurcate at about 44°N in the north Pacific and Atlantic Oceans, near Oregon and northern Spain respectively, and at about 48°S in the South Pacific Ocean. Equatorwards of these points the nearshore wind stress blows towards the equator everywhere. Upwelling of cold subsurface water is to be expected at the coast, to supply the offshore Ekman drift (Figures 4.1 and 8.25). The associated current systems have been described in Chapters 8 and 14 (Figures 8.26, 8.27, 14.16 and 14.17); the surface currents are equatorward, which also tends to bring in cooler water from higher latitudes and further reduce surface temperature.

If air temperature and humidity, wind speed and cloud cover are all regarded as fixed, a reduction of SST reduces the latent and sensible heat loss, and also the net heat loss by longwave radiation, typically by a total of some 35 W m^{-2} per °C of SST change. Upwelling typically produces several degrees of cooling. The large net heat flux into the ocean in coastal upwelling regions is thus readily understood.

Coastal upwelling occurs only in the first few tens of kilometres from the coast, whereas the coastal bands of heat gain in Figure 1.6 are as much as a thousand kilometres or more in width, so coastal upwelling by itself does not explain the existence of these bands of heat flux into the ocean. However, the heat capacity of the surface mixed layer is so great that the upwelled water takes several months to warm to the equilibrium temperature; during these months the water can move 1000 km or so offshore with the Ekman drift. Furthermore, after the water is upwelled, it flows equatorwards with the Sverdrup flow. Consequently the water moves into steadily warmer climates, causing it to continue to warm for substantial periods after it has upwelled. This advective contribution to surface heating is clearly seen by comparing the annual mean SST map (Figure 2.5a) with surface currents (Figures 8.6 and 14.2). In each of the eastern boundary upwelling regions, the surface currents clearly flow from low temperatures towards high.

Equatorial upwelling regions

Strong bands of oceanic heat gain along the equator are found over the entire width of the Atlantic and over the eastern Pacific Oceans; weaker heat gain occurs in the west Pacific Ocean. Comparison with Figure 1.4a shows that there are moderate easterly wind stresses along the equator in both oceans; in relative terms they are strongest on the western side of the Atlantic and in the central Pacific Ocean.

These winds give rise to large poleward Ekman transports on either side of the equator (Figure 4.1). Upwelling occurs to replace the water that is removed by Ekman transport, as discussed in Chapter 8. The upwelled water is in turn supplied by geostrophically balanced flow in the top few hundred meters. A zonal steric height gradient develops to keep these meridional flows in geostrophic balance. (The Coriolis force changes direction across the equator, so the same zonal steric height gradient produces meridional flows of opposite sign on either side of the equator.) This zonal steric height gradient can be seen along the equator in both the Pacific and Atlantic Oceans in Figure 2.8b.

The heat flux maximum in the equatorial Pacific and Atlantic Oceans (Figure 1.6) lies somewhat to the east of the strongest equatorial easterly winds (Figure 1.4a), i.e. to the east of the strongest upwelling. The reason is that the temperature of upwelled water decreases eastwards along the equator, being coldest at the east of each basin. This fact in turn relates to the zonal gradient of steric height along the equator, set up to balance the wind and provide geostrophic inflow. Since steric height is roughly speaking a vertical integral of temperature, the zonal steric height gradient forces near-surface temperatures to be colder at the eastern end of each basin, where the steric height is lower.

As in the case of the eastern boundary heating regions, the width of the equatorial heating region extends about 1000 km on either side of the equator, much wider than the region of actual upwelling. The reason is again the same as in the coastal upwelling regions. After the water upwells it takes several months for it to absorb enough heat to reach thermal equilibrium with the atmosphere, during which time it can travel 1000 km poleward with the Ekman drift.

Two other factors not related to upwelling also influence equatorial SST in the Pacific and Atlantic Oceans. Both are related to current shear. Vertical current shear between the westward surface flow and the eastward flowing Equatorial Undercurrent produces increased turbulence (Figures 8.8 and 14.4) and assists in lowering SST. Horizontal shear between the South Equatorial Current and the North Equatorial Countercurrent gives rise to wave-like instabilities in the central and eastern Pacific Ocean (Figure 8.14). These waves are believed to transport substantial quantities of heat towards the equator, helping to reduce the intensity of the surface heat flux patch in the eastern Pacific Ocean.

Equatorial Indian Ocean heating regions

The pattern of surface heating in the Indian Ocean is unlike the pattern in the other two major ocean basins. The Indian Ocean is also the one where different climatologies show more significant differences in net surface heat flux than in any other ocean; for example, Figure 1.6 indicates an equatorial heating maximum in the western Indian Ocean, a feature weak or absent in other climatologies. However, all climatologies agree that heating is greatest at the *western* side of the basin - a major difference from the pattern in the Atlantic and Pacific Oceans which requires explanation.

Longshore winds along the west coast of Indonesia and the east side of the Bay of Bengal do not strongly favour upwelling at any time, except along the coast of Java in the Asian summer monsoon (Figure 1.4b). Furthermore, as discussed in Chapter 11, the vertical temperature gradient is not quite so sharp near the surface as it is in the eastern Atlantic and Pacific Oceans, so more upward motion is needed before surface cooling can be generated. Upwelling does occur along the south Java coast around August, but when the net heat flux is averaged over the year it does not result in an obvious patch of upwelling-induced heat gain like that off the Atlantic and Pacific east coasts. The net result is that heat gain along the eastern side of the Indian Ocean is somewhat weaker than in the other two oceans. By contrast, annual mean winds favour upwelling along the Arabian, east Indian, and east African coasts in the northern hemisphere (Figure 1.4). This upwelling and associated heat gain along the western boundary, a feature unique to the Indian Ocean, primarily occurs during the summer monsoon when intense southwesterlies blow along all these coasts. As in the *eastern* Pacific and Atlantic Oceans, upwelling and horizontal advection are both essential components in generating the broadscale pattern of heat flux seen in the western equatorial Indian Ocean (McCreary and Kundu, 1989).

As discussed in Chapter 11, the seasonal flow regime that results along the African and Arabian coasts from the monsoon winds is extremely complex, and quite detailed numerical modelling may be necessary even for a rough understanding of the heat budget of the northern Indian Ocean as a whole. Nevertheless, it is worth noting that the annual mean wind stress pattern of Figure 1.4a implies southward Ekman fluxes nearly everywhere north of about 15°S in the Indian Ocean. This must be balanced by an equal and opposite net northward geostrophic flow reaching deeper than the Ekman layer and therefore having an average temperature several degrees lower than that of the surface Ekman drift. The southward heat transport associated with this balanced pair of flows is of the same order of magnitude as the net heat flux through the sea surface (this can be estimated from Figure 1.6 by integrating the fluxes over the Indian Ocean north of 15°S). Thus even in this strongly seasonal region, the amount of heat transported out may be primarily controlled by the annual mean winds.

A notable feature in all climatologies is the region of quite large net heat flux into the ocean in the Indonesian throughflow region. This region does not contain any major upwelling, so the explanation for the large oceanic heat gain must be found somewhere else. In Chapter 13 it was argued that turbulent mixing must be anomalously strong in Indonesian waters and that its effect must be felt to 1000 m depth. This reduces SST and distributes the heat input from the atmosphere over a deeper "slab" then elsewhere in the world ocean. This is the only equatorial region in the world ocean were large heat gain is achieved without upwelling.

The Leeuwin Current

The zero heat flux contour in Figure 1.6 meets the continents near 50°N and S in the eastern Atlantic and Pacific Oceans but closer to 20°S in the Indian Ocean. This major difference between the ocean basins does not reflect a difference in the wind regime (Figure 1.4a) but a difference in the details of the eastern boundary regime. As discussed in Chapter 11, the upwelling that one might expect from the equatorward winds between 20° and 34°S along the western Australian coast is overwhelmed by an onshore geostrophic drift. The meridional pressure gradient needed to maintain the onshore drift is supplied by heat loss near Western Australia (clearly seen in Fig 1.6) which cools water at the southern end of the continent, reducing steric height from the very high levels found off northwestern Australia. This is one rare instance in which the currents carrying heat fluxes in the ocean are created by the heat fluxes themselves, i.e. by thermohaline processes; most surface currents are driven by wind forcing. However, as remarked in Chapter 11, the whole Leeuwin Current system can be regarded as being driven by winds along the equatorial Pacific Ocean, which pile up warm water in the west Pacific region and hence bring very warm water to northwestern Australia.

The subtropical western boundaries

One feature which emerges clearly from Figure 1.6 is that the Kuroshio and Gulf Stream and its extensions release massive amounts of heat to the atmosphere. This occurs mainly in winter when cold winds blow from Siberia and Canada respectively across the warm, poleward flowing waters of these currents; heat loss during summer is low and may turn into heat gain on occasions. On annual average, however, heat is lost from the ocean in the western boundary currents and extension regions. This heat loss results in convective sinking of surface water, so the extension regions of western boundary currents in the northern hemisphere are important regions for water mass formation. The subtropical mode waters have their origin in these regions, from where they are subducted into the subtropical gyres of the north Pacific and Atlantic Oceans.

By contrast, the heat losses from the western boundary currents of the Southern Hemisphere seem very small. In the case of the Brazil and East Australian Currents data are probably quite adequate to yield reasonable heat flux estimates, so that the small heat losses associated with these two currents are probably valid. Furthermore, the result is not unexpected. First, both currents are quite weak, the Brazil Current because it is opposed by the thermal flow towards the north Atlantic Ocean and the East Australian Current because a substantial fraction of its potential transport is drained away by the Indonesian throughflow. Secondly, no analogue of the very cold and dry Canadian and Siberian winter winds blow over either of these two currents.

The apparent weakness of the heat loss from the Agulhas Current and its extension suggested by Figure 1.6 is harder to understand and may be an artefact of the climatology. The Agulhas Current is among the strongest western boundary currents in the world ocean and experiences heat loss in all seasons (Chapter 11). It may be that the observational data used to construct Figure 1.6 are inadequate in the Agulhas Current region; the same is probably true for the western boundary current along New Zealand and its extension.

The subpolar north Atlantic Ocean

A region of heat loss without analogue in the other ocean basins can be seen in the subpolar north Atlantic Ocean. A remarkable aspect of the heat budget in that region is that the ocean currents supplying the surface heat flux are generated to some degree by the heat fluxes themselves, i.e. driven to some degree by thermohaline forcing and not by the wind stress alone. The situation is somewhat similar to that of the Leeuwin Current, but in the north Atlantic Ocean the process is strongly modulated by salinity effects.

It has already been remarked in Chapter 15 that the north Atlantic is substantially saltier than the north Pacific Ocean. Part of the reason for this is that much of the freshwater evaporated from the north Atlantic Ocean is carried over the low mountain ranges of Central America by the Trade Winds and rains out near the terminus of the Pacific North Equatorial Countercurrent. (This is the major reason for the large difference in *P-E* in the Atlantic relative to the Pacific Ocean seen in Figure 1.7). The freshened water in the Pacific Ocean is carried northward into the north Pacific subtropical gyre. Meanwhile the saltier water in the north Atlantic Ocean enters the Gulf Stream and moves northward. Another contribution is thought to come from the Agulhas Current eddies, which carry salt from the Indian into the Atlantic Ocean.

This saltiness of north Atlantic surface water has an important consequence. The salty subtropical water flows northeastward across the Atlantic to the Arctic Ocean in the North Atlantic and Norwegian Currents; it moves fast enough that the net positive *P-E* in the temperate north Atlantic Ocean (Figure 1.7) is not able to freshen this water greatly, so it remains very salty (relative to the north Pacific Ocean) at 60° - 70°N. As was discussed in connection with the formation of North Atlantic Deep Water in Chapter 7, the presence of this high salinity water below the surface water in the Greenland Sea means that little cooling is required to start convective overturn and sinking. This process of North Atlantic Deep Water formation has a self-perpetuating character. Because the water moves through the temperate north Atlantic Ocean so fast it remains salty despite freshwater input into the region and therefore sinks after cooling; in doing so, it pulls more water along behind it, maintaining the rapid flow in the surface layer. The rate of North Atlantic Deep Water formation was estimated in Chapter 7 at about 15 Sv. To put it another way, about 1300 cubic kilometres of water sink from the surface of the north Atlantic Ocean to several thousand metres depth every day, to return southward as North Atlantic Deep Water. If this water is supplied from the upper 500 m of the ocean, this corresponds to an ocean region of 50 km side length - the area of Los Angeles or Greater Tokyo. Not surprisingly, the cooling of this much water from temperatures near 12 - 15°C down to 0 - 4°C implies a major warming of the atmosphere; it is thought to be the reason why northwestern Europe is so markedly warmer than western Canada and Alaska at the same latitudes — water in the upper 500 m there is much too fresh to permit the formation of Deep Water in the north Pacific Ocean.

This process illustrates the subtle nature of the dependence of the earth's climate on apparently minor topographic detail. Central America forms a complete blockage of the Pacific from the Atlantic Ocean; yet it is low enough to permit moisture transport across it in the atmosphere, so that the north Atlantic Ocean becomes salty while the north Pacific Ocean becomes fresh. One may therefore say that northwestern Europe is warmer than Canada and Alaska because there is a low but complete land blockage in Central America.

The global freshwater and salt budgets

As noted earlier, estimates of precipitation and evaporation over the ocean (Figure 1.7) are subject to even greater uncertainties than the net heat flux field. Consequently we will not discuss the freshwater and salt budgets nearly as fully as the heat budget. We begin by noting two things. Firstly, it is evident from our earlier considerations on the transport of heat at the end of Chapter 4 that a net transport of salt or of freshwater can exist even if the net mass transport is zero. Secondly, salt and freshwater can both be transported in the same direction. This fact is somewhat contrary to intuition, since one might think that relative to the oceanic mean, freshwater is negative salt load and should therefore always be flowing in a direction opposite to the direction of the salt flux (readers familiar with estuarine oceanography will recognize this concept). Figure 18.4 shows that this is not so when the $P - E$ difference becomes a major contribution to the budget. The two quantities are then no longer inversely related.

Considering these two facts it turns out that an accurate estimate of the transport through Bering Strait is very important for the correct determination of both the salt and freshwater fluxes. In the discussion of the Arctic Mediterranean Sea (Chapter 7) we argued that mass transport through Bering Strait is less than 1 Sv and can be neglected in the global mass budget; in other words, in any model of the global oceanic circulation Bering Strait could be considered closed. For many years it was taken for granted that the same simplification could be applied in a model of the global salt and freshwater budgets. Such a model produces a net southward transport of salt for the north Atlantic Ocean, as a consequence of high evaporation in the North Atlantic Current and Deep Water formation and return flow (Figure 18.4), and a northward freshwater transport that decreases from large values in the south Atlantic Ocean to zero in the Arctic Mediterranean Sea. Wijffels *et al.* (1991) pointed out only recently that the surface salinities of the north Pacific Ocean are so low (near 32, see Figure 2.5b) that their freshwater content relative to the global mean salinity is of order 10%. In their budget calculations they use a figure of 1.5 Sv for the flow through Bering Strait; this produces a freshwater transport of 150,000 m^3 s^{-1}, several times the mean flow of the Amazon River! Thus the Bering Strait throughflow is a major contributor to the world freshwater balance, and if it is accounted for, freshwater transport in the Atlantic Ocean is southward everywhere. Figure 18.5 shows the resulting global flux distribution for salt and freshwater. Note that unlike

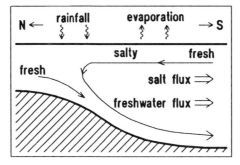

FIG. 18.4. A sketch of salt and freshwater transport in an ocean basin resembling the north Atlantic Ocean. At the surface, water gets saltier as it moves from the tropics through the subpolar gyre to the region where it sinks to carry salt southward in the deeper layers; this produces southward salt transport. Freshwater is imported from the Arctic Mediterranean Sea, in quantities not sufficient to lower the salinity in the deeper layer beyond the tropical surface salinities; while it therefore does not reverse the salt flux, it traverses the ocean from north to south, producing a freshwater flux in the same direction as the salt flux.

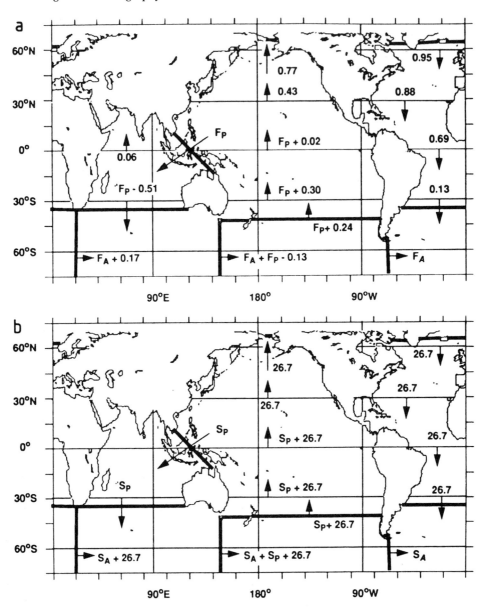

FIG. 18.5. Transports of (a) freshwater (10^9 kg s^{-1}) and (b) salt (10^6 kg s^{-1}) by the oceans, relative to the (undetermined) transports through the passages around Antarctica (index A) and in the Indonesian throughflow (index P). From Wijffels *et al.* (1991).

freshwater, salt does not escape through the sea surface, so the salt transport is the same in each of the major subdivisions of the world ocean.

This discussion illustrates why the north Pacific becomes so much fresher than the north Atlantic Ocean. One can imagine that if the two basins were suddenly forced to have equal salinities, Deep Water might form in both; but to maintain this state of affairs it would be necessary to increase the Central American land barrier sufficiently in order to suppress the flux of moisture from the Atlantic to the Pacific Ocean. With the present topography of Central America and rainfall distribution the surface salinity of the north Atlantic Ocean would gradually increase, while the surface salinity of the north Pacific Ocean would decrease, enhancing Deep Water formation in the Atlantic and retarding it in the Pacific Ocean. Pacific Deep Water formation would eventually cease, and near-surface salinities would continue to decrease in the north Pacific Ocean until they became so low that the small Bering Strait throughflow could drain off a freshwater flux equal to the flux over Central America. It is interesting to speculate what the surface salinity in the north Pacific Ocean must have been during the last Ice Age, when the Bering Strait was blocked.

Model heat flux patterns in the Southern Ocean

The discussion of the last few paragraphs carried us to the limits of what we can achieve with today's data set. Global budgets require global data coverage, which is difficult to achieve even with modern means. Satellites will make a major contribution here, at least for the global heat budget. But we are still desperately short of data particularly in the Southern Ocean (which is a major unknown in Figure 18.4). Recourse to numerical modelling is therefore essential if we want to learn more about the climate without waiting for many more years until the needed data arrive.

Numerical models of the oceanic circulation have the advantage that they do not depend on the details of the heat and evaporation algorithms used in generating Figs. 1.6 and 1.7; their estimates of the net surface heat flux depend only on the velocity, temperature and salinity fields of the model itself. Furthermore, most models force surface temperature and salinity (the best known observational parameters) to stay close to observed patterns. Their usefulness for predicting heat and freshwater fluxes therefore depends mainly on their skill in getting the ocean currents correct. In this section we will compare the surface heat flux field from one such model with the observations of Figures 1.6 and 1.7. The model heat flux field turns out to be in reasonable agreement with observations, over the region where data are adequate; similar results are found in a number of other ocean models. Hence it is reasonable to use such models to give us some ideas on the heat flux pattern in the Southern Ocean, where data are not adequate.

The model used here, like several other models, is driven by the observed wind stresses. Furthermore, and again like many other models, it uses the condition that the surface heat flux at each location is proportional to the departure of model SST from some climatological estimate of mean sea surface temperature (e.g. Figure 2.5a). Similarly, P-E is assumed proportional to the departure of model SSS from some climatological observed SSS (e.g. Figure 2.5b). For climate studies such models are run for several thousand simulation years until the model ocean changes so slowly in time that it can reasonably be treated as being in equilibrium. In the model discussed here, the global mean heat flux into the ocean was balanced to within 0.02 W m^{-2}.

In this equilibrium solution, the current field which develops is primarily set by the surface wind distribution. The winds control the Ekman flow, including the Ekman pumping from the surface layer, and influence the geostrophic flow underneath (through the Sverdrup relation). They also control horizontal variations of the depth-integrated steric height, as discussed in Chapter 4; these are set by the requirement to hold the Sverdrup flow in geostrophic balance. Thermal effects are also important; they control the mean depth of the thermocline (and hence the magnitude and vertical extent of typical surface geostrophic currents), and they also result in some examples of density driven flows roughly where we expect them to occur, on the basis of discussions earlier in this chapter.

Figure 18.6 shows the model's equilibrium heat flux distribution for the world ocean. The observed heat flux pattern (Figure 1.6) is generally reproduced by the model. The heat flux at the eastern Pacific and Atlantic boundaries is not as large as in the observations, but the model ocean gains heat near the coast from at least 40°S to 50°N in the Pacific, and from the Cape of Good Hope to Spain in the Atlantic Ocean. By contrast, heat is lost from the model ocean off western Australia south of 20°S, again in agreement with observations. No heat flux maximum is found in Indonesia because no allowance was made for enhanced vertical mixing in this region. As the model is driven only by annual mean wind stresses, the model heat flux in the western Indian Ocean comes out somewhat smaller than the observed annual mean. Strong heat fluxes out of the ocean occur in the Kuroshio and the Gulf Stream, and weak heat fluxes out of the ocean occur in the Brazil Current and the East Australian Current. All these features are qualitatively much the same as in the observations. Since the forcing of this model contains no seasonal cycle, its broad agreement with reality provides one reason for believing that much of the net heat flux into the ocean is controlled by the annual mean currents.

The similarities between model and observations encourage us to tentatively interpret the model heat fluxes in data-poor regions, as if they were real. It is worth examining the nature of the model surface heat budget in some detail for the Southern Ocean, because the results of this and other models provide the most reliable guide so far as to the nature of heat exchange processes in this very important part of the ocean.

It is evident from Figure 18.6 that two large bands of heat loss from the ocean occur in the model Southern Ocean, stretching eastward and slightly southward from the Agulhas retroflection and from New Zealand. This tends to confirm the suspicion that the observations must be missing a large part of the heat loss from the Agulhas and East Australian Current systems, respectively. Support for the value of the model also comes from the fact that the two heat loss bands coincide with observed locations of Subantarctic Mode Water formation (Chapter 6). Heat loss leads to convective overturn, and the model, too, shows deep mixed layers beneath the heat loss bands.

Two distinct bands of oceanic heat gain are seen south of the two heat loss bands. One occurs in the Atlantic and western Indian Ocean, south of the Agulhas Current; the second occurs south of the New Zealand boundary current. In these regions Ekman transports are northward and increase northwards, so that upwelling occurs, and surface water is moved towards warmer climates; both processes will favour heat gain by the ocean. Further south still more heat loss

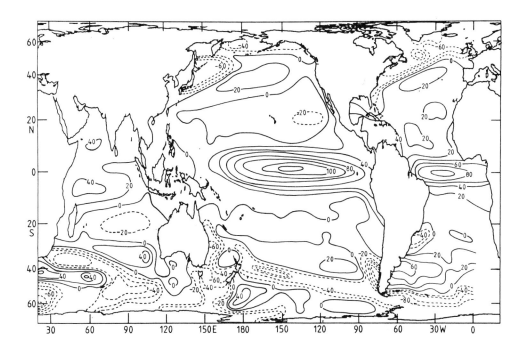

FIG. 18.6. An example of model-derived heat flux (W m^{-2}). After Hirst and Godfrey (1992).

occurs near the Antarctic continent, associated with bottom water formation in the model. However, since the model does not allow for sea ice formation the details cannot be expected to be very realistic.

The model just discussed represents the ocean through a network (or "grid") of data points of between 200 km and 300 km separation. It is therefore unable to resolve the quasigeostrophic eddies which are so ubiquitous in the ocean and in some regions (particularly in the Southern Ocean; see Chapter 6) essential for the transport of heat and salt. Other models which use the most powerful computers presently available use grid representations with mesh sizes as small as 15 km. While we cannot expect these models to reproduce actually observed eddies in size, location, or life span, we can hope that the models reproduce the eddy statistics such as the distribution of eddy kinetic energy (Figure 17.10) and improve our estimates of oceanic heat flux by identifying the relative importance of eddy heat transport in different regions of the world ocean.

SST-based positive feedbacks in the tropical atmosphere

The recourse to a numerical model in the last section completes our discussion of the importance of the sea surface temperature as the coupling agent between ocean and atmosphere for establishing the mean climate. However, sea surface temperature provides the coupling agent on shorter time scales as well, and in the tropics its role for controlling the seasonal cycle is so important that a discussion of the mean climate would be incomplete without consideration of the mean seasonal cycle for the tropics. This also offers a natural introduction to the topic of the next chapter, variations in the seasonal cycle, or interannual variability.

Two prominent features of the atmospheric circulation over the Pacific Ocean are the Intertropical Convergence Zone (ITCZ) and the South Pacific Convergence Zone (SPCZ), which are seen as cloud bands in satellite imagery (Figs. 8.5 and 19.7). They are nearly always located near ridges of high SST, with temperatures of 28°C and above. The clouds are formed when moist near-surface air, transported equatorward and westward by the Trade Winds of both hemispheres, enters the Convergence Zones. The air then rises; moisture condenses into clouds, and intense rainfall occurs in the Convergence Zones (compare Figure 1.7). The condensation process releases huge amounts of latent heat into the surrounding air, which makes the air expand and keep on rising, thereby sucking more moist air in behind it. After reaching the top of the troposphere the warmed, dried air spreads sideways, loses its heat by radiative loss to space, and sinks back towards the sea surface (Figure 1.1), where the Trade Wind cycle begins again.

There are analogues of the ITCZ in the Indian Ocean region, associated with the Asian and Australian monsoons, and also over Africa and South America; but unlike the Pacific convection bands, the locations of these systems are to some extent set by geographic features (the Himalayas and the Andes).

The atmospheric convection cells of the tropics are the most energetic features of the global climate. They compete vigorously with one another, in the sense that a mature convection zone creates wind shears between upper level outflows and near-surface inflows for thousands of kilometres around it, and these shears strongly inhibit the formation of other convection zones. Competition between the two hemispheres produces a strong seasonal cycle, and the mechanism of the SST coupling introduces strong positive feedback, i.e. a tendency for instability, into the world's climate.

This is particularly noticeable each May - June when solar heating in the northern hemisphere starts to favour convection there at the expense of the southern hemisphere. At this time of year (and no other) SST increases to 29°C over most of the tropical Indian Ocean, and a store of moist air builds up over this region. By analogy with the Pacific ITCZ and SPCZ one might expect rainfall to develop over this SST maximum. However, this does not happen, most likely because convection over the Indian Ocean is being suppressed by the southern hemisphere convective systems of the Pacific Ocean. Whether for this or some other reason, convection over the Indian Ocean is weakening south of the equator, and some time around early June convection starts over Asia. The supply of very moist oceanic air is drawn towards the new convection centre which grows very rapidly; the SPCZ weakens rapidly at the same time; one might say that it can no longer compete. As a result of these developments the onset of the Indian summer monsoon is a dramatic and sudden event. For reasons not yet fully understood the onset of the Australian monsoon around November each year is not quite so dramatic; though it can be marked by rainfall that starts within one or two days over a longitude span of about 40° across northern Australia.

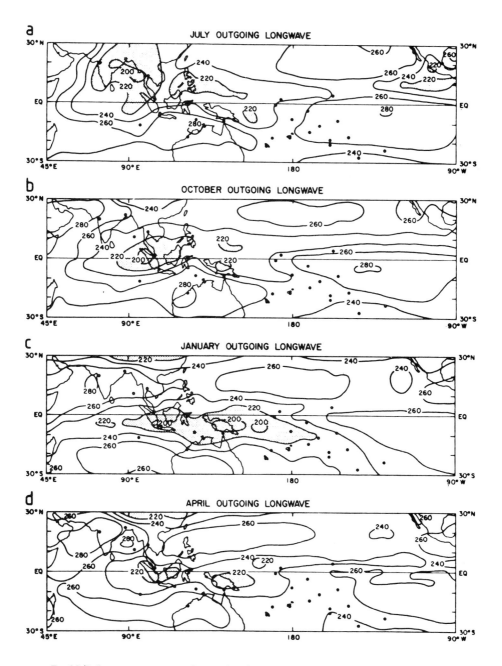

FIG.18.7. Long-term mean of outgoing longwave radiation (OLR, W m⁻²) for the period June 1974 - November 1983 but excluding 1978, for the Indian and Pacific sectors. (a) July, (b) October, (c) January, (d) April. Areas of less than 220 W m⁻² OLR are stippled, indicating greatest tropical convection. Adapted from Janowiak *et al.* (1985).

The clouds associated with the centres of rainfall (which are the driving force for these atmospheric circulations) influence the longwave backradiation of incoming solar energy. Large cumulus towers in high rainfall areas reach very high and have very cold tops; they emit very little heat energy and are therefore seen in maps of outgoing longwave radiation (OLR) as local minima. Seasonal patterns of rainfall can therefore be followed by monitoring OLR from satellites. The observations reveal a prominent difference in the rainfall patterns of October and May (Figure 18.7). Convection crosses the equator rather smoothly in October, following the Indonesian archipelago from Asia to Australia, while in April/May, just before the onset of the Asian summer monsoon, the convection is weak and split between Indonesia and the west Pacific Ocean. This hysteresis or asymmetry in the mean seasonal pattern of the atmospheric convection has the consequence that the biggest interannual changes in the earth's climate system tend to occur in May. The atmospheric circulation is very delicately poised at this time of year, and small changes in SST — and hence of moisture content in the atmosphere — can have disproportionately large effects on the atmospheric circulation. We explore this further in the following chapter.

CHAPTER 19

El Niño and the Southern Oscillation (ENSO)

The previous chapter stressed the importance of the tropics for the coupling between ocean and atmosphere and showed how positive feedback between the sea surface temperature (SST) and the atmospheric convection zones can result in rapid magnifications of initially small disturbances. We now continue this theme and examine in detail the situation in the tropical Pacific Ocean. This region is characterized by very high sea surface temperatures and extremely low zonal temperature gradients in the west (the so-called "warm pool"; see Figure 2.5a); small SST variations in this region can grow into interannual climate variations of global proportions.

We start by looking at some observations. It is well known that climate conditions in the Australian continent vary between the extremes of devastating droughts and equally devastating floods. In eastern Australia, years of severe drought have been documented for nearly two centuries, and it has been noticed that they come at irregular intervals of a few years. We know now that they are part of a global phenomenon known as the El Niño - Southern Oscillation, or ENSO, phenomenon, which manifests itself in fluctuations of rainfall, winds, ocean currents, and sea surface temperature of the tropical oceans, and of the Pacific Ocean in particular. If these fluctuations were strictly periodic people would probably have learnt to plan for their occurrence. What makes them so difficult to cope with is their irregularity. As it is, the ENSO affects national economies often in an unpredictable manner, causing great hardship and social upheaval and at least on one occasion the downfall of a government (Tomczak, 1981a). The data shown in Figure 19.1 are only a poor indicator for this; dollar values and variations of gross national product do not convey the amount of human suffering behind them. But they indicate the magnitude and importance of the task ahead, the development of a reliable forecast of year-to-year climate variations, of which ENSO variations are a major part.

The Southern Oscillation

When Sir Gilbert Walker became Director-General of Observatories in the British colony of India in 1904 he set himself the task of trying to predict variations in the Indian monsoons and related droughts. To this end he started a project to examine global records of sea-level pressure, temperature, rainfall, and other variables from around the world. These records had been collected in the colonies of the major European powers and accumulated over some decades. Within each year of record, Walker calculated the seasonal averages for pressure and rainfall at each station. The averages would differ from year to year; but the patterns of the differences turned out to be similar over wide regions of the globe. An example is shown in Figure 19.2, which applies the technique of a twelve-month running mean to observations of air pressure at Darwin and of rainfall over the equatorial mid-Pacific Ocean (thus giving values for every month rather than seasonal values only, as in Walker's original method). The similarity in pattern comes out clearly.

Walker plotted world maps of these differences and found that they were dominated by a single spatial pattern. Figure 19.3 shows a modern compilation of this spatial pattern, using objective mathematical methods. The "Southern Oscillation Index" used for the figure is a composite number derived from

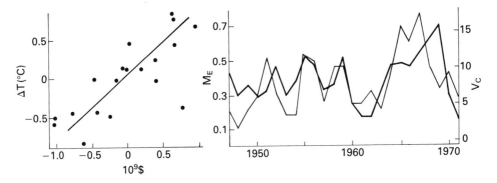

FIG. 19.1. Examples of the impact of climate variability on economic activity. (a) Anomalies of sea surface temperature (SST) in northern Australian winter (June-August, 5° - 15°S, 120 - 160°E) against variations in aggregate value of five Australian crops (wheat, barley, oats, sugar cane, and potatoes) in the subsequent summer, after removal of long-term trends due to productivity. Low SST is associated with drought; the regression line indicates a decrease of 1 billion $ for a 1°C decrease in SST. (b) Variations in wind intensity along the coast of Oregon, as measured by the mean offshore Ekman transport M_E during April to September (m² s⁻¹ per m of coastline, left scale and thin line), and variations in the catch of Dungeness crab (V_c, in million pounds, right scale and thick line) eighteen months later. The total catch reflects upwelling conditions remarkably well. (The time scale is correct for M_E but shifted by 18 months for V_c.)

observations of air pressure at sea level for Cape Town, Bombay, Djakarta, Darwin, Adelaide, Apia, Honolulu, and Santiago de Chile. (This is not the only definition of the Index in use; a more commonly used simpler version uses the difference in air pressure at sea level between Tahiti and Darwin). Its time history during the present century is given in Figure 19.4 (which, by comparison with Figure 19.2, shows that air pressure at Darwin contributes to the Southern Oscillation Index (SOI) in the inverse sense: Darwin air pressure is low when the SOI is high, and high when the SOI is low). The correlation map indicates a cellular structure of the air pressure field in the tropics, with high pressure in the central South Pacific Ocean and low air pressure over Australia, south-east Asia and India, central and southern Africa, and South America when the SOI is high. As the

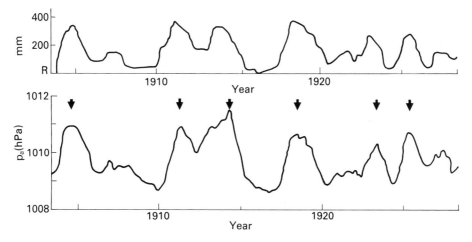

FIG. 19.2. Twelve-month running mean of air pressure p_a for Darwin (bottom) and of average rainfall R at a series of islands in the central equatorial Pacific Ocean between 160°E and 150°W (top). Arrows identify ENSO events for comparison with Fig. 19.4.

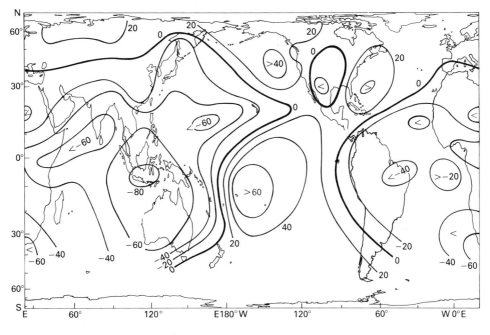

FIG. 19.3. Correlation (%) between air pressure at sea level and the Southern Oscillation Index (for explanation of the index, see text) for December - February. Adapted from Wright (1977).

SOI reverses from positive to negative (as occurred, for example, from 1971 to 1972; see Figure 19.4), so does the spatial pattern, the highs turning into lows and the lows into highs. When searching for a name for the phenomenon, Walker compared it with the more regular seasonal variations of air pressure over the North Atlantic Ocean (the natural point of reference for a colonial officer of the British Empire) and chose the term "Southern Oscillation" for what is essentially a phenomenon of the tropics. His choice of name has now been accepted by meteorologists in both hemispheres.

El Niño

One of the richest fishing regions of the World Ocean, the South Pacific coastal upwelling region along the coast of Peru, Chile, and Ecuador, occasionally experiences an influx of warm tropical water which suppresses the upwelling of nutrients. The *anchoveta*, which inhabit these waters in their millions forming the nutritional basis for a huge bird population and the stock for an important fish meal industry, depend on the supply of nutrients into the surface layer. They avoid the warm nutrient-poor water, which causes mass mortality amongst the birds (Figure 19.5). If the extent of the tropical influx is very severe, mass mortality can occur among the fish as well; hydrogen sulphide from decaying fish has been known to blacken the paint on ships in Callao harbour. The high temperatures along the South American coast last for about a year or more before conditions return to those which prevailed before the influx of tropical water.

The phenomenon has become known as El Niño, a term originally used by the fishermen of the Peruvian port of Paita to describe an influx of warm but nutrient-rich coastal water from the Gulf of Guayaquil (Tomczak, 1981b; Philander, 1990). This influx, which heralded good catches, usually occurs in December, which the fishermen (and millions of people in Christian communities around the world) associate with Christmas. The choice of "el niño" (the child) for an oceanic phenomenon as welcome and awaited as the birth of Christ appears sensible. Unfortunately, the influx of tropical nutrient-poor oceanic water associated with the suppression of the upwelling also manifests itself in a rise of surface temperature in December. When oceanographers began studying the phenomenon they failed to differentiate between the two different advective processes and adopted the local term El Niño for an oceanic phenomenon of much larger scale and of devastating effects for the fishermen.

Although oceanographers have been aware of the phenomenon for many decades (e.g. Sverdrup *et al.*, 1942), it was not until 1966 that Bjerknes (1966, 1969), a meteorologist who had become interested in ocean dynamics, pointed out the close relation between El Niño and the Southern Oscillation. As an illustration from modern data, Figure19.6 shows correlation coefficients between sea surface temperature and the Southern Oscillation Index in December - February. It is seen that sea surface temperature is low when the SOI is high (negative correlation) in a broad region of the east Pacific Ocean surrounding Peru; but it is also negative (at this time of year) in the far western Pacific, most of the Indian, and the central South Atlantic Ocean. It is evident that the phenomenon is not restricted to the south Pacific coastal upwelling region but is of global scale.

The combined process of El Niño and the Southern Oscillation has become known as ENSO, and the suppression of upwelling in the east accompanied by a drought in the west is now called an *ENSO event*. During the last two decades, with the availability of rainfall estimates over many years (derived from the outgoing long wave radiation measurements; see Chapter 18) and with improved

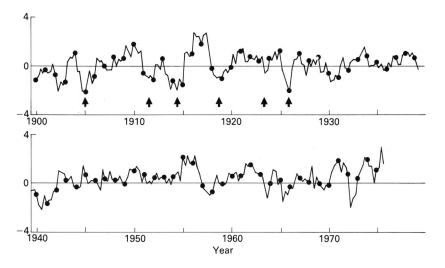

Fig. 19.4. Time series of seasonal values of the Southern Oscillation Index in units of one standard deviation from the mean. Seasons are defined February - April (identified by the dots on the curve), May - July, August - October, and November - January. For an explanation of the index, see text. Arrows identify ENSO events for comparison with Fig. 19.2.

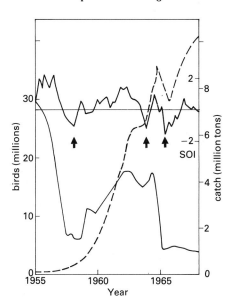

FIG. 19.5. An example of the effect of El Niño on the biosystem in the upwelling region along the coast of Chile and Peru. Annual catch of *anchoveta* (dotted line), number of sea birds, and Southern Oscillation Index (SOI). El Niño events are indicated by arrows on the SOI curve. The 1957 El Niño decimated the bird population. The subsequent build-up of an *anchoveta* fishery resulted in a new competitor and prevented the bird population from recovering to pre-1957 levels. The 1963 and 1965 El Niños affected both competitors, though the bird population suffered more severely. The fishery eventually collapsed to below 3 million tons as a result of overfishing and the 1973 El Niño (see Tomczak (1981a) for more details).

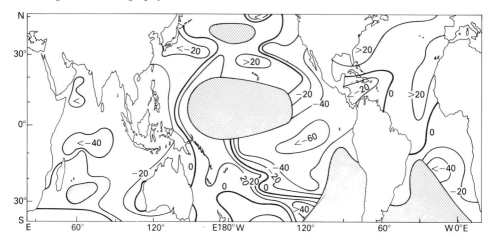

Fɪɢ. 19.6. Map of the correlation (%) between sea surface temperature and the Southern Oscillation Index for the December - February season. Data were obained from merchant vessels along the major shipping routes. Shading indicates regions with no data due to a lack of shipping activity.

knowledge of the SST distribution, it has become clear that ENSO is an instability of the coupled ocean - atmosphere system in the tropics. An indication of the effectiveness of the coupling can be seen with the intense rainfall bands of the ITCZ and SPCZ and the associated SST maxima; both are consistently found in close proximity. In an ENSO event, the entire Pacific air - sea system - rain bands and their associated winds, wind-driven currents and SST patterns - all move eastwards together, and the apex formed by the ITCZ and SPCZ in the west, which before the ENSO event was located at point *A* of Figure 19.7, moves to the dateline (point *B*). On average, the rainfall and SST maxima remain so close to one another that it is not possible to tell through the noise of shorter timescale rainfall variations whether the SST changes are causing the changes in rainfall or vice versa. The rainfall in the central Pacific Ocean strengthens, so that the convection system there competes successfully with the neighbouring convection systems for a while, suppressing their rainfall. Severe drought in Australia, Indonesia and to a lesser extent South Asia results. The centre of convection stays near the central equatorial Pacific Ocean (point *B* in Figure 19.7) for a year or more, before returning to the more common location in the far west (point *A*), bringing the ENSO event to an end. More detailed analysis of the time development of an ENSO event requires some elementary knowledge of tropical ocean dynamics, which we introduce in the next few paragraphs.

Some aspects of ENSO dynamics

Understanding the evolution of an ENSO event begins with an understanding of the evolution of the SST field. Many factors can influence the sea surface temperature. A change in wind speed affects evaporative heat loss; wind stresses create Ekman drifts, which advect surface water horizontally, and also create Ekman pumping, which changes the deeper density field and therefore the

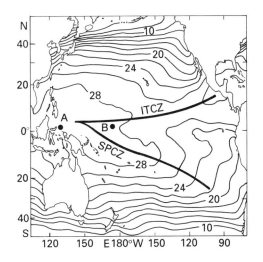

FIG. 19.7. The positions of the Intertropical Convergence Zone (ITCZ) and the South Pacific Convergence Zone (SPCZ) in relation to the annual mean sea surface temperature (°C), based on Levitus (1982). See text for the significance of locations *A* and *B*.

temperature of the water available for upwelling (a particularly important mechanism for SST change in the equatorial eastern Pacific Ocean, as we shall see). The changed density field alters the geostrophic flow, which also contributes to advection; and finally winds also provide the mechanical energy for stirring deeper water into the mixed layer. A seventh important mechanism for SST change is that due to changes in cloud cover. This plethora of different mechanisms for SST change has meant that, despite considerable progress in recent years, it has not yet been possible to identify a clearly-defined, single mechanism as the trigger for ENSO events. Indeed, there may not be a single dominant mechanism.

However, there is general agreement on one point: Westerly wind bursts in the western Pacific Ocean, i.e. reversals of the general Trade Wind pattern, seem to be a necessary ingredient of the initialization process for an ENSO event. The winds in the equatorial western Pacific Ocean are usually very light; but occasionally an outbreak of westerly winds occurs, perhaps for a week or more at a time, over thousands of kilometres - sometimes along the entire 4000 km stretch from Indonesia to about 170°W. These bursts are linked at their eastward end with pairs of low pressure cells that eventually grow into tropical cyclones (or typhoons, as they are known in the northern hemisphere). Figure 19.8 shows an example of such a situation. The two low pressure cells of 18 May 1986 later separated to lead independent lives as tropical cyclones on either side of the equator.

These westerly wind bursts are important for setting in train wave motions characteristic for the equatorial region (Figure 19.9). They literally blow surface water eastward along the equator. Because during westerly winds the Ekman transports are directed towards the equator, these wind bursts also deepen the thermocline on the equator. The equatorward Ekman transports are generally

Fig. 19.8. Cloud cover over the western Pacific Ocean as observed by satellite on 18 May 1986, indicating a tropical cyclone pair in formation near 160°E. (The latitude/longitude grid shows every 10°; the centre longitude is 140°E.) Note the westerly winds at the equator between the two centres of the cyclonic rotation.

confined to the band between about 5°N and 5°S, so the thermocline shallows near 5° - 7° N or S. A strong westerly burst can create thermocline disturbances such as sketched in Figure 19.9b within a few days. These disturbances then begin to move, through the action of equatorial wave dynamics.

The regions of shallowed thermocline near 5-7°N and 5 - 7°S at first move westward, by the Rossby wave propagation mechanism discussed in Chapter 3. It can be shown that eqn (3.11), which gave us the Rossby wave propagation speed as a function of latitude in a 1 1/2 layer ocean, is no longer valid at these short distances from the equator; nevertheless, it gives a useful first approximation. For typical thermocline depths H of 150 m and a density ratio $\Delta\rho/\rho$ = 0.004 the shallow thermocline regions move west at 6°N or S at a speed of about 0.3 m s^{-1}, taking a few months to reach the western boundary from the dateline, as sketched in Figure 19.9c.

Figure 19.9c shows also that the equatorial thermocline depression has moved rapidly eastward. This movement occurs at a speed of $c_g = (g\,\Delta\rho\,H\,/\,\rho)^{1/2}$, or about 2.5 m s^{-1}, and is due to the action of equatorial *Kelvin waves*. The principle of a Kelvin wave propagating along the equator is illustrated in Figure 19.10a. Note that an equatorial Kelvin wave can only move eastwards. Several clear examples of Kelvin wave generation by westerly wind bursts have been seen on Pacific tide gauge records. They travel with little dissipation over the entire width of the Pacific Ocean, about 10 - 20% faster than the predicted speed; the excess speed is due to advection by the Equatorial Undercurrent.

After another month or so the patch of shallow thermocline has reached the western boundary and has started to produce a disturbance in the western boundary currents there. This occurs primarily equatorward of the original patch of shallow thermocline (Figure 19.10b). These equatorial disturbances in turn

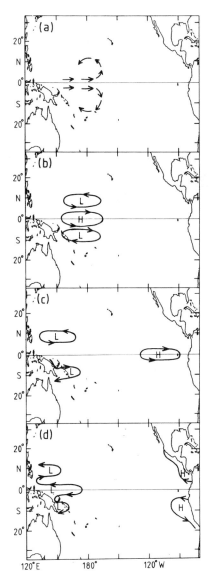

FIG. 19.9. Sketch of wave propagation during an ENSO event. (a) Wind stress *anomalies* associated with a typical westerly wind burst, (b) distribution of thermocline depth anomalies a few days after the westerly wind burst; Ekman convergence has piled thermocline water up near the equator, at the expense of off-equatorial regions, (c) distribution of thermocline depth anomalies about a month after a westerly wind burst; the off-equatorial regions of shallowed thermocline have moved slowly westward as Rossby waves, while the equatorial region of deepened thermocline has moved rapidly eastwards as an equatorial Kelvin wave, (d) distribution of thermocline depth anomalies about two months after a westerly wind burst; the Rossby waves have reached the western boundary, propagated equatorward and created a new (upwelling) equatorial Kelvin wave emanating from the western boundary; the first equatorial Kelvin wave has reached the eastern boundary, spread poleward and created new Rossby waves that are starting to propagate slowly back into the Pacific Ocean.

generate new equatorial Kelvin waves which propagate rapidly eastward, this time involving an uplifting of the thermocline (the "nose" of shallow thermocline emanating from the west in Figure 19.9d). The reflected Kelvin waves have longer period than the westerly wind burst that gave rise to them, and correspondingly lower amplitude. Meanwhile the first equatorial Kelvin wave has reached the east Pacific coast and propagated poleward as two coastal Kelvin waves. These in turn generate new - though diffuse - Rossby waves that radiate away from the eastern boundary (Figure 19.10b).

It can be imagined that these changes in thermocline depth induced by a westerly wind burst will affect SST in rather complex ways, far beyond the winds that produce them. The initial downwelling Kelvin waves of Figure 19.9b have two effects in the east Pacific Ocean. First they depress the thermocline, so that — even though upwelling favourable winds are still active — the upwelled water is substantially warmer than before. (This effect is not as strong in the central west Pacific Ocean, where the upwelled water is quite warm both before and during the passage of the Kelvin wave.) Secondly, the associated eastward currents point down the mean zonal SST gradient in the equatorial Pacific Ocean; i.e.

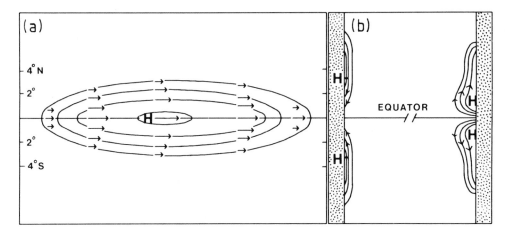

FIG. 19.10. Sketch of Kelvin wave dynamics. (a) For an internal equatorial Kelvin wave in a $1^1/_2$ layer ocean; contours indicate upper layer pressure or thermocline depth, arrows show flow direction. An isolated thermocline depression spans the equator; flow is purely zonal, in geostrophic balance and thus eastward on both sides of the equator. This removes thermocline water from the western end of the region and deposits it at the eastern end, resulting in eastward movement. (An isolated patch of *shallow* thermocline also moves eastward, though the currents and pressure gradients both have the opposite signs); (b) for internal Kelvin waves along meridional coastlines. The pressure gradients in the offshore direction are in geostrophic balance; on the western side this implies equatorward flow. This removes thermocline water from the poleward end of the region and deposits it at the equatorward end, resulting in equatorward movement. Rossby wave action keeps these waves tightly confined to the western boundary, i.e. they are disturbances of the western boundary current. Similarly, along the eastern boundary flow is polewards; this removes thermocline water from the equatorward end of the region and deposits it at the poleward end, resulting in poleward movement. In this case the patches can also propagate slowly westward through Rossby wave propagation, resulting in broadening of the pattern, especially near the equator.

horizontal advection also results in warming. As described in Chapter 18, Ekman drifts carry the warmer upwelled water polewards, so the effects on SST extend substantially further from the equator than the 300 km width of the original Kelvin wave. When the coastal Kelvin waves pass along the region of very shallow thermocline next to the Peru coast, they cause the increases in SST associated with an El Niño. The Rossby waves reflected off the eastern boundary in Figure 19.9c probably also play a role in SST change.

Anatomy of an ENSO event

This very brief summary of equatorial dynamics and our qualitative understanding of the competition between tropical convection cells from the last chapter allows us to investigate the development of a typical ENSO event in some detail. Before entering the discussion it is useful to define some terms. The time history of the Southern Oscillation Index (Figure 19.4), which is low during ENSO events, tells us that ENSO years are significantly less frequent than non-ENSO years. Distributions of oceanic or atmospheric parameters drawn from long-term annual means (such as the SST map of Figure 19.7) therefore reflect conditions during non-ENSO years. Some scientific publications and many press and television reports therefore often refer to the non-ENSO situation as the "normal" situation. From the previous discussion it should be clear that ENSO events are not abnormalities; they are basic elements of the coupled ocean - atmosphere system. Labelling a particular set of years as normal is therefore not justified. This fact has become more and more accepted in recent years, and definitions such as non-ENSO mean, ENSO-mean, pre-ENSO mean, and post-ENSO mean have found more widespread use. They refer to average conditions during years with comparable SOI values. An ENSO-mean, for example, is computed from data for all years with an SOI minimum; the non-ENSO mean would then be computed from data for all remaining years. A pre-ENSO mean uses only data from years preceeding ENSO years, while a post-ENSO mean results from data for years following ENSO years. More recently, years with unusually high SOI values, which represent a particularly strong "run" of the feedback loop when the centre of high SST is at point A of Figure 19.7, have become known as "La Niña" years (the girl, as opposed to the strict meaning of El Niño, the boy). Whether this term will become generally accepted remains to be seen.

The existence of a positive feedback loop infers that the steady state which corresponds to the long-term mean distributions of the oceanic and atmospheric parameters is rarely, if ever, realized. The combined ocean - atmosphere system is continuously changing, in response to the positive feedback described in Chapter 18. What remains to be discussed is how the system changes from one operational state of the feedback to the other, and why the ENSO state is less frequent than the non-ENSO state. The answer to this question lies in a closer study of the time evolution of an ENSO event. Because individual ENSO events vary widely in intensity and duration, the best data set to produce a somewhat general answer is a "composite" event, i.e. the pattern which shows up in the pre-ENSO, ENSO, post-ENSO, and non-ENSO means. The data base required for such an undertaking has become available over the last three or four decades and was strengthened significantly through an international research programme designed to study ENSO dynamics. The programme, known as TOGA (Tropical Ocean, Global Atmosphere), began in 1985 and will continue into 1994 as part of the World Climate Research Programme (WCRP).

For the purpose of this description, we follow Rasmusson and Carpenter and divide the composite ENSO event into five "phases", the antecedent, onset, peak, transition, and mature phases. Figure 19.11 identifies the phases in relation to the composite ENSO year and the rainfall history at two island locations. It should be noted that there is much variability between different ENSO events, and the composites discussed below are not very useful as a practical forecasting tool. Nevertheless, they provide a frame of reference for studying in broad outline what might be happening during an ENSO event.

Figure 19.12 shows *anomalies* of near-surface wind vectors and SST for the antecedent phase, in August - October preceding ENSO. The Southwest Monsoon is still active at this time of year, and is stronger than usual, drawing moist air from the Pacific Ocean and thus causing the particularly strong Trades seen in the western equatorial Pacific region. SST values are slightly below the non-ENSO average across most of the equatorial Pacific Ocean, but slightly higher SST occurs near Indonesia and Papua New Guinea, (A in Figure 19.7). Because the mean SST maximum in Figure 19.7 is so broad, this implies an absolute SST maximum near A. Figure 19.11 shows that the high SST near A (Indonesia) is accompanied by high rainfall anomalies there; rainfall is below average at B (near Nauru, 170°E; Figure 19.11) at this time.

The onset phase, seen in Figure 19.13, corresponds to November - January preceding the ENSO event. The sun has crossed the equator, and the Australian summer monsoon has started. Wind anomalies have reversed in the equatorial western Pacific region and just north of Australia. Because the monsoon winds

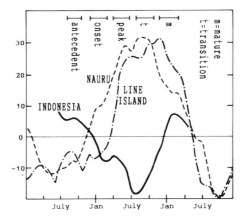

Fig. 19.11. Time development of rainfall, expressed in a rainfall index, in Indonesia, near Nauru (167°E), and in the Line Islands (near 160°W), for the composite ENSO event. Antecedent: August - October of pre-ENSO year; onset: November - January; peak: March - May; transition: August - October; mature: December - February of post-ENSO year. All values are three month running means. Note the out-of-phase relationship between Indonesian and Nauru rainfall anomalies, and the eastward propagation of the rain anomaly from Nauru to the Line Islands.

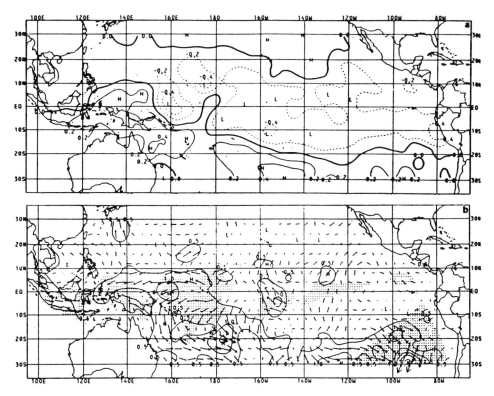

Fɪɢ. 19.12. SST anomaly (°C) and wind anomaly (m s⁻¹), during the antecedent phase of ENSO (August - October preceding the ENSO year). The magnitude of the wind anomaly is indicated by arrow length and also contoured in 0.5 m s⁻¹ intervals. Shading indicates regions where fewer than 10 observations were available in a 2° square. From Rasmusson and Carpenter (1982).

reverse seasonally, this in fact represents a strengthening of the Australian summer monsoon winds. This is not unusual, given that the preceding Asian summer monsoon was strong; a strong Asian monsoon is usually followed by a strong Australian monsoon (Meehl, 1987). The slight cooling of SST of about 0.4°C for the Indonesian region is probably due to excess evaporation, caused by the stronger than usual Australian monsoon winds. There is also a clear warming of about the same amount near 170°E (point *B* in Figure 19.7), and strong rainfall has started here (Figure19.11). Rain has correspondingly decreased near *A*. Once again, inspection of Figure 19.7 shows that these small changes in SST imply substantial shifts of the position of the absolute SST maximum eastward towards *B*. The strengthening of the westerlies east of Papua New Guinea is in fact due to an increased frequency of westerly wind bursts. An indication of these bursts can be seen in the composite mean as well (Figure 19.13): the wind anomalies near 170°E show a tendency for tropical cyclone pair formation near 170°E (anti-clockwise north, clockwise south of the equator).

What appears to be happening at this time of development of an ENSO event is that, associated with the strong Australian monsoon, a new convection centre forms near *B*. It competes with the more usual convection centre near *A*, sucking

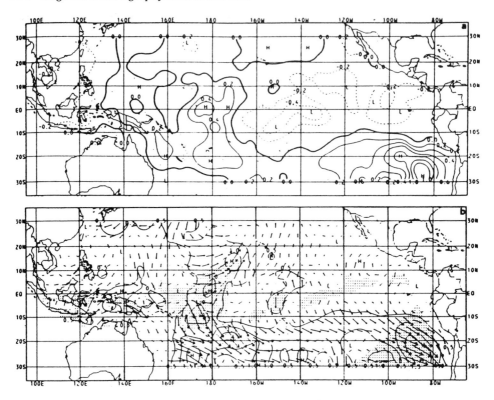

Fig. 19.13. SST anomaly (°C) and wind anomaly (m/s⁻¹), during the onset phase
of ENSO (November - January). For details see Fig. 19.12. From Rasmusson and
Carpenter (1982).

westerly winds towards it. We have noted that the very small SST changes from
Figure 19.12 to Figure 19.13 are in fact enough to displace the very broad SST
maximum from near *A* in Figure 19.7 to near *B*, so the new convection centre
near *B* is consistent with the principle that convection over the Pacific Ocean
follows SST maxima. Note that mean rainfall west of the date line is uniformly
large (about 3 m per year from Indonesia to Nauru), so a modest fractional
change in rainfall at either place is a very big change in absolute terms. The
reduction of rain over Indonesia and accompanying increase over Nauru is
apparent in Figure 19.11.

The westerly wind anomalies in the western Pacific region of Figure 19.13
continue weakly into the peak phase of March - May, and they evidently drive
downwelling equatorial Kelvin waves. The effect of these is evident in Figure 19.14;
a marked warm SST anomaly has developed near South America. As explained
in the preceding section, the Kelvin waves have a stronger effect on SST in the
shallow thermocline of the east Pacific than in the central and west Pacific Ocean.
In contrast, the small warming in the central Pacific Ocean in Figure 19.13 is
probably largely due to horizontal advection.

On the basis of competition between convection centres, one might expect
the centre near *B* of Figure 19.7 to "run away" after its formation. Curiously,
however, this does not happen between November - January and March - May.

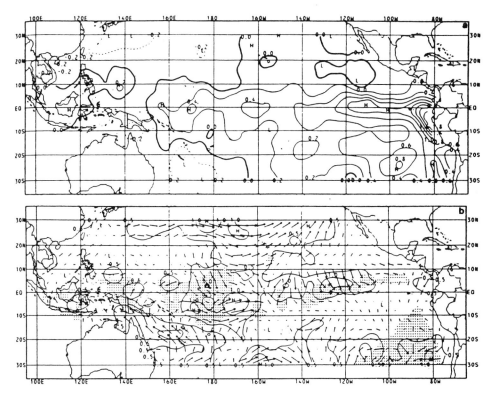

FIG. 19.14. SST anomaly (°C) and wind anomaly (m/s⁻¹), during the peak phase of ENSO (March - May). For details see Fig. 19.12. From Rasmusson and Carpenter (1982).

Perhaps the reason is that the most active convection has moved with the SST to be well south of the equator during late southern summer, i.e. from February through April.The rainfall maximum near Nauru moves further east, to the Line Islands, during this period.

However, around May - June of an ENSO year when the Southern Hemisphere convection dies and a new Southwest Monsoon begins, drastic changes occur in the Pacific circulation (Figure 19.15). By August - October, violent westerly wind bursts are occurring, and SST increases throughout the east and central Pacific; the SST is reduced near Indonesia, and Indonesia and Australia experience their greatest drought intensity.

The ENSO event usually does not break until the next change of season (December - February following El Niño), when the winds are disturbed throughout most of the Northern Hemisphere over the Pacific Ocean (Figure 19.16). A significant positive SST anomaly develops in the South China Sea and the Indonesian waters, attracting the winds from the far western Pacific Ocean which begin to blow strongly towards it, breaking the drought there, and the Trades are strong again in the west Pacific region. The east Pacific SST anomaly dies shortly thereafter.

The above description of an ENSO event places strong emphasis on the small SST anomalies in the western Pacific rather than the much more dramatic ones

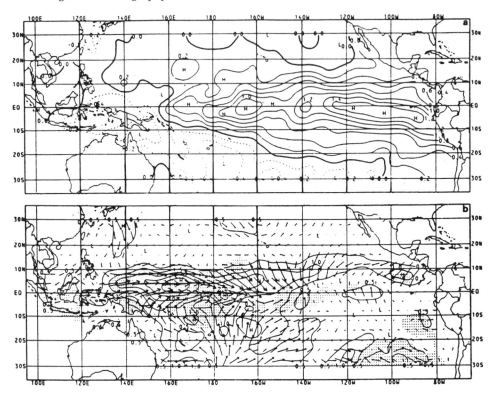

Fig. 19.15. SST anomaly (°C) and wind anomaly (m/s⁻¹), during the transition phase
of ENSO (August - October). For details see Fig. 19.12. From Rasmusson and
Carpenter (1982).

in the eastern Pacific Ocean, because they shift the SST maximum and hence
the convection patterns to the central Pacific region. Strong support for the
relative importance of the small SST anomalies in the west for ENSO comes from
sensitivity tests with atmospheric models which show that a change of SST in
the east affects the wind field much less than a corresponding SST change in
the west. This suggests that our skills in predicting ENSO events should be closely
tied to our ability to forecast very small SST changes in the western Pacific Ocean.
In this region, as we have seen, the equatorial Kelvin waves do not play a
dominant role in SST change; local changes in surface heat fluxes — solar
radiation and evaporative heat loss — are sufficient to account for the observed
SST changes in the west Pacific Ocean. However, we are then confronted with
a data quality problem. For example, the amount of heat needed to generate
the 0.4°C warming in the top 50 m or so near B at the start of an ENSO event
is only about 10 W m⁻². As was noted in the last chapter, the algorithms used
to estimate heat fluxes currently have errors of substantially more than this. Hence
improvement of heat flux algorithms has become a primary goal of ENSO
research. It has already been shown that latent heat loss at low wind speeds has
been seriously underestimated by some commonly used algorithms (Bradley *et.
al.*, 1991) and that inclusion of low-wind evaporation in some atmospheric

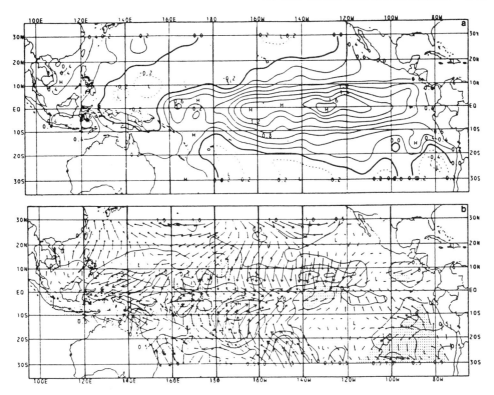

FIG. 19.16. SST anomaly (°C) and wind anomaly (ms^{-1}), during the mature phase of ENSO (December - February of the year following ENSO). For details see Fig. 19.12. From Rasmusson and Carpenter (1982).

circulation models radically improves their representation of the monsoons. The mixing of cool water into the surface mixed layer is certainly crucial in the east Pacific and can also easily make significant contributions to SST change in the west Pacific Ocean, so improvement of mixing algorithms is also a high priority for ENSO research.

The possibility that ENSO events and other movements of tropical convection systems might be accessible to reliable forecasting one or more seasons in advance has led to new demands on the climate observation network. This need led to the TOGA programme already mentioned earlier, which set itself the goal to "gain a description of the tropical oceans and the global atmosphere as a time dependent system in order to determine the extent to which the system is predictable on time scales of months to years and to understand the mechanisms and processes underlying its predictability". This aim is tackled with a variety of instrumentation. Figure 19.17 shows some components of the station network in the Pacific Ocean; others include island tide gauge installations and oceanographic satellites. TOGA is increasingly seen as a forerunner of a permanent oceanic observation network analogous to the network of meteorological observation stations on land. As in meteorology, success in forecasting climate variability will be achieved by transmitting the data to

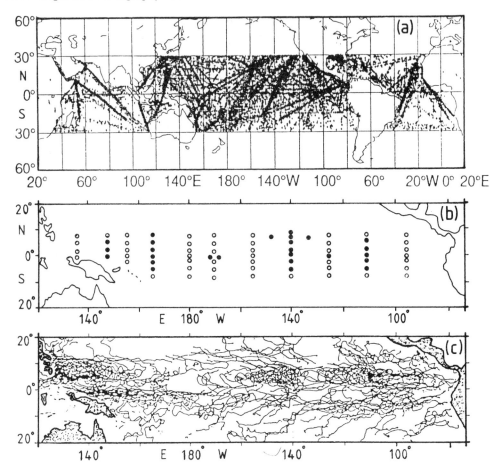

Fig. 19.17. Some of the observation components of TOGA. (a) Station positions where temperature profiles for 0 - 450 m depth, using expendable instrumentation, were obtained by merchant vessels during 1987; (b) existing (dots) and planned (circles) arrays of current meter moorings; (c) trajectories of TOGA surface drifters for July 1988 - February 1990.

information processing centres in real time and assimilating them into numerical models of the oceanic and atmospheric circulation. The planning for the Global Ocean Observing System (GOOS) began in 1992.

Interannual variability of the equatorial Atlantic Ocean

In contrast to the Pacific Ocean, interannual variability in the equatorial Atlantic Ocean is much weaker than its strong seasonal changes. The sea surface temperature along the equator is mainly controlled by advection, and seasonal changes in the current field result in a temperature change of 6 - 8°C at the surface. In comparison, the largest documented interannual temperature change did not exceed 4°C. Nevertheless, when it is recalled that the seasonal upwelling

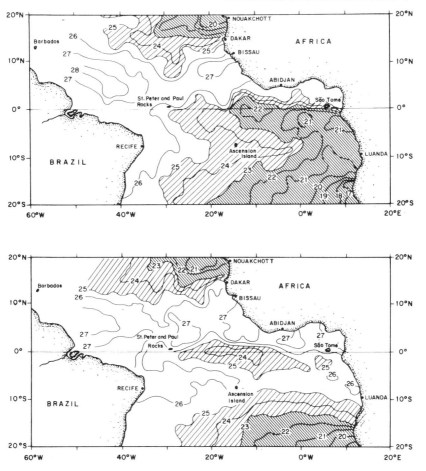

FIG. 19.18. Sea surface temperature (°C) in the tropical Atlantic Ocean. (top) June 1983, (bottom) June 1984. From Philander (1990).

along the coast of Ghana and Ivory Coast is possibly the result of remote wind forcing and Kelvin wave propagation along the equator (Figure 14.20), it is seen that the El Niño mechanism may be equally important to the Atlantic as to the Pacific Ocean. It plays a major role in the seasonal behaviour of the tropical ocean, and it appears to be responsible for occasional anomalous warmings of the water along the equator.

Figure 19.18 shows the sea surface temperature for June 1983 and twelve months later. While temperatures in the two upwelling regions outside the 20°S - 20°N equatorial belt showed little change over the period, the water within 500 km either side of the equator east of 15°W was anomalously cold in 1983 and anomalously warm in 1984. The 1983 situation was accompanied by dry conditions in northeastern Brazil, strong equatorial Trades, and a continuation of strong upwelling in the northern part of the Benguela Current upwelling system (which ususally retreats southward during this time of year). In the following year, westward flow in the equatorial current system was nearly halted, and an intrusion of tropical water similar to that observed along the coast of Peru during El Niño years dominated the northern part of the Benguela Current upwelling system, leading to a temperature increase of 3°C in the upper 50 m of water near 23°S. The causes for such extreme situations in the Atlantic Ocean are not clear. There is also uncertainty about the frequency of their occurrence; only two intrusions of tropical water have been documented for the 40 year period between 1950 and present, although historical records of coastal water temperatures from Namibia seem to indicate one intrusion for every ten years.

CHAPTER 20

The ocean and climate change

In this final chapter we turn from the short-term climate variability of ENSO to consider variations with timescales of several decades or more. Small but persistent changes in climate can have a huge impact on the ecosphere (the finely tuned interplay between living matter and its environment). Such changes can occur naturally, as they have done in the past, experienced for example as ice ages and interglacial periods. This century is witnessing the possibility of climate changes through human activity. The last few decades have seen the introduction of the so-called greenhouse gases (CO_2, CH_4 and others) into the atmosphere on an ever increasing scale; this alters the radiation balance and may result in climate changes of magnitudes comparable to those which occurred naturally in the past. Many countries therefore have made research into these longer-term variations a national priority, and oceanographers are asked to quantify the role of the oceans in long-term climate change. However, when oceanographers try to come to grips with longer-term changes in the ocean, they immediately find themselves facing a major problem - the very patchy nature of the observational record. There have been large fluctuations in the number and spatial distribution of observations made over the decades, as shipping routes changed and wars interrupted continuous data collection and research plans. In many cases the technology used changed drastically over time, and data in the same series are not directly comparable. There is considerable risk that changes in observation density and technique may introduce bias, and great care has to be taken in interpreting long time series of ocean data. Nevertheless, as we shall see, there are several incontrovertible examples of changes in the ocean over decades, and some of these lend themselves to preliminary interpretation in terms of large-scale changes in the ocean circulation patterns.

Given the problems with the data, and the rapidly increasing sophistication of ocean numerical models, it is not surprising that climatologists have recently come to rely rather heavily on numerical models to explore the ways in which changes in ocean circulation are likely to occur. These models, and in particular those which couple the ocean with the atmosphere, have generated some valuable ideas about the likely nature of some kinds of longer-term climate variations. Some of these new ideas and their implications are discussed in this chapter.

Using models to extend our limited observational data base into the future is of course risky business, since few of the model predictions can be verified

381

with existing data. Modellers and observationalists are both well aware of the dangers of predicting long-term climate variations in this way. One of the major goals of the international climate research community is therefore directed towards building an observing network that will keep track of climate changes throughout the world ocean, as well as in the atmosphere. How this can be achieved and what is on the drawing board is discussed at the end of this chapter.

The observations

Three major sources are available for inferring changes in ocean circulation over the last hundred years or so. The first is the record of sea surface temperature from merchant ships. Some 150 years ago Captain Matthew Fontaine Maury of the U.S. Navy introduced an international program of data collection at sea, which has been gaining impetus ever since. Ships' officers have always recognized the value of good marine weather data (which was even more important in the days of sailing ships than today), and so the quality of the merchant ship record (though by no means perfect, as we shall see) is remarkably good. The extremely laborious task of entering all these decades of data from around the world into computer-compatible form is also now largely complete (there were about 63 million SST observations alone, to 1979); various versions of the world record of marine observations, known as the Comprehensive Ocean-Atmosphere Data Set or COADS, are now readily available for interested users, in various stages of editing and compression. In addition to sea surface temperature, this data set contains surface winds, air pressure, temperature and humidity, cloud cover and rainfall, which also provide valuable information from which - given our present understanding of the ocean as a dynamical system - changes in ocean circulation can be inferred. However, most of their value is in meteorology, and detailed discussion is beyond the scope of this book.

The second data set consists of a few long-term tide-gauge records, mostly from Europe and North America. These have been much studied in connection with the possibility of long-term sea-level rise as a result of increasing CO_2 levels. We will address both their limitations and the things that have been learned from them.

Thirdly, accurate profiles of ocean temperature and salinity throughout depth have been undertaken since the introduction of the Nansen bottle and the reversing thermometer, the precursors of the modern CTD, some 110 years ago. However, reliable sections across ocean basins do not go back further than about 70 years, and examples of high-quality sections that have been repeated a few decades apart are quite rare; thus the interpretation problem in this case is usually one of the representativeness of the data rather than one of data quality.

Sea surface temperature

In the early days of the merchant ship measurement program, sea surface temperature (SST) was measured by picking up a water sample in a canvas bucket and measuring its temperature when it reached the ship's deck. Evaporation from the bucket's walls will generally result in cooling, depending on weather conditions and the time taken to make the measurement. By contrast, for the last forty years SST has mostly been measured by devices installed in the intake for the engine cooling system; in this case water is usually warmed during its transit from the ocean to the thermometer. The difference between uninsulated bucket temperatures and engine room intake temperatures is generally in the range 0.3-0.7°C. Since the data forms filled out by the officers usually did not include a space for recording the device used for measurement (uninsulated bucket, insulated bucket or engine room thermograph), there is an inherent uncertainty in the long-term SST records. Reasonable assumptions have been made about the change of instrumentation through time, but these cannot now be directly checked. Similarly, air temperatures are often measured in rather sheltered locations that may be subject to deck heating during daytime, and such details are not recorded. Nevertheless, it has been found that when reasonable correction procedures are applied to each data set, the magnitude of the corrections turns out to be extremely consistent between, for example, the northern and southern hemisphere, and the variations in corrected air and sea temperature track one another quite closely. This gives one some confidence that along major shipping routes averages of each time series over a 10°x10° square for a decade will be reliable to one or two tenths of a °C.

When processed in such a way, i.e. compressed into mean annual values for 10°x10° squares, sea surface temperature is seen to be a function f of space and time defined at constant space and time intervals: SST = $f(x,y,t)$, where x, y represent the longitude and latitude of the centre of the 10°x10° square and t is the year. Several mathematical techniques are available to analyze space and time trends in such data sets. One method regularly used in meteorology and oceanography is a technique known as Empirical Orthogonal Function (EOF) analysis. The method represents the data as a sum of products of functions: $f(x,y,t) = \Sigma\, F_i(x,y)\, G_i(t)$, where the F_i express the data distribution in space and the G_i give the contribution of the respective space distribution to the observed SST field at any given time. Theory shows that an infinite sum of function products can reproduce the observations to any required accuracy and that many such representations of the data are possible. In practical applications the summation is truncated after the first few terms. The strength of EOF analysis lies in the fact that it arranges the contributions of the sum in such a way that the first term ($i = 1$) accounts for more of the variance found in the observations than any other term; the second term ($i = 2$) then accounts for more of the variance found in the difference between the observations and the first term than any of the following terms, and so on. This allows one to extract the dominant spatial and temporal signals with the help of very few function products.

Figures 20.1 - 20.3 show results of an EOF analysis of SST based on a data set similar to COADS but going back to "only" 1900. $F_1(x,y)$, the space function for the first EOF, is shown in Figure 20.1a, while Figure 20.1b shows the corresponding time function $G_1(t)$. The space function is seen to be positive

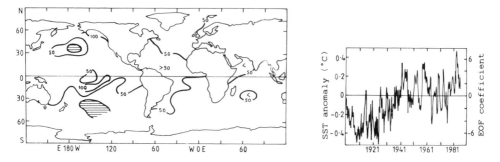

FIG. 20.1. Empirical Orthogonal Eigenfunction analysis of sea surface temperature.
First Eigen function: (a) spatial pattern (arbitrary units, negative values hatched),
(b) temporal amplitude. From Folland *et al.* (1986a)

nearly everywhere, and the time function is increasing irregularly with time. The
net increase in $G_1(t)$ over the last 80 years is about 0.5°C. The net contribution
of the first EOF to SST is given by the product of both functions, which is negative
everywhere before 1940 and turns positive from 1975 or so.

This provides fairly convincing evidence that - despite the uncertainties discussed
earlier - the sea surface temperature has indeed risen over the present century
by about 0.5°C on global average. However, it should be noted that this
interpretation relies in part on the particular presentation of the first EOF; a
similar analysis for the COADS data set which goes back another fifty years shows
the same trend but cooling between 1860 and 1910 (Folland *et al.*, 1984).

As outlined above, the first EOF can be subtracted from SST(x,y,t) and the
same analysis applied to the resulting difference, to give the second EOF which
describes most of the remaining SST variation. The spatial pattern $F_2(x,y)$ is seen
in Figure 20.2a and the corresponding time amplitude $G_2(t)$ in Figure 20.2b.
In this case the spatial pattern has a maximum in the eastern Pacific Ocean and
is in general closely reminiscent of the ENSO pattern of SST variability seen in
Figure 19.6. Furthermore, the time amplitude $G_2(t)$ has maxima at each of the
ENSO events. It follows that during ENSO years the product $F_2(x,y) G_2(t)$ has
positive values (positive SST anomalies) in the eastern Pacific and negative values
(negative SST anomalies) in the western Pacific Ocean, while the reverse is true
during anti-ENSO years. In other words, the second EOF can be identified with
the ENSO signal in SST discussed in Chapter 19. (EOF analysis also reveals that
ENSO events are associated with a *global* warming of the ocean surface; this is
seen in the first EOF which shows maxima in $G_1(t)$ during ENSO years.)

Subtracting the second EOF and applying the procedure once again produces
the third EOF shown in Figure 20.3. $F_3(x,y)$ shows positive values in the north
Atlantic and north Pacific Oceans and negative values through most of the
Southern Hemisphere, while the time function shows slow variation over several
decades. It might be thought that after such mathematical manipulations the
result would be more noise than signal; however, there is such strong spatial

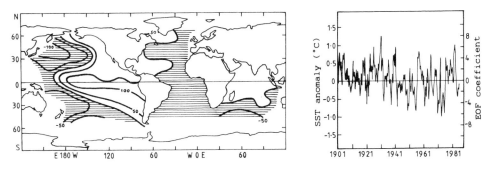

FIG. 20.2. As for Fig. 20.1 but for the second Empirical Orthogonal Eigenfunction.

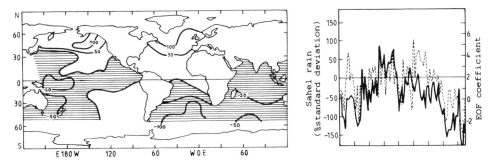

FIG. 20.3. As for Fig. 20.1 but for the third Empirical Orthogonal Eigenfunction. The dashed line in Fig. 20.3b shows an area average of rainfall in the Sahel region.

and temporal coherence in F_3 and G_3 that this is unlikely. Furthermore, both functions seem to be associated with significant changes in the world's climate. An idea of the magnitude of these effects can be gained from the fact that between 1950 and 1980, the temperature of the southern hemisphere oceans (plus the northern Indian Ocean) increased relative to the rest of the northern hemisphere oceans by about 0.4°C.

Figure 20.3b includes a measure of average rainfall in the Sahel region at the southern edge of the Sahara Desert. There is evidently a fair degree of correlation between the rainfall time series and the third EOF time function $G_3(t)$ of SST. To test whether this correlation is a coincidence or is based in physics, atmospheric numerical models have been run with the SST anomalies of Figure 20.3a superimposed on a mean SST climatology. These generate rainfall anomalies over tropical Africa fairly similar to observation. A plausible explanation is that addition of the SST pattern of Figure 20.3a to climatology shifts the tropical SST maximum northwards. As discussed in the previous two chapters, this tends to move the rainfall maximum north with it. However, the true explanation may be more complex than this, and some time will pass before climatologists will be able to assist the people of the Sahel region to avoid the hardship and suffering which they experience during the present series of droughts.

These examples show that observed changes in SST can be related to observed variations in climate. It is therefore realistic to hope that with our increasing data base we will be able one day to go beyond mere description and come to conclusions about causes and effects.

Sea level measurements and sea level rise

Unlike the merchant ship data for which results are usually averaged over observations from a large number of ships, each tide gauge is an individual instrument. Its reliability over decades depends on the care taken by its operators in preventing fouling and damage, in meticulously recording any shifts in the tide staff fixed to a wall next to the gauge, and in taking accurate surveys every few years to relate the height of the staff to stable bench marks on the shore. 179 stations exist with records of more than 30 years; only 22 of these have records of more than 80 years, and of these only 3 are located outside the northeast Atlantic Ocean. Figure 20.4 shows the locations of sites with records of more than 10 years.

Many of these 179 records have to be rejected for use in long-term climate studies. For example, much of the coast of Japan and western North America is tectonically active, so that all the bench marks to which the tide gauge height has been measured may have shifted by unknown amounts. In fact it is now recognized that *all* tide gauges are subject to slow rises and falls of land level, because the magma beneath the earth's crust is still slowly flowing back towards the regions occupied by thick ice sheets only about 10,000 years ago; however, outside tectonically active areas recent numerical models of this process appear to be successfully capturing the main features of this "postglacial rebound" (Peltier and Tushingham, 1991).

Using these data, Gornitz and Lebedeff (1987) found that much of the variations from region to region could be removed if the tide gauge records were corrected for post glacial rebound. After correcting for this effect, Gornitz and Lebedeff found a global mean sea level rise of 1.2 ± 0.3 mm/year. Two more recent estimates of global mean sea level rise in the last century are 1.8 ± 0.1 mm/year (Douglas, 1991) and 2.4 ± 0.9 mm/year (Peltier and Tushingham, 1991). Both of the new estimates rely on a global model of postglacial rebound, which indicates that far away from the previously glaciated regions, coastal lands are rising relative to the sea. It is apparently the correction of this effect which leads to the discrepancy between Gornitz and Lebedeff's(1987) estimate and the two newer estimates. The sharpening of the estimate of sea level trends achieved by correcting for postglacial rebound is illustrated in Figure 20.5; note that the corrected trend is clearly greater than zero. However, Peltier and Tushingham(1991) report that their estimate of global mean sea level rise is "extremely sensitive to relatively modest alterations to the analysis procedure" which no doubt applies also to Douglas' estimate. All estimates are of necessity biased by the heavy concentration of available sea level records in Europe and North America.

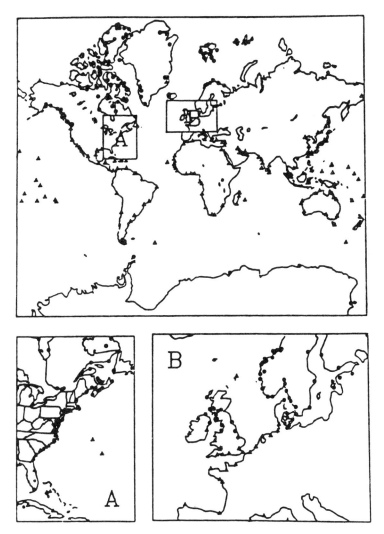

FIG. 20.4. Locations of validated tide gauges from which records are available that are of length greater than 10 years. The data are archived with the Permanent Service for Mean Sea Level at Bidston, Merseyside, United Kingdom. From Peltier and Tushingham (1991).

Much of this observed sea level rise can be accounted for by thermal expansion of seawater. As seen in Figure 20.1, the ocean has certainly warmed at the surface in the last century, and water expands upon warming. Convergence of Ekman transports and the convective overturn of water after surface cooling results in downward motions in certain parts of the ocean. Both processes provide the principal means by which warmed water is carried below the ocean surface; it is an advective rather than a diffusive process, so its magnitude can be directly estimated from large-scale ocean observations of currents. This makes it somewhat easier to assess thermal expansion rates for a given history of global mean surface

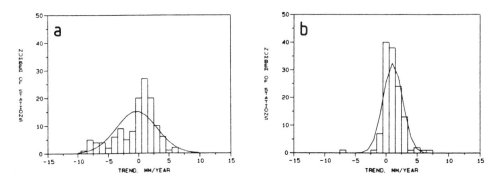

FIG. 20.5. Histograms of sea level trends from all tide gauge records with length of 50 years or longer, (a) before and (b) after correction for postglacial rebound. From Douglas (1991).

temperature rise. The subducted water tends to accumulate in the subtropical gyres. However, Kelvin and Rossby waves tend to redistribute the warming over the rest of the globe; if this process went to completion, and the wind stress field did not change, the increase in depth-integrated steric height would be uniform over the world. This results in a rise in surface steric height (i.e. sea level) that is fairly spatially uniform, though thermal expansion rates are predicted to be somewhat greater in the tropics than near the poles (Figure 20.6). For the global

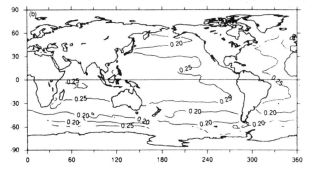

FIG. 20.6. Estimated sea level rise (m) by 2050 caused by thermal expansion, from the model of Church *et al.* (1991), assuming a global averaged temperature increase of 3°C by 2050.

mean SST rise of 0.4° - 0.6°C per century inferred from Figure 20.1, the model yields a global mean thermal expansion of 0.8 ± 0.2 mm/year.

A complete assessment of sea level rise has to include a number of other factors. The volume of water generated by the observed melting of non-polar glaciers over the last century is equivalent to a sea level rise of 0.46 ± 0.26 mm/year (Meier, 1984). It is seen that the two effects together, thermal expansion and non-polar glacier melt, yield between them sea level rise estimates of 1.3 ± 0.5 mm/year, somewhat below the most recent estimates quoted above ($1.8 \pm .1$ mm/yea $\cdot$ 2.4 $\pm$ .9 mm/year). The uncertainties in the remaining contributions are certainly large enough to account for the discrepancy. The effects of a warming on Greenland and Antarctica are believed to be of opposite sign, and neither is well known — the large ice sheets on Greenland's flanks are thought to have retreated under warmer conditions, whereas Antarctica (and inner Greenland) are so cold that the warming should have produced little or no melting over the last century. Instead, increased sea temperatures should have generated increased snowfall over Antarctica and inner Greenland. The net contribution from both polar icesheets is believed to be near zero. However, this can evidently only be a very approximate figure, and the uncertainty of our estimate of the total sea level rise in the last century is substantially greater than the figure of ± 0.5 mm/year from thermal expansion and non-polar ice melt alone.

Contributions of both signs arise also in consideration of groundwater, the water trapped on or under land masses. Artesian bores have removed substantial quantities of groundwater over the last century, while large water storages for hydroelectric dams etc. have been created. Newman and Fairbridge (1986) estimate that the latter effect would have reduced sea level by up to 0.75mm/year from 1957-1980, mostly in small water storages. The USSR Committee for the International Hydrologic Decade (Korzun, 1978) estimated a sea level rise of 0.8 mm/year due to reduction of groundwater storage.

In summary, our best estimate at present is that the combined contribution from groundwater and ice storage to sea level rise has been positive over the last century, and perhaps of the same order of magnitude as the combined contribution from thermal expansion and non-polar ice melt. However, in the absence of better information, the most recent projections of future sea level rise assume that there will be no net contribution from polar ice and groundwater.

Regional variations of sea level on decadal time scales

Before leaving the topic of long-term sea level change it is worth noting that after correction for postglacial rebound, other observations of interest from the point of view of decadal sea level change can be extracted from the available sea level records on the eastern coast of the USA and Canada.

Significant differences in the rate of sea level rise do occur from place to place, that probably originate in oceanographic effects. When sea level, corrected for postglacial rebound, from tide gauges along the North American east coast is examined for a trend over the period 1930-1980, it is found that over the 50-year interval sea level rose about 0.05 m more at gauges south of 38°N than at gauges to the north of 38°N (Figure 20.7). No tectonic explanation for this feature is known. The most likely cause is that the Gulf Stream altered its strength over the period. A very strong mean drop in steric sea level of order 0.7 m, about the strongest in the world ocean, is associated with the separation of the Gulf Stream from the coast near 35-38°N. The break in sea level trends over the past

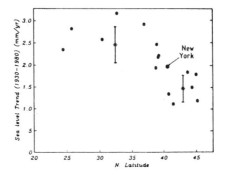

FIG. 20.7. Trends of sea level (corrected for postglacial rebound) on the North American east coast, 1930-1980, plotted as a function of latitude. From Douglas (1991).

50 years seen in Figure 20.7 occurs quite close to this mean sea level drop. It is rather reasonable to suppose that the 0.7 m sea level drop might have changed by 0.05 m, or 7%, over the last 50 years due to changes in climate.

Ocean hydrology

The third source of long-term data for the ocean comes from accurate hydrological observations made by ocean research vessels and Ocean Weather Stations. The latter are vessels that have been stationed at fixed locations for some decades to supplement the land-based meteorological observation network and provide advance warning of weather events approaching the continents. Figure 20.8 shows a time series of monthly average salinity profiles based on data collected daily from Ocean Weather Station Bravo (56°30'N, 51°00'W) near the centre of the Labrador Sea. Freshening in late summer is evident at 10 m each year; in winter, salinity increases due to sea-ice formation. In most winters, a slight freshening can be seen at and below 100 m, bringing winter salinities close together over the top several hundred meters; this is due to convective overturn of the relatively fresh but cold waters above. Maximum mixed layer depths reached

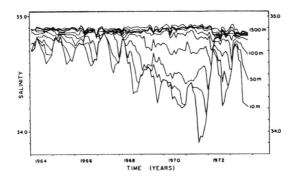

FIG. 20.8. Time series of monthly mean salinities at Ocean Weather Station Bravo, (56°30'N, 51°W), for 1964-1974. From Lazier (1980).

to over 1000 m in the winters of 1964-65, 1966-67, 1971-72 and 1972-73. However, from 1967-1971 the surface salinity was significantly lower than in other years, and mixed layers in these winters only penetrated to 200 m. Lazier (1980) suggested that abnormal northerly winds during these years blew sea ice southward, leading to greater ice melt in summer.

Brewer *et al.* (1983) complemented Lazier's work by examining two salinity sections across the Atlantic near 57°N, one taken in 1962 and the other in 1981. They found a systematic salinity decrease of about 0.02 between the two cruises. A plot of mean and standard deviation of salinity as a function of σ_{1000} from each cruise is seen in Figure 20.9 (σ_{1000} is the density the water would have if it was brought to 1000 m without changing salinity or potential temperature). The freshening is particularly evident when σ_{1000} is greater than 37.1 and for the range $36.8 < \sigma_{1000} < 36.9$. The first of these water mass modifications is thought to originate north of Denmark Strait, the latter in the Labrador Sea. These studies show that widespread changes in deep water can occur quite rapidly in response to rather modest changes in atmospheric conditions.

These deep water effects are not the only large-scale changes that have been observed in the north Atlantic Ocean over the last few decades. Levitus (1989) undertook a statistical study of the changes in the north Atlantic circulation from 1955-59 to 1970-74, using the data bank of all historical hydrographic observations. Unfortunately the north Atlantic Ocean is the only region where sufficient data exist to make such an analysis possible on a basin-wide scale. Figure 20.10b shows the difference in the depth of the 26.5 σ_θ surface for 1970-74 against 1955-59. Evidently, the thermocline shallowed significantly in most of the subtropical gyre,

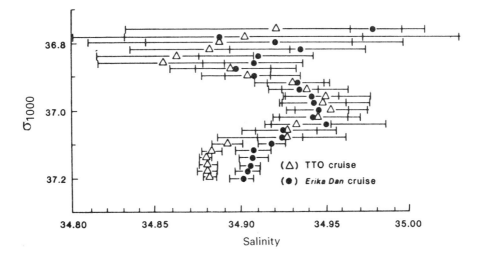

Fig. 20.9. Salinity means and standard deviations (horizontal bars) from two cruises across the North Atlantic near 57°N (dots for the 1962 cruise, triangles for the 1981 cruise) as functions of density σ_{1000} (for explanation see text). Note the clear freshening of the later curve, below $\sigma_{1000} = 37.1$ (corresponding to potential temperatures of about 2.5°C) and for σ_{1000} near 36.8 - 36.9. From Brewer *et al.* (1983) .

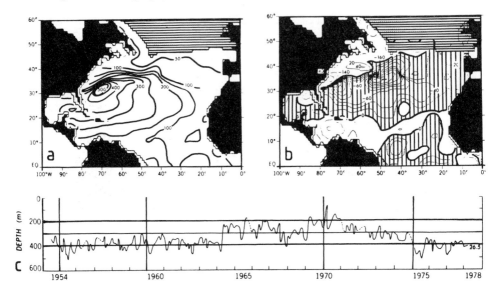

FIG. 20.10. Changes in thermocline depth in the North Atlantic Ocean. (a) Depth
(m) of the 26.5 σ_θ surface for 1955-1959; crosses indicate regions where the surface
does not exist, (b) depth difference (m) of the 26.5 σ_θ surface 1970-74 minus
1955-59, (c) depth of the 26.5 σ_θ surface at 32°N, 64°W for 1954-1978. Adapted
from Levitus (1989).

with some deepening on the inshore edge of the Gulf Stream. The net
appearance is of significant weakening of the density gradients across the Gulf
Stream over this period, and hence (through geostrophy) a weakening of the
near-surface Gulf Stream itself. However, a time series (Figure 20.10c) of the
depth of the 26.5 σ_θ surface at 32°N, 64°W near the maximum depth of this
isopycnal surface suggests that the shallowing near this location is not part of
a longer-term trend but is associated with a minimum depth of the ispycnal surface
in the early 1970s. Much more systematic monitoring of the oceans is required
to link these observations with possible climate trends.

These broad-scale changes in the Atlantic subtropical gyre seem to involve
changes in isopycnal depths without much clear signal in the water mass structure.
Changes in water mass properties can be investigated by monitoring changes
of potential temperature for given densities. As an example, Figure 20.11 shows
the change in potential temperature on potential density surfaces along 49.5°W.
A systematic cooling and therefore freshening is apparent north of 45°N; though
it should be noted that this is a near-surface effect (water with $\sigma_\theta < 27.5$ is
confined to the top 200 m at these latitudes).

It would be useful to extend such studies to the southern hemisphere, but
opportunities for doing so are unfortunately rare. One recent example uses
hydrographic sections at 43°S and 28°S in the western South Pacific Ocean taken
in 1967 and repeated in 1989-90. Over the 22 years between these sections, there
has been a depth-averaged warming at most depths below the surface mixed layer
(Bindoff and Church, 1992). As in the case of the Atlantic Ocean, these changes

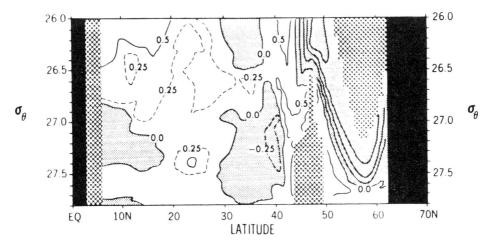

Fɪɢ. 20.11. Change of potential temperature 1970-1974 minus 1955-1959 (°C) on potential density surfaces along 49.5°W, as a function of potential density. Cross-hatching indicates regions where the corresponding potential density does not exist. From Levitus (1989).

are mostly due to changes in isopycnal depth with rather little change in water mass properties on surfaces of constant density. A slight freshening for water of temperature 8°C or higher can account for a rise in sea level of about 2-3 cm, roughly equal to the observed rate of sea level rise over 20 years.

In conclusion, it will be evident from these examples that, while interdecadal variations definitely have occurred on basinwide scales in the ocean, our ability of keeping track of these interdecadal changes observationally is extremely sketchy. This needs to be borne in mind when considering results from numerical model studies.

Model results, salinity and climate

In an attempt to extend our understanding of long-term climate change beyond the limits posed by the available data, several research groups have begun the task of developing numerical models of the coupled ocean/atmosphere system. Such models require massive computing power, which is becoming available now. When applied to a simulation of our present climate, most models give satisfactory results when the ocean or the atmosphere are modelled in isolation but develop unreasonable climate trends (e.g. a rapid warming of the ocean surface) when the two components are treated as a coupled system. The salt budget proves to be particularly difficult, most probably because we do not yet understand how to incorporate the process of tropical rainfall correctly. A number of more or less empirical methods have been developed to prevent the models from diverging from the known climate trend of the last decades. Our hope is that by applying these methods to simulations into the future we can get reasonably accurate estimates of future climate trends.

One of the most intriguing results from these models is the role of sea surface salinity. Models which simulate the oceanic and atmospheric circulation for several thousand years have revealed the existence of (at least) two stable steady states. One of these steady states corresponds to the circulation system we observe today, with North Atlantic Deep Water formation and recirculation through all ocean basins. The other steady state does not have North Atlantic Deep Water formation; it shows a very much colder and fresher north Atlantic Ocean and very little Deep Water exchange between the three major oceans. Figure 20.12 shows the SST difference between the two possible steady solutions. Compared to the present situation, the solution without NADW formation shows the north Atlantic Ocean colder by as much as 7°C and the north Pacific Ocean colder by some 2°C. The qualitative resemblance between Figure 20.12 and the third EOF derived from SST data (Figure 20.3a) is striking; however the amplitude of the SST differences of Figure 20.12 is larger than the largest differences associated with the third EOF by a factor of 5. One way of interpreting this is to say that the data support the possibility of an alternative steady state of the oceanic circulation, in the sense that the time function $G_3(t)$ can be seen as an indicator for the speed with which the ocean circulation may be changing from one steady state to the other.

The existence of two steady states may seem to contradict our argument from Chapter 18 that a cold but fresh north Atlantic Ocean will eventually return to its present state by increasing its salinity through water vapour export across Central America. However, if the change in north Atlantic SST is large enough (and a 7°C SST difference is an enormous change) it might be expected to have a major effect on the atmospheric circulation and associated rainfall which might

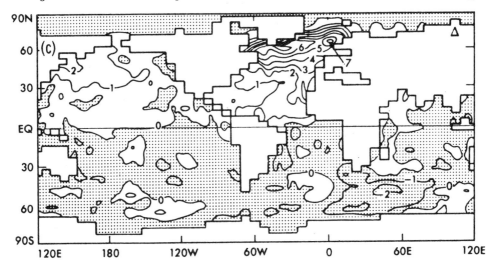

FIG. 20.12. Difference in sea surface temperature, $T_1 - T_2$ (°C), between two stable steady states of the oceanic circulation found in a coupled ocean/atmosphere model. T_1 is SST in the model with NADW formation, T_2 is SST in the model without NADW formation. From Manabe and Stouffer (1988).

result in a suppression of water vapour export from the Atlantic Ocean. A large shift in the distribution of tropical rainfall was indeed found in the coupled model; however, as already mentioned, modelling tropical rainfall is one of the weak points of all models at present.

Whether the ocean circulation (within the confines of the present world topography) can, or did in the past, have more than one stable steady state as indicated by models, is a subject for paleoceanography. Whatever the answer will be, there is no doubt that the north Atlantic Ocean is the major determinant of the process. Consider the sketch shown in Figure 20.13. Deep convection in the Greenland Sea will be inhibited if the salinity in the northward flow of water from the subtropics is reduced; this is the mechanism behind Figure 20.12. The same effect would be observed if the amount of freshwater and ice exported from the Arctic Ocean is increased to cover the surface layer of the Greenland and Iceland Seas. This would force the warm, salty water from the south to submerge below the fresher surface water well before entering the Greenland Sea (in today's climate it does not submerge until it reaches the Greenland Sea; see Chapter 7). The fresh surface layer would insulate the underlying subtropical water and prevent it from cooling, stopping the deep convection in the Greenland Sea. Formation of North Atlantic Deep Water can thus be inhibited by various means, and it seems more and more likely that circulation patterns without NADW formation did exist in the past. To give just one example, sediment cores from the Antarctic Circumpolar Current region show large variations in carbon isotope composition between periods of glaciation and interglacial periods, which can be related to changes in the NADW contribution to Circumpolar Water (Charles and Fairbanks, 1992) and suggest that NADW formation was much reduced during the last ice age. This of course means that a circulation with very little or no NADW formation can develop again. Does the introduction of greenhouse gases

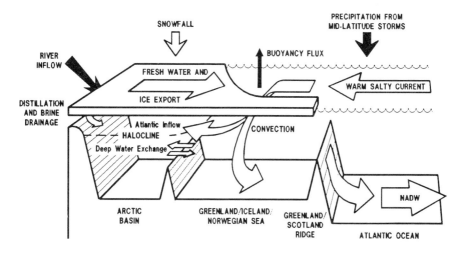

FIG. 20.13. Sketch of the circulation in the north Atlantic Ocean.

into the atmosphere promote the change to such a pattern, or does it stabilize the existing circulation? The answer, although impossible to give today, is of tremendous interest to the people of Europe, which would see drastic changes if the north Atlantic sea surface temperature were to fall by several °C.

While the model results and possible scenarios are extremely interesting, they are only indicative of the types of phenomena that may occur in the real coupled ocean-atmosphere system. They give us no indication, for example, whether the fluctuations of the north Atlantic gyre discussed above are associated with some coupled ocean-atmosphere fluctuation or merely a response to random fluctuations in atmospheric forcing; nor do we have the data to know if such phenomena are occurring in most of the other basins. Regional oceanographers and modellers agree that there is a lamentable gap between the comprehensive global coverage of the ocean apparently provided by models and the extraordinarily sparse nature of the data base with which to verify them. We will certainly be living with this problem for some decades. However, oceanographers are devoting considerable efforts to improving the situation. The World Ocean Circulation Experiment (WOCE) introduced in Chapter 2 is one component of the strategy, TOGA (see Chapter 19) another. Out of these efforts will evolve the Global Ocean Observation System (GOOS) as the equivalent of the network of meteorological observations supported by all member countries of the World Meteorological Organization (WMO). Although it will have a very similar function, namely the provision of real-time data for the forecasting of the oceanic circulation, it will have to be based on very different technology. Factors affecting its design are the lack of availability of voluntary observers for many ocean regions and the - in comparison to atmospheric space scales - much smaller scales of oceanic eddies and frontal variability. The increased demand posed by small space scales is partly compensated by the much longer oceanic time scales which allow less frequent sampling than in the atmosphere. Scientists involved in the design of GOOS are actively working on questions of data resolution in space and time required for forecasting the variability of the oceanic circulation.

References

Aagaard, K. and P. Greisman (1975) Towards new mass and heat budgets for the Arctic ocean. *Journal of Geophysical Research* 80, 3821-3827.

Aagaard, K., L. K. Coachman and E. C. Carmack (1981) On the halocline of the Arctic Ocean. *Deep-Sea Research* 28, 529-545.

Aagaard, K., A. T. Roach and J. D. Schumacher (1985a) On the wind-driven variability of the flow through Bering Strait. *Journal of Geophysical Research* 90, 7213-7221.

Aagaard, K., J. H. Swift and E. C. Carmack (1985b) Thermohaline circulation in the Arctic Mediterranean Seas. *Journal of Geophysical Research* 90, 4833-4846.

Armi, L. and H. Stommel (1983) Four views of a portion of the north Atlantic subtropical gyre. *Journal of Physical Oceanography* 13, 828-857.

Arnault, S. (1987) Tropical Atlantic geostrophic currents at the equator. *Journal of Geophysical Research* 6, 287-297.

Arnone, R. A., D. E. Wiesenburg and K. D. Saunders (1990) The origin and characteristics of the Algerian Current. *Journal of Geophysical Research* 95, 1587-1598.

Bainbridge, A. E. (editor) (1980) *GEOSECS Atlantic Expedition, volume 2: sections and profiles.* U.S. Government Printing Office, Washington.

Bang, N. D. (1971) The southern Benugela Current region in February, 1966: part II. Bathythermography and air-sea interactions. *Deep-Sea Research* 18, 209-224.

Bang, N. D. and W. R. H. Andrews (1974) Direct current measurements of a shelf-edge frontal jet in the southern Benguela system. *Journal of Marine Research* 32, 405-417.

Bindoff, N. L. and J. A. Church (1992) Warming of the water column in the southwest Pacific Ocean *Nature* 357, 59-62.

Bjerknes, J. (1966) A possible response of the atmospheric Hadley circulation to equatorial anomalies of ocean temperature. *Tellus* 18, 820-829.

Bjerknes, J. (1969) Atmospheric teleconnections from the equatorial Pacific. *Monthly Weather Review* 97, 163-172.

Börngen, M., P. Hupfer and M. Olberg (1990) Occurrence and absence of strong salt influxes into the Baltic Sea. *Beiträge zur Meereskunde* 61, 11-19.

Bourke, R. H., V. G. Addison and R. G. Paquette (1989) Oceanography of Nares Strait and northern Baffin Bay in 1986 with emphasis on deep and bottom water formation. *Journal of Geophysical Research* 94, 8289-8302.

Brewer, P. G., C. D. Densmore, R. Munns and R. J. Stanley (1969) Hydrography of the Red Sea brines. In E. T. Degens and D. A. Ross (editors): *Hot brines and recent heavy metal deposits in the Red Sea.* Springer-Verlag Berlin, 138-147.

Brewer, P. G., W. S. Broecker, W. J. Jenkins, P. B. Rhines, C. G. Rooth, J. H. Swift, T. Takahashi and R. T. Williams (1983) A climatic freshening of the deep North Atlantic North of 50°N over the past 20 years. *Science* 222, 1237-1239.

397

Brink, K.-H., D. Halpern, A. Huyer and R. L. Smith (1983) The physical environment of the Peruvian upwelling system. *Progress in Oceanography* 12, 285-305.

Bruce, J. G., D. R. Quadfasel and J. C. Swallow (1980) Somali eddy formation during the commencement of the Southwest Monsoon, 1978. *Journal of Geophysical Research* 85, 6654-6660.

Bruun, A. F., S. V. Greve, H. Mielche and R. Spärck (editors) (1956) *The Galathea deep sea expedition 1950-1952.* George Allen and Unwin, London.

Bryden, H. L. and R. D. Pillsbury (1977) Variability of deep flow in the Drake Passage from year-long current measurements. *Journal of Physical Oceanography* 7, 803-810.

Camden-Smith, F., L.-A. Perrins, R. V. Dingle and G. B. Brundit (1981) A preliminary report on long-term bottom-current measurements and sediment transport/erosion in the Agulhas Passage, southwest Indian Ocean. *Marine Geology* 39, M81-M88.

Carissimo, B. C., A. H. Oort and T. H. Vonder Haar (1985) Estimating the meridional energy transports in the atmosphere and ocean. *Journal of Physical Oceanography* 15, 82-91.

Cember, R. P. (1988) On the sources, formation, and circulation of Red Sea deep water. *Journal of Geophysical Research* 93, 8175-8191.

Charles, C. D. and R. G. Fairbanks (1992) Evidence from Southern Ocean sediments for the effect of North Atlantic deep-water flux on climate. *Nature* 355, 416-419.

Church, J. A., J. S. Godfrey, D. R. Jackett and T. J. McDougall (1991) A model of sea level rise caused by ocean thermal expansion. *Journal of Climate* 4, 438-455.

Clarke, A. J. (1991) On the reflection and transmission of low-frequency energy at the irregular western Pacific Ocean boundary. *Journal of Geophysical Research* 96, 3289-3305.

Clarke, R. A. and J.-C. Gascard (1983) The formation of Labrador Sea Water. part I: large-scale processes. *Journal of Physical Oceanography* 13, 1764-1778.

Coachman, L. K. (1986) Circulation, water masses, and fluxes on the southeastern Bering Sea shelf. *Continental Shelf Research* 5, 23-108.

Coachman, L. K. and K. Aagaard (1974) Physical oceanography of Arctic and Subartic seas. In Y. Herman(Editor): *Marine geology and oceanography of the Arctic seas.* Springer, Berlin, 1-72.

Coachman, L. K. and K. Aagaard (1988) Transports through Bering Strait: annual and interannual variability. *Journal of Geophysical Research* 93, 15535-15539.

Collin, A. E. (1966) Canadian Arctic Archipelago and Baffin Bay. In R. W. Fairbridge(editor): *The Encyclopedia of Oceanography.* Reinhold, New York, 157-160.

Craig, H., W. S. Broecker and D. Spencer (1981) *GEOSECS Pacific Expedition, volume 4.* U.S. Government Printing Office, Washington.

Cresswell, G. R. and T. J. Golding (1980) Observations of a southflowing current in the southeastern Indian Ocean. *Deep-Sea Research* 27, 449-466.

Cresswell, G. R., T. J. Golding and F. M. Boland (1978) A buoy and ship examination of the Subtropical Convergence south of Western Australia. *Journal of Physical Oceanography* 8, 315-320.

Cresswell, G. R. and R. Legeckis (1986) Eddies off southeastern Australia. Deep-Sea *Research* 33, 1527-1562.

Currie, R. I., A. E. Fisher and P. M. Hargreaves (1973) Arabian Sea upwelling. In B. Zeitzschel(editor): *The biology of the Indian Ocean* Springer, Berlin, 37-52.

Cutler, A. N. and J. C. Swallow (1984) Surface currents of the Indian Ocean. *Institute of Oceanographic Sciences Report* 187, 8pp. & 36 charts.

Davis, R.E. (1976) Predictability of sea surface temperature and sea level pressure anomalies over the North Pacific Ocean. *Journal of Physical Oceanography* 6, 249-266.

Delcroix, T. and C. Henin (1988) Observations of the Equatorial Intermediate Current in the western Pacific Ocean (165°E). *Journal of Physical Oceanography* 18, 363-366.

Dietrich, G., K. Kalle, W. Krauss and G. Siedler (1980) *General Oceanography* 2nd ed. Wiley-Interscience, New York.

Donguy, J. and C. Henin (1975) Evidence of the South Tropical Countercurrent in the Coral Sea. *Australian Journal of Marine and Freshwater Research* 26, 405-409.

Douglas, B. C. (1991) Global sea level rise. *Journal of Geophysical Research* 96, 6981-6992.

Duncan, C. P., S. G. Schladow and W. G. Williams (1982) Surface currents near the Greater and Lesser Antilles. *International Hydrographic Review* 59(2), 67-78.

Edwards, R. J. and W. J. Emery (1982) Australasian Southern Ocean frontal structure during summer 1976-77. *Australian Journal of Marine and Freshwater Research* 33, 3-22.

Emery, W. J. and J. S. Dewar (1982) Mean temperature-salinity, salinity- depth and temperature-depth curves for the North Atlantic and the North Pacific. *Progress in Oceanography* 11, 219-305.

England, M. H., J. S. Godfrey, A. C. Hirst and M. Tomczak (in press) The mechanism for Antarctic Intermediate Water renewal in a world ocean model. *Journal of Physical Oceanography*

Etter, P. C., P. J. Lamb and D. H. Portis (1987) Heat and freshwater budgets of the Caribbean Sea with revised estimates for the Central Amercian Seas. *Journal of Physical Oceanography* 17, 1232-1248.

Evans, R. H. and O. B. Brown (1981) Propagation of thermal fronts in the Somali Current system. *Deep-Sea Research* 28, 521-527.

Fein, J. S. and P. L. Stephens (1987) *Monsoons.* John Wiley and Sons, New York.

Ffield, A. and A. Gordon (1992) Vertical mixing in the Indonesian thermocline. *Journal of Physical Oceanography* 22, 184-198.

Fine, R. A., W. H. Peterson and H. G. Ostlund (1987) The penetration of tritium into the tropical Pacific. *Journal of Physical Oceanography* 17, 553-564.

Firing, E. (1987) Deep zonal currents in the central equatorial Pacific. *Journal of Marine Research* 45, 791-812.

Foldvik, A., K. Aagaard and T. Tørresen (1988) On the velocity field of the East Greenland Current. *Deep-Sea Research* 35, 1335-1354.

Folland, C.K., D.E. Parker and F.E. Kates (1984) Worldwide marine temperature fluctuations 1856-1981. *Nature* 310, 670-673.

Folland, C. K., D. E. Parker, M. N. Ward and A. W. Coleman (1986a) Sahel rainfall, northern hemisphere circulation anomalies and worldwide sea temperature changes. Meteorological Office London (amended July 1987).

Folland, C. K., T. N. Palmer and D. E. Parker (1986b) Sahel rainfall and worldwide sea surface temperatures, 1901-85. *Nature* 320, 602-606.

Fu, L.-L., D. B. Chelton and V. Zlotnicki (1988) Satellite altimetry: observing ocean variability from space. *Oceanography* 1(2), 4-11 and 58.

Gamberoni, L., J. Geronimi, P. F. Jeannin and J. F. Murail (1982) Study of frontal zones in the Crozet-Kerguelen region. *Oceanologica Acta* 5, 289-299.

Gamo, T., Y. Nozaki, H. Sakai, T. Nakai and H. Tsubota (1986) Spacial and temporal variations of water characteristics in the Japan Sea bottom layer. *Journal of Marine Research* 44, 781-793.

Gill, A. E. (1982) *Atmosphere-Ocean Dynamics*. Academic Press, New York.

Gill, A. E. and K. Bryan (1971) Effects of geometry on the circulation of a three-dimensional southern-hemisphere ocean model. *Deep-Sea Research* 18, 685-721.

Gloersen, P. and W. J. Campbell (1988) Variations in the Arctic, Antarctic, and global sea ice covers during 1978-1987 as observed with the Nimbus 7 scanning multichannel microwave radiometer. *Journal of Geophysical Research* 93, 10666-10674.

Godfrey, J. S. (1973) Comparison of the East Australian Current with the western boundary flow in BRYAN and COX's (1968) numerical model ocean. *Deep-Sea Research* 20, 1059-1076.

Godfrey, J. S. (1989) A Sverdrup model of the depth-integrated flow for the world ocean allowing for island circulations. *Geophysical and Astrophysical Fluid Dynamics* 45, 89-112.

Godfrey, J. S., M. Nunez, E. F. Bradley, P.A. Coppin and E. J. Lindstrom (1991) On the net surface heat flux into the Western equatorial Pacific. *Journal of Geophysical Research* 96, 3391-3400.

Godfrey, J. S. and A. J. Weaver (1991) Is the Leeuwin Current driven by Pacific heating and winds? *Progress in Oceanography* 27, 225-272.

Goedecke, E., J. Smed and G. Tomczak (1967) Monatskarten des Salzgehaltes der Nordsee dargestellt für verschiedene Tiefenhorizonte. *Deutsche Hydrographische Zeitschrift* B9, 1-110.

Gordon, A. L. (1981) South Atlantic thermocline ventilation. *Deep-Sea Research* 28, 1239-1264.

Gordon, A. L. (1982) Weddell Deep Water variability. *Journal of Marine Research* 40 suppl., 199-217.

Gordon, A. L. (1985) Indian-Atlantic transfer of thermocline water at the Agulhas retroflection. *Science* 227, 1030-1033.

Gordon, A. L. (1986a) Interocean exchange of thermocline water. *Journal of Geophysical Research* 91, 5037-5046.

Gordon, A. L. (1986b) *Southern Ocean Atlas*. A. A. Balkema, Rotterdam.

Gordon, A. L. and W. B. Owens (1987) Polar oceans. *Review of Geophysics* 25, 227-233.

Gornitz, V. and S. Lebedeff (1987) Global sea-level changes during the past century. In: D. Nummedal, O.H. Pilkley and J.D. Howard (editors): *Sea level-change and coastal evolution*. Society of Economic Palaeontologists and Mineralogists Special Publication 41.

Gründlingh, M. L. (1985a) An intense cyclonic eddy east of the Mozambique Ridge. *Journal of Geophysical Research* 90, 7163-7167.

Gründlingh, M. L. (1985b) Occurrence of Red Sea Water in the southwestern Indian Ocean, 1981. *Journal of Physical Oceanography* 15, 207-212.

Gründlingh, M. L. (1987) Cyclogenesis in the Mozambique Ridge. *Deep-Sea Research* 34, 89-103.

Guan, B. (1986) Evidence for a counter-wind current in winter off the southeast coast of China. *Chinese Journal of Oceanology and Limnology* 4, 319 - 332.

Halpern, D. A. (1980) A Pacific equatorial temperature section from 172 E to 110 W during winter and spring 1979. *Deep-Sea Research* 27, 931-940.

Halpern, D. A., R. A. Knox and D. S. Luther (1988) Observations of 20-day period meridional current oscillations in the upper ocean along the Pacific equator. *Journal of Physical Oceanography* 18, 1514-1534.

Hamon, B. V., J. S. Godfrey and M. A. Greig (1975) Relation between mean sea level, current and wind stress on the east coast of Australia. *Australian Journal of Marine and Freshwater Research* 26, 389-403.

Hanawa, K. and S. Kizu (1990) In situ measurement of solar radiation over the sea south of Japan. *Journal of the Meteorological Society of Japan* 68, 607-611.

Hastenrath, S. and P. J. Lamb (1979) *Climatic Atlas of the Indian Ocean. Part II: The Oceanic Heat Budget.* The University of Wisconsin Press. 93 pp.

Hellerman, S. and M. Rosenstein (1983) *Journal of Physical Oceanography* 13, 1093-1104.

Hendon, H. H., N. E. Davidson and B. Gunn (1989) Australian summer monsoon onset during AMEX 1987. *Monthly Weather Review* 117, 370-390

Herman, Y. (1974) *Marine geology and oceanography of the Arctic seas.* Springer, Berlin.

Hirst, A. C. and J. S. Godfrey (in press) The role of the Indonesian throughflow in a global ocean GCM. *Journal of Physical Oceanography*

Hsiung, J. (1985) Estimates of global oceanic meridional heat transport. *Journal of Physical Oceanography* 15,1405-1413.

Hughes, P. and E. D. Barton (1974) Stratification and water mass structure in the upwelling area off NW Africa in April/May 1969. *Deep-Sea Research* 21, 611-628.

Huyer, A. (1976) A comparison of upwelling events in two locations: Oregon and northwest Africa. *Journal of Marine Research* 34, 531-546.

IHO/IOC/CHS (1984) *GEBCO - General Bathymetric Chart of the Oceans* (5th ed.). International Hydrographic Organization/Intergovernmental Oceanographic Commission/Canadian Hydrographic Service, Ottawa. 74 pp. and 19 charts.

Inue, N., T. Miita and S. Tawara (1985) Tsuchima Strait II: physics. In H. Kunishi *et al.* (editors): *Coastal Oceanography of Japanese Islands.* Tokai University Press, 914-933 (in Japanese).

Janowiak, J. E., A. F. Krueger, P. A. Arkin and A. Gruber (1985) *Atlas of outgoing longwave radiation derived from NOAA satellite data.* NOAA Atlas No. 6, U.S. Dept. of Commerce, Silver Spring, Md., 44 pp.

Jones, P. D., T. M. L. Wigley and P.B. Wright (1986) Global temperature variations between 1861 and 1984. *Nature* 332, 430-434.

Joyce, T. M. (1977) A note on the lateral mixing of water masses. *Journal of Physical Oceanography* 7, 626-629.

Joyce, T. M. (1987) Hydrographic sections across the Kuroshio Extension at 165 E and 175 W. *Deep-Sea Research* 34, 1331-1352.

Joyce, T. M., B. A. Warren and L. D. Talley (1986) The geothermal heating of the abyssal subarctic Pacific Ocean. *Deep-Sea Research* 33, 1003-1015.

Käse, R. H., W. Zenk, T. B. Sanford and W. Hiller (1985) Currents, fronts and eddy fluxes in the Canary Basin. *Progress in Oceanography* 14, 231-257.

Kawai, H. (1972) Hydrography of the Kuroshio Extension. In H. Stommel and K. Yoshida (editors): *Kuroshio, Physical Aspects of the Japan Current.* University of Washington Press, Seattle, 235-352.

Kim, K. and R. Legeckis (1986) Branching of the Tsushima Current in 1981-83. *Progress in Oceanography* 17, 265-276.

Kinder, T. H. and G. Parrilla (1987) Yes, some of the Mediterranean outflow does come from great depth. *Journal of Geophysical Research* 92, 2901-2906.

Kinder, T. H., G. W. Heburn and A. W. Green (1985) Some aspects of the Caribbean circulation. *Marine Geology* 68, 25-52.

Knox, R.A. and D.Halpern (1982) Long range Kelvin wave propagation of transport variations in Pacific Ocean equatorial currents. *Journal of Marine Research* 40,suppl., 329-339.

Korzun, V. I. (ed) (1978) *World water balance and resources of the world.* UNESCO, Moscow.

Krauss, W. (1986) The North Atlantic Current. *Journal of Geophysical Research* 91, 5061-5074.

Krauss, W. and R. H. Käse (1984) Mean circulation and eddy kinetic energy in the eastern North Atlantic. *Journal of Geophysical Research* 89, 3407-3415.

Lazier, J. R. N. (1980) Oceanographic conditions at Ocean Weather Ship Bravo, 1964-1974. *Atmosphere-Ocean* 18, 227-238.

Leaman, K. D., E. Johns and T. Rossby (1989) The average distribution of volume transport and potential vorticity with temperature at three sections across the Gulf Stream. *Journal of Physical Oceanography* 19, 36-51.

Legeckis, R. (1977) Long waves in the eastern equatorial Pacific Ocean: a view from a geostationary satellite. *Science* 197, 1179-1181.

Legeckis, R. (1986) A satellite time series of sea surface temperatures in the eastern equatorial ocean. *Journal of Geophysical Research* 91, 12879-12886.

Legeckis, R. and A. L. Gordon (1982) Satellite observations of the Brazil and Falkland currents - 1975 to 1976 and 1978. *Deep-Sea Research* 29, 375-401.

Levitus, S. (1982). *Climatological Atlas of the World Ocean.* NOAA Professional Paper 13.

Levitus, S. (1989) Interpentadal variability of temperature and salinity at intermediate depths of the North Atlantic Ocean, 1970-1974 versus 1955-1959. *Journal of Geophysical Research* 94, 6091-6131.

Lewis, J. K. and A. D. Kirwan jr (1987) Genesis of a Gulf of Mexico ring as determined from kinematic analyses. *Journal of Geophysical Research* 92, 11727-11740.

Lukas, R. (1986) The termination region of the Equatorial Undercurrent in the eastern Pacific. *Progress in Oceanography* 16, 63-90.

Lukas, R. and E. Firing (1984) The geostrophic balance of the Pacific Equatorial Undercurrent. *Deep-Sea Research* 31, 61-66.

Lukas, R. and E. Lindstrom (1991) The mixed layer of the western equatorial Pacific Ocean. *Journal of Geophysical Research* 96, 3343-3357.

Lutjeharms, J. R. E. (1988) Remote sensing corroboration of retroflection of the East Madagascar Current. *Deep-Sea Research* 35, 2045-2050.

Lutjeharms, J. R. E., N. D. Bang and C. P. Duncan (1981) Characteristics of the currents east and south of Madagascar. *Deep-Sea Research* 28, 879-899.

Lutjeharms, J. R. E. and R. C. van Ballegooyen (1988) The retroflection of the Agulhas Current. *Journal of Physical Oceanography* 18, 1570-1583.

Luyten, J. R. and D. H. Roemmich (1982) Equatorial currents at semi-annual period in the Indian Ocean. *Journal of Physical Oceanography* 12, 406-413.

Luyten, J. R. and J. C. Swallow (1976) Equatorial undercurrents. *Deep-Sea Research* 23, 999-1001.

Manabe, S. and R. J. Stouffer (1988) Two stable equilibria of a coupled ocean-atmosphere model. *Journal of Climate* 1, 841-866

Mantyla, A. W. and J. C. Reid (1983) Abyssal characteristics of the world ocean waters. *Deep-Sea Research* 30, 805-833.

Matthäus, W. and H. Franck (1990) The water volume penetrating into the Baltic Sea in connection with major Baltic inflows. *Gerlands Beiträge zur Geophysik* 99, 377-386.

Maul, G. A. and A. Hermann (1985) Mean dynamic topography of the Gulf of Mexico with application to satellite altimetry. *Marine Geodesy* 9, 27-44.

McCartney, M. S. (1977) Subantarctic Mode Water. *Deep-Sea Research* 24 suppl., 103-119.

McCartney, M. S. (1982) The subtropical recirculation of mode waters. *Journal of Marine Research* 40 suppl., 427-464.

McCreary, J. P. and P. K. Kundu (1989) A numerical investigation of sea surface temperature variability in the Arabian Sea. *Journal of Geophysical Research* 94, 16,097-16,114.

McPhaden, M. J. (1982) Variability in the central equatorial Indian Ocean part I: Ocean dynamics. *Journal of Marine Research* 40, 157-176.

Meehl, G.A. (1987) The annual cycle and interannual variability in the tropical Pacific and Indian Ocean regions. *Monthly Weather Review* 115, 27-49.

Meier, M.F. (1984) Contribution of small glaciers to global sea level. *Science* 226, 1418-1421.

Meincke, J. (1978) On the distribution of low salinity intermediate waters around the Faroes. *Deutsche Hydrographische Zeitschrift* 31, 50-64.

Millero, F. J. and A. Poisson (1981) International one-atmosphere equation of state of sea-water. *Deep-Sea Research* 28, 625-629.

Mizuno, K. and W. B. White (1983) Annual and interannual variability in the Kuroshio current system. *Journal of Physical Oceanography* 13, 1847-1867.

Moore, D., P. Hisard, J. McCreary, J. Merle, J. O'Brien, J. Picaut, J.-M. Verstraete and C. Wunsch (1978) Equatorial adjustment in the eastern Atlantic. *Geophysical Research Letters* 5, 637-640.

Moore, W. S., J. L. Sarmiento and R. M. Key (1986) Tracing the Amazon component of surface Atlantic water using ^{228}Ra, salinity and silica. *Journal of Geophysical Research* 91, 2574-2580.

Muench, R. D., J. D. Schumacher and S. A. Salo (1988) Winter currents and hydrographic conditions on the northern central Bering Sea shelf. *Journal of Geophysical Research* 93, 516-526.

Mulhearn, P. J. (1987) The Tasman Front: a study using satellite infrared imagery. *Journal of Physical Oceanography* 17, 1148- 1155.

Murray, J. W. (1991) The 1988 Black Sea oceanographic expedition: introduction and summary. *Deep-Sea Research* 38 suppl., S655 - S661.

Murray, S. P. and D. Arief (1988) Throughflow into the Indian Ocean through the Lombok Strait, January 1985 - January 1986. *Nature* 333, 444-447.

Murty, A. V. S. (1987) Characteristics of neritic waters along the west coast of India with respect to upwelling, dissolved oxygen & zooplankton biomass. *Indian Journal of Marine Sciences* 16, 129-131.

Neal, V. T., S. Neshyba and W. Denner (1969) Thermal stratification in the Arctic Ocean. *Science* 166, 373-374.

Nehring, D. (1990) Die hydrographisch-chemischen Bedingungen in der westlichen und zentralen Ostsee von 1979 bis 1988 - ein Vergleich. *Meereswissenschaftliche Berichte Warnemünde* 2, 1-45.

Nehring, D. and W. Matthäus (1990) Aktuelle Trends hydrographischer und chemischer Parameter in der Ostsee, 1958 - 1989. *Meereswissenschaftliche Berichte Warnemünde* 2, 46-79.

Nelson, G. (1989) Poleward motion in the Benguela area. In Barber (editors): *Poleward flows along eastern ocean boundaries.* Springer, New York, 110-130.

Neumann, G. and W. J. Pierson jr. (1966) Pri*nciples of Physical Oceanography*. Prentice-Hall, Englewood Cliffs N.J.

Newman, W. S. and R. W. Fairbridge (1986) The management of sea level rise. *Nature* 320, 319-321.

Nilsson, C. S. and G. R. Cresswell (1981) The formation and evolution of East Australian Current warm-core eddies. *Progress in Oceanography* 9, 133-183.

Nosaki, Y., V. Kasemsupaya and H. Tsubota (1989) Mean residence time of the shelf water in the East China and the Yellow Seas determined by ^{228}Ra/^{226}Ra measurements. *Geophysical Research Letters* 16, 1297-1300.

Nowlin, W. D. jr and W. Zenk (1988) Westward bottom currents along the margin of the South Shetland Island Arc. *Deep-Sea Research* 35, 269-301.

Oberhuber, J. M. (1988) An atlas based on the COADS data set: the budget of heat, buoyancy and turbulent kinetic energy at the surface of the global ocean. *Max-Planck-Istitut für Meteorologie Report* 15.

Östlund, H. G., G. Possnert and J. H. Swift (1987) Ventilation rate of the deep Arctic ocean from carbon 14 data. *Journal of Geophysical Research* 92, 3769-3771.

Osborne, J., P. Rhines and J. Swift (1991) *OceanAtlas for Macintosh, a microcomputer application for examining oceanographic data, version 1.0.* Scripps Institute of Oceanography, La Jolla Calif. S.I.O. ref. #91-5.

Peltier, W. R. and A. M. Tushingham (1991) Influence of isostatic adjustment on tide gauge measurements of secular sea level change. *Journal of Geophysical Research* 96, 6779-6796

Peters, A. (1989) *Peters' Atlas of the World*. Longman Press, Harlow.

Peters, H. (1976) The spreading of the water masses of the Banc d'Arguin in the upwelling area off the northern Mauritanian coast. *"Meteor" Forschungs-Berichte* A18, 78-100.

Peters, H., M. C. Gregg and J. M. Toole(1988) On the parameterization of equatorial turbulence. *Journal of Geophysical Research* 93, 1199-1218.

Peterson, W. H. and C. G. H. Rooth (1976) Formation and exchange of deep water in the Greenland and Norwegian Seas. *Deep-Sea Research* 23, 273-283.

Peterson, R. G. and L. Stramma (1991) Upper-level circulation in the South Atlantic Ocean. *Progress in Oceanography* 26, 1-73.

Philander, S. G. (1990) *El Niño, La Niña, and the Southern Oscillation*. Academic Press, San Diego.

Pickard, G. L. and W. J. Emery (1990) *Descriptive Physical Oceanography* 5th ed. Pergamon Press, Oxford.

Pillsbury, R. D. and J. S. Bottero (1984) Observations of current rings in the Antarctic zone at Drake Passage. *Journal of Marine Research* 42, 853-874.

Piola, A. R. and D. T. Georgi (1982) Circumpolar properties of Antarctic Intermediate Water and Subantarctic Mode Water. *Deep-Sea Research* 29, 687-711.

Pond, S. and G. L. Pickard (1983) *Introductory Dynamical Oceanography* 2nd ed. Pergamon Press, Oxford.

Quadfasel, D. R. and F. Schott (1982) Water-mass distributions at intermediate layers off the Somali coast during the onset of the Southwest Monsoon, 1979. *Journal of Physical Oceanography* 12, 1358-1372.

Quadfasel, D. R. and F. Schott (1983) Southward subsurface flow below the Somali Current. *Journal of Geophysical Research* 88, 5973-5979.

Rasmusson, E. M. and T. H. Carpenter (1982) Variations in tropical sea surface temperature and surface wind fields associated with the Southern Oscillation/ El Niño. *Monthly Weather Review* 110, 354-384.

Reid, J. L. (1986) On the total geostrophic circulation of the South Pacific Ocean: flow patterns, tracers, and transports. *Progress in Oceanography* 16, 1-61.

Reid, J. L. (1989) On the total geostrophic circulation of the South Atlantic Ocean: flow patterns, tracers, and transports. *Progress in Oceanography* 23, 149-244.

Reid, J. L. and R. J. Lynn (1971) On the influence of the Norwegian-Greenland and Weddell seas upon the bottom waters of the Indian and Pacific oceans. *Deep-Sea Research* 18, 1063-1088.

Richardson, P. L. (1983a) Eddy kinetic energy in the North Atlantic from surface drifters. *Journal of Geophysical Research* 88, 4355-4367.

Richardson, P. L. (1983b) Gulf Stream rings. In A. R. Robinson (editor): *Eddies in Marine Science.* Springer, Berlin, 19-45.

Richardson, P. L. (1985) Average velocity and transport of the Gulf Stream near 55W. Journal of *Marine Research* 43, 83-111.

Richardson, P. L. and D. Walsh (1986) Mapping climatological seasonal variations of surface currents in the tropical Atlantic using ship drifts. *Journal of Geophysical Research* 91, 10537-10550.

Rintoul, S. R. (1991) South Atlantic interbasin exchange. *Journal of Geophysical Research* 96, 2675-2692.

Robinson, M. K., R. A. Bauer and E. H. Schroeder (1979) *Atlas of North Atlantic-Indian Ocean monthly mean temperatures and mean salinities of the surface layer.* U.S. Naval Oceanographic Office ref. pub. 18, Washington D.C.

Roemmich, D. and B. Cornuelle (1990) Observing the fluctuations of gyre-scale ocean circulation: a study of the subtropical South Pacific. *Journal of Physical Oceanography* 20, 1919-1934

Rodman, M. R. and A. L. Gordon (1982) Southern Ocean bottom water of the Australian-New Zealand sector. *Journal of Geophysical Research* 87, 5771-5778.

Royer, T. C. (1982) Coastal fresh water discharge in the Northeast Pacific. *Journal of Geophysical Research* 87, 2017-2021.

Royer, T. C. and W. J. Emery (1987) Circulation in the Gulf of Alaska, 1981. *Deep-Sea Research* 34, 1361-1377.

Rudels, B. and D. Quadfasel (1991) Convection and deep water formation in the Arctic Ocean - Greenland Sea system. *Journal of Marine Systems* 2, 435-450.

Sælen, O. H. (1988) On the exchange of bottom water between the Greenland and Norwegian Seas. *Geophysical Institute, Div. A., Physical Oceanography, University of Bergen, report* 67.

Sankey, T. (1973) The formation of deep water in the northwestern Mediterranean. *Progress in Oceanography* 6, 159-188.

Schemainda, R., D. Nehring and S. Schulz (1975) Ozeanologische Untersuchungen zum biologischen Produktionspotential der nordwestafrikanischen Wasserauftriebsregion 1970 - 1973. *Geodätische und Geophysikalische Veröffentlichungen* IV 16, 1-88.

Schmitz, W. J. jr (1987) Observations of new, large and stable abyssal currents atmidlatitudes along 165°E. *Journal of Physical Oceanography* 17, 1309-1315.

Schott, G. (1912) *Geographie des Atlantischen Ozeans.* C. Boysen, Hamburg.

Schott, G. (1935) *Geographie des Indischen und Stillen Ozeans.* C. Boysen, Hamburg.

Schott, F., J. C. Swallow and M. Fieux (1990) The Somali Current at the equator: annual cycle of currents and transports in the upper 1000 m and connection to neighbouring latitudes. *Deep-Sea Research* 37, 1825-1990.

Shetye, S. R., A. D. Gouveia, S. S. C. Shenoi, G. S. Michael, D. Sundar, A. M. Almeida and K. Santanam (1991) The coastal current off western India during the northeast monsoon. *Deep-Sea Research* 38, 1517-1529.

Siedler, G., A. Kuhl and W. Zenk (1987) The Madeira Mode Water. *Journal of Physical Oceanography* 17, 1561-1570.

Simpson, J. H. (1981) The shelf-sea fronts: implications of their existence and behaviour. *Philosophical Transactions of the Royal Society of London* A302, 531-546.

Smethie, W. M. jr, D. W. Chipman, J. H. Swift and K. P. Koltermann (1988) Chlorofluoromethanes in the Arctic Mediterranean Seas: evidence for formation of bottom water in the Eurasian Basin and deep-water exchange through Fram Strait. *Deep-Sea Research* 35, 347-369.

Smith, O. P. and J. M. Morrison (1989) Shipboard acoustic doppler current profiling in the eastern Caribbean Sea, 1985-1986. *Journal of Geophysical Research* 94, 9713-9719.

Sprintall, J. and M. Tomczak (1990) Salinity considerations in the oceanic surface mixed layer. *Ocean Sciences Institute (the University of Sydney) Report* 36.

Sprintall, J. and M. Tomczak (1992) Evidence of the barrier layer in the surface layer of the tropics. *Journal of Geophysical Research* 97, 7305-7316.

Sprintall, J. and M. Tomczak (1993) On the formation of Central Water in the southern hemisphere. *Deep-Sea Research* 40, 827-848.

Stanton, B. R. (1981) An oceanographic survey of the Tasman Front. *New Zealand Journal of Marine and Freshwater Research* 15, 289-297.

Stommel, H. and K. Yoshida (1972) *Kuroshio, Physical Aspects of the Japan Current.* University of Washington Press, Seattle. 517 pp.

Stramma, L. (1992) The South Indian Ocean Current. *Journal of Physical Oceanography* 22, 421-430.

Stramma, L., Y. Ikeda and R. G. Peterson (1990) Geostrophic transport in the Brazil Current region north of 20°S. *Deep-Sea Research* 37, 1875-1886.

Stramma, L., R. G. Peterson and M. Tomczak (in press) The South Pacific Current. *Journal of Physical Oceanography*

Streten, N. A. (1980) Some synoptic indices of the Southern Hemisphere mean sea level circulation 1972-77. *Monthly Weather Review* 108, 18- 36.

Sverdrup, H. U., M. W. Johnson and R. H. Fleming (1942) *The oceans: their physics, chemistry and general biology.* Prentice-Hall, Englewood Cliffs.

Swallow, J. C. and M. Fieux (1982) Historical evidence for two gyres in the Somali Current. *Journal of Marine Research* 40 suppl., 747-755.

Swallow, J. C. and R. T. Pollard (1988) Flow of bottom water through the Madagascar Basin. *Deep-Sea Research* 35, 1437-1440.

Swallow, J., M. Fieux and F. Schott (1988) The boundary currents east and north of Madagascar 1. Geostrophic currents and transports. *Journal of Geophysical Research* 93, 4951-4962.

Tabata, S. (1982) The anticyclonic, baroclinic eddy off Sitka, Alaska, in the northeast Pacific Ocean. *Journal of Physical Oceanography* 12, 1260-1282.

Taljaard, J. J., H. van Loon, H. L. Crutcher and R. J. Lenne (1969) *Climate of the upper air part 1 - southern hemisphere; volume 1: temperatures, dew points, and heights at selected levels.* Naval Weather Service Command, Washington D.C.

Talley, L. D. and R. A. deSzoeke (1986) Spatial fluctuations north of the Hawaiian Ridge. *Journal of Physical Oceanography* 16, 981-984.

Tchernia, P. (1980) *Descriptive Regional Oceanography*. Pergamon Press, Oxford.

Thompson, K. R., J. R. N. Lazier and B. Taylor (1986) Wind-forced changes in Labrador Current transport. *Journal of Geophysical Research* 91, 14261-14268.

Thompson, R. O. R. Y. (1984) Observations of the Leeuwin Current off western Australia. *Journal of Physical Oceanography* 14, 623-628.

Tintore, J., P. E. La Violette, I. Blade and A. Cruzado (1988) A study of an intense density front in the eastern Alboran Sea: the Almeria-Oran front. *Journal of Physical Oceanography* 18, 1384-1397.

Tolmazin, D. (1985a) Changing coastal oceanography of the Black Sea. I: northwest shelf. *Progress in Oceanography* 15, 217-276.

Tolmazin, D. (1985b) Changing coastal oceanography of the Black Sea. II: Mediterranean effluent. *Progress in Oceanography* 15, 277-316.

Tomczak, G. and E. Goedecke (1964) Die thermische Schichtung der Nordsee auf Grund des mittleren Jahresganges der Temperatur in $^1/_2$°- und 1° - Feldern. *Deutsche Hydrographische Zeitschrift* B8, 1-182.

Tomczak, M. (1980) A review of Wüst's classification of the major deep-sea expeditions 1873-1960 and its extension to recent oceanographic research programmes. In M. Sears and D. Merriman (editors): *Oceanography: the past*. Springer, New York, 188-194.

Tomczak, M. (1981a) Prediction of environment changes and the struggle of the Third World for national independence: the case of the Peruvian fisheries. In M. H. Ghantz and J. D. Thompson(editors): *Resources Management and Environmental Uncertainty: Lessons from Coastal Upwelling Fisheries*. John Wiley & Sons, New York, 401-435.

Tomczak, M. (1981b) Coastal upwelling systems and eastern boundary currents: a review of terminology. *Geoforum* 12, 179-191.

Tomczak, M. (1981d) Bass Strait Water intrusions in the Tasman Sea and mean temperature-salinity curves. *Australian Journal of Marine and Freswater Research* 32, 699-708.

Tomczak, M. (1981e) Longshore advection during an upwelling event in the Canary Current area as detected by airborne radiometer. *Oceanologica Acta* 4, 161-169.

Tomczak, M. (1985) The Bass Strait Water cascade during winter 1981. *Continental Shelf Research* 4, 225-278.

Tomczak, M. and D. Hao (1989) Water masses in the thermocline of the Coral Sea. *Deep-Sea Research* 36, 1503-1514.

Tomczak, M. and P. Hughes (1980) Three dimensional variability of water masses and currents in the Canary Current upwelling region. *"Meteor" Forschungs-Ergebnisse* A 21, 1-24.

Tomczak, M. and D. G. B. Large (1989) Optimum multiparameter analysis of mixing in the thermocline of the eastern Indian Ocean. *Journal of Geophysical Research* 94, 16141-16149.

Tomczak, M. and G. Miosga (1976) The sea surface temperature as detected by airborne radiometer in the upwelling region off Cap Blanc, NW-Africa. *"Meteor" Forschungs-Ergebnisse* A17, 1-20.

Trenberth, K. E., J. G. Olson and W. G. Large (1989) *A global ocean wind stress climatology based on ECMWF analyses*. NCAR Technical Note NCAR/TN-338+STR.

Tsuchiya, M. (1986) Thermostads and circulation in the upper layer of the Atlantic Ocean.

Progress in Oceanography 16, 235-237.

Tsuchiya, M., R. Lukas, R. A. Fine, E. Firing and E. Lindstrom (1989) Source waters of the Pacific Equatorial Undercurrent. *Progress in Oceanography* 23, 101-147.

Unesco (1981) Tenth Report of the Joint Panel on Oceanographic Tables and Standards. *Unesco Technical Papers in Marine Science* 36.

University of East Anglia, Climatic Research Unit (1992) *World Climate Disc* (CD-ROM and software). Chadwick-Healey, Cambridge.

U.S. Navy (1981) *Marine climatic atlas of the world IX; world-wide means and standard deviations.* U.S. Naval Oceanography Command Detachment, Asheville N.C.

van Aken, H. M., J. Punjanan and S. Saimima (1988) Physical aspects of the flushing of the east Indonesian Basins. *Netherlands Journal of Sea Research* 22, 315-339.

van Bennekom, A. J. (1988) Deep-water transit times in the eastern Indonesian basins, calculated from dissolved silica in deep and interstitial waters. *Netherlands Journal of Sea Research* 22, 341-354.

Vastano, A. C. and R. L. Bernstein (1984) Mesoscale features along the First Oyashio Intrusion. *Journal of Geophysical Research* 89, 587-596.

Villanoy, C. L. and M. Tomczak (1991) Influence of Bass Strait Water on the Tasman Sea thermocline. *Australian Journal of Marine and Freshwater Research* 42, 451-464.

Voituriez, B. (1981) Les sous-courants équatoriaux nord et sud et la formation des dômes thermiques tropicaux. *Oceanologica Acta* 4, 497-506.

Vukovich, F. M. (1988) Loop Current boundary variations. *Journal of Geophyical Research* 93, 15,585-15,591.

Wakasutchi, M. and K. I. Ohshima (1990) Observations of ice-ocean eddy streets in the Sea of Okhotsk off the Hokkaido coast using radar images. *Journal of Physical Oceanography* 20, 585-599.

Walker, N. D. (1986) Satellite observation of the Agulhas Current and episodic upwelling south of Africa. *Deep-Sea Research* 33, 1083-1106.

Wang, J. and Ching-Sheng Chern (1988) On the Kuroshio branch in the Taiwan Strait durin wintertime. *Progress in Oceanography* 21, 469 - 491.

Warren, B. A. (1981a) Deep circulation of the world ocean. In B. A. Warren and C. Wunsch (editors): *Evolution of Physical Oceanography.* MIT Press, Cambridge (Massachusetts), 6-42.

Warren, B. A. (1981b) Transindian hydrographic section at lat. 18 S: Property distributions and circulation in the south Indian Ocean. *Deep-Sea Research* 28, 759-788.

Warren, B. A. (1982) The deep water of the Central Indian Basin. *Journal of Marine Research* 40 suppl., 823-860.

Warren, B. A. (1983) Why is no deep water formed in the North Pacific? *Journal of Marine Research* 41, 327-347.

Weng, X. And C. Wang (1988) On the Taiwan Warm Current water. *Chinese Journal of Oceanology and Limnology* 6, 320-329.

White, W. B. and A. E. Walker (1985) The influence of the Hawaiian archipelago upon the wind-driven subtropical gyre in the western north Pacific. *Journal of Geophysical Research* 90, 7061-7064.

Whitworth, T. III and R. G. Peterson (1985) Volume transport of the Antarctic Circumpolar Current from bottom pressure measurements. *Journal of Physical Oceanography* 15, 810-816.

Wijffels, S. E., R. W. Schmitt, H. L. Bryden and A. Stigebrandt (1992) On the transport of freshwater by the oceans. *Journal of Physical Oceanography* 22, 155-162.

World Climate Research Programme (1985) *International Implementation Plan for TOGA*, ITPO Document °1 (first edition), World Meteorological Office, Geneva.

World Climate Research Programme (1988) *World Ocean Circulation Experiment Implementation Plan, Vol. II: Scientific Background.* WCRP-12.

World Climate Research Programme (1990) *JSC/CCCO TOGA Scientific Steering Group: Report of the Ninth Session.* Kona, Hawaii, U.S.A. WCRP-47.

Worthington, L. V. (1969) An attempt to measure the volume transport of Norwegian Sea overflow water through the Denmark Strait. *Deep-Sea Research* 16 suppl., 421-432.

Worthington, L. V. (1981) The water masses of the world ocean: some results of a fine-scale census. In B. A. Warren and C. Wunsch (editors): *Evolution of Physical Oceanography.* MIT Press, Cambridge (Massachusetts), 43-69.

Wright, P. B. (1977) *Southern oscillation; patterns and mechanisms of the teleconnections and the persistence.* Hawaii University Institute of Geophysics Report HIG-77-13.

Wright, P. W. (1988) An atlas based on the COADS data set: fields of mean wind, cloudiness and humidity at the surface of the global ocean. *Max-Planck-Institut für Meteorologie Report* 14.

Wüst, G. (1936) Schichtung und Zirkulation des Atlantischen Ozeans: Das Bodenwasser und die Gliederung der Atlantischen Tiefsee. *Wissenschaftliche Ergebnisse der Deutschen Atlantischen Expedition "Meteor" 1925-1927* vol. VI part 1, 3-107.

Wüst, G. (1961) Das Bodenwasser und die Vertikalzirkulation des Mittelländischen Meeres. *Deutsche Hydrographische Zeitschrift* 14, 81-92

Wüst, G. (1963) On the stratification and the circulation in the cold water sphere of the Antillean-Caribbean basins. *Deep-Sea Research* 10, 165-187.

.Wüst, G. (1964) *Stratification and Circulation of the Antilles and Caribbean Basins.* Columbia University Press, 201 pp

Wyrtki, K. (1961) Physical oceanography of the southeast Asian waters. Scientific Results of Maritime Investigations of the South China Sea and the Gulf of Thailand 1959-1961. *Naga Report,* 2. Scripps Institution of Oceanography, La Jolla California.

Wyrtki, K. (1971) *Oceanographic Atlas of the International Indian Ocean Expedition.* National Science Foundation, Washington D.C. 531 pp. Reprinted 1988 by A.A. Balkema, Rotterdam.

Wyrtki, K. (1973a) An equatorial jet in the Indian Ocean. *Science* 181, 262-264.

Wyrtki, K. (1973b) Teleconnections in the equatorial PacificOcean. *Science* 180, 66-68.

Wyrtki, K. (1977) Sea level during the 1972 El Niño. *Journal of Physical Oceanography* 7, 779-787.

Yasunari, T. (1990) Impact of Indian monsoon on the coupled ocean/atmosphere system in the tropical Pacific. *Meteorology and Atmospheric Physics* 44, 29-41.

You, Y. and M. Tomczak (1993) Thermocline circulation and ventilation in the Indian Ocean derived from water mass analysis. *Deep-Sea Research* 40, 13-56.

Zahn, W. (1984) Eine Abschätzung des Volumentransportes im Kanal von Moçambique während des Zeitraumes Oktober-November 1957. *Beiträge zur Meereskunde* 51, 67-74.

Zimmerman, J. T. F. (1984) Windscale effluent as a tracer for continental shelf circulation. *Nature* 311, 102-103.

Glossary

anti-cyclonic sense of rotation around a centre of high pressure (clockwise in the northern hemisphere, anti-clockwise in the southern hemisphere); see also cyclonic

barrier layer the depth range between the bottom of the mixed layer and the seasonal thermocline

cast (also hydrographic cast or hydrographic station) the measurement of temperature, salinity and other properties using either a series of water sampling devices attached to a wire ("bottle cast") or a CTD mounted in a rack ("rosette") holding such devices ("CTD cast"), lowered into the ocean from a ship; also a set of data (usually depth, temperature, salinity, oxygen, and nutrients) collected in that way

convection vertical movement produced by increasing the density of a fluid at the upper surface of a volume or by decreasing the density at the bottom

convergence horizontal movement through a volume of fluid in which more fluid enters the volume than leaves it horizontally, resulting in vertical movement out of the volume

cyclonic sense of rotation around a centre of low pressure (anti-clockwise in the northern hemisphere, clockwise in the southern hemisphere); derived from the circulation around tropical cyclones

diapycnal directed across surfaces of constant density

divergence horizontal movement through a volume of fluid in which less fluid enters the volume than leaves it horizontally, resulting in vertical movement into the volume

downwelling downward vertical movement of water through the bottom of the surface layer produced by a convergence at the surface

eddy circulation system in which the water follows closed circular or elliptic paths; can be cyclonic or anti-cyclonic

entrainment movement of mass from one layer of a fluid into another layer without compensatory movement of fluid in the opposite direction

finestructure variability of a property in space on scales of a metre or less *haline* related to salinity

halocline the layer where salinity changes most rapidly with depth

interleaving a process where fluid with given properties moves laterally into a region occupied by fluid with different properties; as a result, layers of the first type of fluid form within the second type of fluid

isobars contours of constant pressure

isohalines contours of constant salinity

isopycnals contours of constant density

isotherms contours of constant temperature

latitude the north-south co-ordinate of a position on the earth's surface expressed in degrees, from 90°S (-90°) at the south pole to 0° at the equator and 90°N (+90°) at the north pole

longitude the east-west co-ordinate of a position on the earth's surface expressed in degrees, from 0° at the longitude of Greenwich to 180° at the date line in the Pacific Ocean, positive or °W to the west of 0° longitude, negative or °E to the east of 0° longitude

meridian a line of constant longitude

meridional in the direction of meridians, i.e. north-south

nautical mile unit of length used in navigation; for oceanographic purposes (taking the earth as perfectly spherical) the nautical mile can be defined as one minute of arc along the equator or along any meridian. One degree of arc has sixty minutes, so one degree of latitude corresponds to 60 nautical miles, which is very close to 111 km

nutrients in oceanography the name given to the group of dissolved mineral salts most important for marine life, usually comprising anorganic phosphate, nitrate, and silicate; sometimes nitrite and organic and particulate phosphate are included as well

oxygen in oceanography the amount of oxygen dissolved in seawater, given in millilitres per litre (ml/l) or in micromols per kilogram (μmol kg^{-1}); an approximate conversion, exact near a temperature of 5°C and 34.45 salinity, is 1 ml/l = 44.66 μmol kg^{-1}

polar pertaining to the regions under the influence of the easterly winds of very high latitudes

potential temperature temperature of a water particle, found at some depth, after it is moved adiabatically (i.e. without exchange of heat with its surroundings) to the surface

pycnocline the layer where density changes most rapidly with depth *pycnostad* a layer where the vertical change of density is very small and displays a local minimum

ring an eddy formed by separation of part of a strong current (such as a western boundary current); it is characterized by a current band of roughly the width of the parent current and uniform large velocity, and by the trapping of water with properties different from the properties found outside the ring

subpolar pertaining to the regions between the polar and temperate climate zones

subtropical pertaining to the regions under the influence of the Trade Winds

temperate pertaining to the regions under the influence of the Westerlies

thermal relating to temperature

thermocline the layer where temperature changes most rapidly with depth during summer (the seasonal thermocline); the depth range where temperature changes rapidly with depth throughout the year (the permanent or oceanic thermocline). Consult chapter 5 for a full explanation of terms

thermohaline relating to temperature and salinity

thermostad a layer where the vertical change of temperature is very small and displays a local minimum

tropical pertaining to the regions between the Trade Winds of the two hemispheres (the Doldrums)

upwelling upward vertical movement of water through the bottom of the surface layer produced by a divergence at the surface

water mass a body of water with a common formation history

water type a set of parameter values to describe water with the corresponding properties

subduction sinking of water through movement on inclined isopycnal surfaces

source water type a set of parameter values to describe the properties of a newly formed water mass

tracers a common name for properties which do not affect the density of seawater and therefore have no impact on water movement but can be used to indicate water movement; in addition to the classical tracers (oxygen and nutrients) oceanography now uses tracers introduced or enriched by human activity such as carbon, cesium, the chlorofluorocarbons (CFCs or freons), plutonium, strontium, tritium, and others

tritium radioactive isotope of hydrogen with mass number 3; naturally found in seawater at low concentration levels, during the last decades found at elevated concentration levels as a result of fallout from atmospheric bomb testing

zonal in the direction parallel to the equator, i.e. east-west

units and conversions:

property	unit	derived units		
distance	metre (m)	1 nautical mile	=	1853.2 m
			=	1.8532km
velocity	metres per second (m s^{-1})	1 knot	=	1 nautical mile per hour
			=	0.515 m s^{-1}
			=	44.5 km/day
			=	16 234 km/year
transport	cubic metres per second (m^3 s^{-1})	1 Sverdrup (Sv)	=	10^6 m^3 s^{-1}
			=	3.6 km^3/hour
pressure	Pascal (Pa; 1 Pa = 1 kg m^{-1} s^{-2})	1 dbar	=	10 kPa (equivalent to 1 m depth increase)

Wind velocity is related to wind force, expressed in Beaufort, through the following table:

Beaufort force	knots	m s^{-1}	km/hour
0	under 1	0.0 - 0.2	under 1
1	1 - 3	0.3 - 1.5	1 - 5
2	4 - 6	1.6 - 3.3	6 - 11
3	7 - 10	3.4 - 5.4	12 - 19
4	11 - 16	5.5 - 7.9	20 - 28
5	17 - 21	8.0 - 10.7	29 - 38
6	22 - 27	10.8 - 13.8	39 - 49
7	28 - 33	13.9 - 17.1	50 - 61
8	34 - 40	17.2 - 20.7	62 - 74
9	41 - 47	20.8 - 24.4	75 - 88
10	48 - 55	24.5 - 28.4	89 - 102
11	56 - 63	28.5 - 32.6	103 - 117
12	over 63	over 32.6	over 117

INDEX

The following abbreviations are used in this index: WO = World Ocean, AO = Atlantic Ocean, IO = Indian Ocean, PO = Pacific Ocean, Med = Mediterranean, C = Current.